Invertebrate
Tissue Culture
Research Applications

Contributors

NORIAKI AGUI
CHRISTOPHER J. BAYNE
MARION A. BROOKS
SONJA M. BUCKLEY
HARUO CHINO
JAMES W. FRISTROM
EDER L. HANSEN
DONALD HEYNEMAN
W. FRED HINK

HIROYUKI HIRUMI
TIMOTHY J. KURTTI
ARTHUR H. McINTOSH
KARL MARAMOROSCH
EDWIN P. MARKS
JUN MITSUHASHI
J. D. PASCHKE
JAMES L. VAUGHN
S. R. WEBB

MARY ALICE YUND

Invertebrate Tissue Culture

Research Applications

EDITED BY

Karl Maramorosch

Waksman Institute of Microbiology
Rutgers University
The State University of New Jersey
New Brunswick, New Jersey

ACADEMIC PRESS   New York / San Francisco / London   1976

A Subsidiary of Harcourt Brace Jovanovich, Publishers

ACADEMIC PRESS, INC.
111 Fifth Avenue, New York, New York 10003

United Kingdom Edition published by
ACADEMIC PRESS, INC. (LONDON) LTD.
24/28 Oval Road, London NW1

Library of Congress Cataloging in Publication Data

Main entry under title:

Invertebrate tissue culture.

    Bibliography: p.
    Includes index.
    1.   Invertebrates—Cultures and culture media.
2.   Tissue culture.   I.   Maramorosch, Karl.
[DNLM: 1.   Invertebrates.   2.   Tissue culture. QS530
I94]
QL362.8.I58   592'.08'2028   75-26351
ISBN 0–12–470270–8

Contents

List of Contributors

Numbers in parentheses indicate the pages on which the authors' contributions begin.

NORIAKI AGUI (133), Laboratory of Applied Zoology, Faculty of Agriculture, Tokyo University of Education, Komaba, Meguro-ku, Tokyo, Japan

CHRISTOPHER J. BAYNE (61), Department of Zoology, Oregon State University, Corvallis, Oregon

MARION A. BROOKS (181), Department of Entomology, Fisheries, and Wildlife, University of Minnesota, Saint Paul, Minnesota

SONJA M. BUCKLEY* (201), Yale Arbovirus Research Unit, Department of Epidemiology and Public Health, Yale University School of Medicine, New Haven, Connecticut

HARUO CHINO (103), Biochemistry Division, The Institute of Low Temperature Science, Hokkaido University, Sapporo, Japan

JAMES W. FRISTROM (161), Department of Genetics, University of California, Berkeley, California

EDER L. HANSEN† (75), Clinical Pharmacology Research Institute, Berkeley, California

DONALD HEYNEMAN (57), Hooper Foundation, University of California, San Francisco, California

W. FRED HINK (319), Department of Entomology, Ohio State University, Columbus, Ohio

HIROYUKI HIRUMI‡ (233), Boyce Thompson Institute, Yonkers, New York

TIMOTHY J. KURTTI (39), Department of Entomology, Fisheries, and Wildlife, University of Minnesota, Saint Paul, Minnesota

ARTHUR H. McINTOSH (3), Waksman Institute of Microbiology,

* Present address: 60 College Street, New Haven, Connecticut.
† Present address: 561 Santa Barbara Road, Berkeley, California.
‡ Present address: International Laboratory for Research on Animal Diseases, P.O. Box 30709, Nairobi, Kenya.

Rutgers University, The State University of New Jersey, New
Brunswick, New Jersey
KARL MARAMOROSCH (305), Waksman Institute of Microbiology,
Rutgers University, The State University of New Jersey, New
Brunswick, New Jersey
EDWIN P. MARKS (117), Metabolism and Radiation Research Labo-
ratory, Agricultural Research Service, United States Department of
Agriculture, State University Station, Fargo, North Carolina
JUN MITSUHASHI (13), Division of Entomology, National Institute
of Agricultural Sciences, Nishigahara, Kita-ku, Tokyo, Japan
J. D. PASCHKE (269), Department of Entomology, Purdue University,
West Lafayette, Indiana
JAMES L. VAUGHN (295), Insect Pathology Laboratory, Plant Pro-
tection Institute, Agricultural Research Service, United States De-
partment of Agriculture, Beltsville, Maryland
S. R. WEBB* (269), Department of Entomology, Purdue University,
West Lafayette, Indiana
MARY ALICE YUND (161), Department of Genetics, University of
California, Berkeley, California

* Present address: Department of Microbiology, Medical College of Virginia,
Richmond, Virginia.

Preface

In the past decade there has been a tremendous surge in the applications of invertebrate tissue culture to various aspects of biomedical as well as agricultural research. The sudden growth of interest in the use of insect and snail cells coupled with the progress in techniques and new areas of investigation motivated a United States–Japan seminar in Tokyo, December 9–13, 1974, on the current status of invertebrate tissue culture in research. The seminar proved helpful in stimulating further research in several fields as well as in promoting the wider use of invertebrate tissue culture. The Tissue Culture Association (United States) enthusiastically endorsed the seminar, and financial support was obtained from the National Science Foundation, the Japan Society for the Promotion of Science, U.S. Department of Agriculture, and WRARI (Walter Reed).

The conference concerned itself with an inquiry into diverse areas of basic science linked by a common tool: the *in vitro* culturing of tissues and cells of invertebrate animals. The program covered areas of particular importance to genetics, embryology, endocrinology, parasitology, virology, plant pathology, entomology, and neurophysiology. The United States–Japan Cooperative Science Program should be congratulated for effectively serving a very useful purpose: promoting cooperation between outstanding scientists of the two countries. A small multidisciplinary group of researchers in such diverse fields as biology, medicine, biochemistry, agriculture, botany, entomology, genetics, insect pathology, plant pathology, and zoology could hardly have been assembled by any other organization. The goals of the seminar, to stimulate interdisciplinary research and to promote international cooperation, have been achieved to the satisfaction of the participants.

After the seminar, several participants from the United States and Japan were invited to contribute extended chapters, comprising their own as well as work done by others, to a book that would cover the broad field and current status of basic research in invertebrate tissue culture. The invited authors were requested to combine authoritative reviews of research areas with a discussion of the latest developments in each field. Many of the authors pioneered the development of their areas of expertise. The contributors were requested to discuss in depth such topics as cell growth and differentiation, cloning of established cell lines, the breakthrough in molluscan tissue culture and the establishment of the first snail line, invertebrate endocrinology, ecdysone biosynthesis *in vitro*, the identification of distinct juvenile hormones from corpora allata and the production of peptide neurohormones by cultured insect brains, the use of *Drosophila* discs *in vitro* to study gene activity sites, and the applications of insect tissue culture to the study of intracellular parasites, symbionts, and arboviruses. The topics were chosen because of their particular importance in biomedical fields and because of their rapid current development. Insect pathogenic viruses in insect cell lines, extranous contaminants in invertebrate cell cultures, the uses of invertebrate cells in plant pathology, and a description of invertebrate cell lines complete this volume.

The text is quite detailed. It is aimed at an audience with at least an introductory knowledge of tissue culture. The presentation of the most recent results of original research, interpretations, and original conclusions makes this book a unique body of information and brings into sharp focus current findings and new dimensions of invertebrate tissue culture. An attempt was made to have all chapters well documented by tables, photographs, and up-to-date and complete bibliographies. By combining diverse areas currently under investigation, this book should be of interest to microbiologists, parasitologists, virologists, entomologists, geneticists, and medical researchers working in the field and to graduate students in related fields of biomedical research.

I am deeply indebted to the contributors for the effort and care with which they have prepared their chapters and to the National Science Foundation and the Japan Society for the Promotion of Science for supporting the seminar that brought together several authors of this book. Finally I would like to mention the important roles played by the staff of Academic Press in their cooperation and assistance throughout the planning and completion of this volume. It is my hope that this book will contribute both to basic scientific and medical problems and will benefit all who are interested in invertebrate tissue culture.

KARL MARAMOROSCH

PART A

Cell Growth and Differentiation

1

Agar Suspension Culture for the Cloning of Invertebrate Cells

ARTHUR H. McINTOSH

I. Introduction

Established insect cell lines, like their vertebrate counterparts, usually consist of a heterogeneous population of cell types resulting from selective influences of the particular cell culture system. As more invertebrate cell lines become established and available to investigators, the need for the successful cloning of such cells becomes essential. The main advantage of working with cloned lines is that the variability due to the presence of diverse cell types would be avoided, although it is recognized that wide variations in clonal populations themselves do occur. It also should be emphasized that cloning should be done as early as possible, even before the cultures become established cell lines. If performed later it is possible that only a few cell types might remain.

The many techniques employed in the cloning of vertebrate cells, such as the capillary method (Sanford *et al.*, 1948), the microdrop technique (Lwoff *et al.*, 1955), the feeder layer method (Puck and Marcus, 1955), and the dilution method described by Paul (1961), could be applied to invertebrate cell lines. However, despite these available techniques there have been relatively few reports regarding the successful cloning of insect cells. In this regard, Suitor *et al.* (1966) and Grace (1968) reported the successful cloning of *Aedes aegypti* and *Antheraea eucalypti* cell lines utilizing the capillary and dilution methods, respectively. The *A. aegypti* cloned line has been shown to be a moth line (Greene *et al.*, 1972). More recently McIntosh and Rechtoris (1974) have cloned the insect-established *Trichoplusia ni* cell line (TN-368) of Hink (1970) by an agar suspension culture technique. Nakajima and Miyake (1974) have reported the succesful cloning of several *Drosophila* cell lines using spent medium.

This chapter will be concerned with the agar suspension culture technique as a means of isolating clones from several established insect cell lines. In addition some of the criteria used in assessing transformation of vertebrate cells will be discussed with respect to invertebrate cell cultures.

II. Materials and Methods

1. *Cell Cultures and Media*

A description of the insect cell lines and media employed in this study is presented in Table I. The cultivation of the lepidopteran and homop-

TABLE I
Insect-Established Cell Lines

Cell line	Medium	Source
Trichoplusia ni (cabbage looper)	TC199-MK[a]	Hink (1970)
Carpocapsa pomonella 169 (codling moth)	TC199-MK	Hink and Ellis (1971)
Spodoptera frugiperda (fall army worm)	TC199-MK	Goodwin *et al.* (1970)
Agallia constricta (leafhopper)	TC199-MK	Chiu and Black (1967)
Drosophila melanogaster Line 2 (fruit fly)	TC199-MK	Schneider (1972)
Aedes albopictus (ATCC) (mosquito)	M and M[b]	Singh (1967)
Aedes albopictus (Webb)	M and M	Singh (1967)
Aedes aegypti	M and M	Peleg (1968)

[a] McIntosh *et al.* (1973).
[b] Mitsuhashi and Maramorosch (1964).

teran lines have been previously reported (McIntosh *et al.*, 1973; McIntosh and Rechtoris, 1974). The *Aedes albopictus* cell lines were obtained from Dr. Sonja Buckley, Yale Arbovirus Research Unit, Yale University School of Medicine, New Haven, Connecticut; *A. aegypti* from Dr. Victor Stollar, Rutgers Medical School, New Brunswick, New Jersey; and *Drosophila melanogaster* from D. B. Skowronski, State University of New York, Purchase, New York. The *D. melanogaster* line has been adpated to grow in TC199-MK and is now in its tenth passage in this medium.

2. Virus

Chikungunya virus was obtained from Dr. Sonja Buckley and was used to challenge *A. albopictus* (ATCC). Approximately 2×10^6 cells in a Falcon T-flask (25 cm²) were inoculated with 10^4 PFU/ml. Cells were subcultured every 3–4 days by making a 1:5 split.

3. Agar Suspension Culture Technique

This technique has been described in a previous report (McIntosh and Rechtoris, 1974). Briefly the procedure consists of preparing base layers comprised of heart infusion agar and TC199-MK to which is added the cell suspension in 0.33% Special Agar-Noble as overlay. Plates are incubated at 28°–30°C for 10–14 days and colonies counted.

4. Agglutination with Concanavalin A

Seventy-two hour cultures of the insect cell lines were employed in these studies. The medium was removed from the flask and the cell surface washed twice with Hanks' balanced salt solution (HBSS). Cells were collected in 3–4 ml of HBSS by pipetting back and forth and counted in a hemacytometer (McIntosh *et al.*, 1973). The cell density was adjusted to 1×10^6 cells/ml with HBSS and 0.5 ml cells mixed with 0.5 ml of concanavalin A (Con A) (200 μg/ml) in a petri plate (60 × 15 mm). For controls, 0.5 ml of HBSS was substituted for Con A. The reaction was observed with an inverted microscope at final magnifications of ×60 and ×100.

5. Mycoplasma

All cultures were tested for the presence of mycoplasma by a previously described method (McIntosh and Rechtoris, 1974) with the exception that the American Type Culture Collection (ATCC) medium **27563** was substituted for heart infusion agar.

III. Results

A. FORMATION OF COLONIES IN SOFT AGAR BY CELLS FROM ESTABLISHED CELL LINES

Insect cell lines belonging to the orders Lepidoptera, Homoptera, and Diptera are capable of forming colonies in soft agar medium. Of the five established cell lines tested, four demonstrated the ability to form such colonies (Table II). Figure 1 illustrates the formation of colonies from *T. ni* in soft agar medium; the more spread out colony (indicated by the arrow in Fig. 1) is probably a result of growth occurring between the overlay and base surfaces. The colony-forming efficiency was approximately 1%.

B. PROPERTIES OF CLONES

The cloning efficiency of *T. ni* cells is 25%. Of the 10 clones isolated, 9 have been passaged at least 20 times and preserved in liquid nitrogen. One clone was lost by contamination.

Some of the clones resembled the parent line morphologically in size and shape, whereas other clones appeared to be comprised of much smaller or larger cells. The doubling times of several of these clones as

TABLE II

Some Biological Characteristics of Insect-Established Cell Lines

Cell line	Formation of colonies in agar	Agglutination with Con A	Maximum cell density[a]
T. ni	+	+	10×10^6
S. frugiperda	NT[b]	−	16×10^6
C. pomonella	−	+	17×10^6
A. constricta	+	+	5×10^6
D. melanogaster	NT	−	17×10^6
A. albopictus (ATCC)	+	+	—
A. abopictus (Webb)	+	−	90×10^6
A. aegypti	NT	−	36×10^6

[a] Cells harvested from T-flasks after 1-week incubation at 28°C.
[b] NT = not tested.

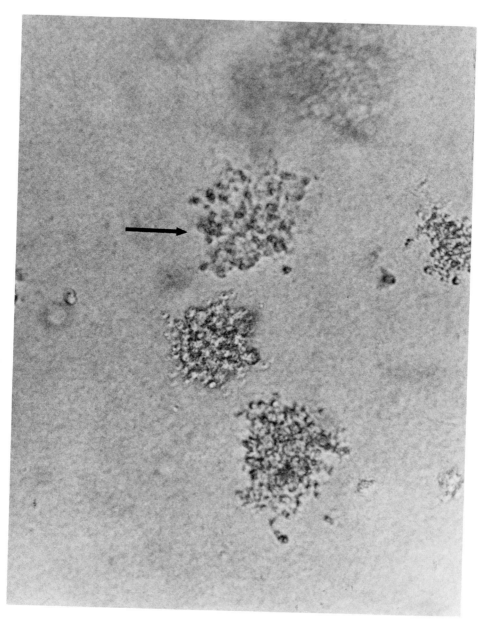

Fig. 1. The appearance of *T. ni* colonies in soft agar. ×200.

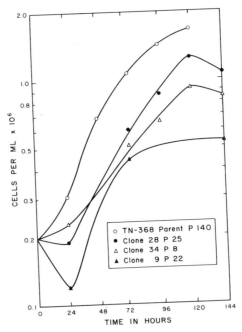

Fig. 2. Growth curves of *T. ni* (TN-368) established cell line and three clones. Reproduced with permission of the Tissue Cuture Association, Inc.

well as the parent line were estimated from the growth curves illustrated in Fig. 2. The *T. ni* parent line (TN-368) had the shortest doubling time of 30 hours, whereas the doubling times of the other clones were 63 (clone 9), 53 (clone 34), and 45 hours (clone 28). For some clones the initial cell density of 2×10^5 cells/ml was too low to establish successful growth. This difference in the cell doubling time of clones has also been observed for *T. ni* cells by Brown and Faulkner (1975), and for *Antheraea eucalypti* cells by Grace (1968).

C. AGGLUTINATION WITH CONCANAVALIN A

Of the eight established cell lines tested for their ability to agglutinate with Con A, four gave positive agglutination. The extensive clumping that occurs within a few minutes following the addition of Con A is quite noticeable visually without the aid of a microscope. An example of such clumping using *T. ni* cells is depicted in Fig. 3a. Figure 3b is the control showing the attachment of cells onto the surface of the petri dish. In contrast Fig. 3c illustrates the failure of *A. albopictus* (Webb) cells to clump when treated with Con A. Figure 3d is the control.

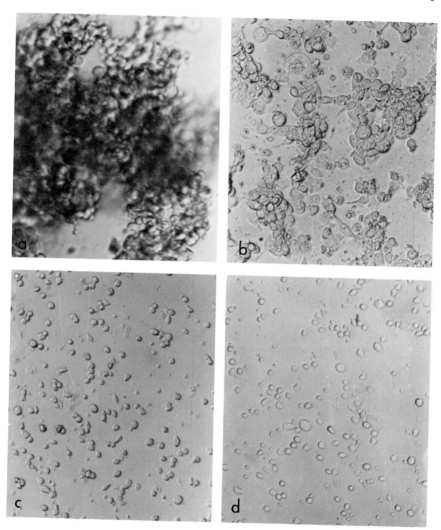

Fig. 3. (a) The clumping of *T. ni* cells with Con A. (b) Control *T. ni* cells show no clumping. (c) *A. albopictus* (Webb) cells treated with Con A. Note absence of clumping. (d) *A. albopictus* (Webb) control. ×112.

It is interesting to note that viruses have been found in cell lines of *A. albopictus* (Webb) (Cunningham *et al.*, 1975), *D. melanogaster* (Williamson and Kernaghan, 1972), and *A. aegypti* (Stollar *et al.*, 1975). All three of these cell lines showed no clumping with Con A. Furthermore, when *A. albopictus* (ATCC) was infected with Chikungunya virus, this infected line showed little or no clumping of cells when reacted with Con A.

D. CELL DENSITY

Table II illustrates the maximal cell density of seven established cell lines after 1-week incubation at 28°C. The cell density represents the total number of cells present in a 25-cm^2 T-flask when resuspended in 5 ml of medium. As is readily apparent, the highest density occurred with *A. albopictus* (Webb) cell line in the Mitsuhashi and Maramorosch medium (1964) containing 20% fetal bovine serum.

IV. Discussion

Successful use of the agar cloning suspension technique is largely dependent on the ability of cells to form colonies in soft agar. Results of the present study show that cells of several invertebrate cell lines can form such colonies. The colony-forming efficiency of cells could conceivably be increased by modification of the present method, e.g., by using a conditioned medium. Since the presence of mycoplasma in cell cultures has been shown to influence the formation of colonies in soft agar (Macpherson and Russell, 1966), all invertebrate cultures employed in this study were screened for these agents and found to be free of them. In addition karyotype analysis of these cell lines confirmed them to be of insect origin. It is, therefore, concluded that the ability of cells to form colonies in agar is not unique to vertebrate or plant cells.

The agglutination of vertebrate cells with Con A has been shown to occur in both normal and transformed cells (Sivak and Wolman, 1972), although this property has been commonly used as a criterion of transformation. It would, therefore, appear that such agglutination is a reflection of the surface property of cells and that some invertebrate cells also possess this property of agglutination with Con A. Indications are that the presence of virus in some of the invertebrate lines may interfere with the agglutination of cells with Con A. This conclusion is based on the findings that three of the insect lines which failed to agglutinate with Con A are known to be infected with viruses, and that *A. albopictus* (ATCC) when inoculated with Chikungunya virus lost its ability to agglutinate. The mechanism of this inhibition is not known but could be due to alteration in the surface membrane properties or blockage of the Con A binding sites. Before any affirmative conclusion can be drawn, further investigation is necessary.

The attainment of high cell densities due to loss of contact inhibition is another criterion commonly employed in assessing transformation of vertebrate cells. Although this phenomenon has not been described for invertebrate cell cultures, the high cell density achieved by *A. albopictus* (Webb) is suggestive of the lack of regulation of cell division.

The results of the present study indicate that a number of properties commonly employed in assessing transformation of vertebrate cells, such as formation of colonies in agar, agglutination with Con A, and attainment of high cell densities, are also possessed by invertebrate cells.

V. Summary

Invertebrate cells possess a number of biological properties which are commonly employed as criteria for the transformation of vertebrate cells. These properties include the formation of colonies in agar, agglutination with Con A, and the achievement of high cell densities. The formation of colonies in soft agar has been utilized as a method for the cloning of invertebrate cells.

Acknowledgments

The author thanks Dr. Karl Maramorosch for reviewing this manuscript, Mr. Neil Goldstein for his generous donation of Con A, and Mrs. Rebecca Shamy for her technical assistance. This research was supported in part by NSF Grant GB 41997.

References

Brown, M., and Faulkner, P. (1975). *J. Invertebr. Pathol.* **26,** 251–257.
Chiu, R.-J., and L. M. Black. (1967). *Nature (London)* **215,** 1076–1078.
Cunningham, A., Buckley, S. M., Casals, J., and Webb, S. R. (1975). *J. Gen. Virol.* **27,** 97–100.
Goodwin, R. H., Vaughn, J. L., Adams, J. R., and Kouloudes, S. J. (1970). *J. Invertebr. Pathol.* **16,** 284–288.
Grace, T. D. C. (1968). *Exp. Cell Res.* **52,** 451–458.
Greene, A. E., Charney, J., Nichols, W. W., and Coriell, L. L. (1972). *In Vitro* **7,** 313–322.
Hink, W. F. (1970). *Nature (London)* **226,** 466–467.
Hink, W. F., and Ellis, B. J. (1971). *Curr. Top. Microbiol. Immunol.* **55,** 19–28.
Lwoff, A., Dulbecco, R., Vogt, M., and Lwoff, M. (1955). *Virology* **1,** 128–139.
McIntosh, A. H., and Rechtoris, C. (1974). *In Vitro* **10,** 1–5.
McIntosh, A. H., Maramorosch, K., and Rechtoris, C. (1973). *In Vitro* **8,** 375–378.
Macpherson, I., and Russell, W. (1966). *Nature (London)* **210,** 1343–1345.
Mitsuhashi, J., and Maramorosch, K. (1964). *Contrib. Boyce Thompson Inst.* **22,** 435–460.
Nakajima, S., and Miyake, T. (1976). *In* "Invertebrate Tissue Culture: Applications in Medicine, Biology and Agriculture" (E. Kurstak and K. Maramorosch, eds.). Academic Press, New York (in press).

Paul, J. (1961). "Cell and Tissue Culture," 2nd ed., 312 pp. Williams & Wilkins, Baltimore, Maryland.

Peleg, J. (1968). *Virology* **35**, 617–619.

Puck, T. T., and Marcus, P. I. (1955). *Proc. Nat. Acad. Sci. U.S.* **41**, 432–437.

Sanford, K. K., Earle, W. R., and Likely, G. D. (1948). *J. Nat. Cancer Inst.* **9**, 229–246.

Schneider, I. (1972). *J. Embryol. Exp. Morphol.* **27**, 353–365.

Singh, K. R. P. (1967). *Curr. Sci.* **36**, 506–508.

Sivak, A., and Wolman, S. (1972). *In Vitro* **8**, 1–6.

Stollar, V., and Thomas, V. L. (1975). *Virology* **64**, 367–377.

Suitor, E. C., Chang, L. L., and Liu, H. H. (1966). *Exp. Cell Res.* **44**, 572–578.

Williamson, D. L., and Kernaghan, R. P. (1972). *Drosophila Inform. Serv.* **48**, 58–59.

2

Establishment and Characterization of Cell Lines from the Pupal Ovaries of *Papilio xuthus*

JUN MITSUHASHI

I. Introduction

Since Grace (1962) established the cell lines from the ovarian tissues of *Antheraea eucalypti* pupae, more than 40 insect cell lines have been established. Among them, four cell lines have been reported to originate from the ovaries of lepidopterous insects (Grace, 1962, 1967; Hink, 1970; Hink and Ignoffo, 1970). In addition to these cell lines, several other lepidopterous cell lines have been established from embryos and blood cells. The lepidopterous cell lines may be useful for the study of insect viruses and probably for the mass production of insect viruses as micro-

14 JUN MITSUHASHI

bial insecticides. Therefore, it is desirable to establish more cell lines from the lepidopterous insects.

This article will describe the establishment of two cell lines from the ovarian tissues of the swallow tail butterfly, *Papilio xuthus*, which is a pest of citrus trees, and some characteristics of the established cell lines.

II. Insects

The *P. xuthus* is a common butterfly in Japan. Numerous eggs and young larvae of this butterfly were easily collected from citrus trees on the campus of the National Institute of Agricultural Sciences. The collected materials were reared on citrus leaves in the laboratory at 25°C until the time of pupation. By rearing the larvae under long-day photoperiodism (16 hours of light per day), nondiapausing pupae were obtained. The pupae were usually stored in a refrigerator and incubated at 25°C for several days before use in order to initiate adult development. When it was necessary to store the pupae for long periods, the larvae were reared under short photoperiodism (12 hours of light per day) so as to obtain diapausing pupae. The diapausing pupae could be stored for more than 1 year in a refrigerator. The diapausing pupae could start adult development when they were incubated at 25°C, if they were stored in a refrigerator long enough to terminate diapause. Usually the ovary to be cultivated was obtained from a pupa at the middle stage of adult development because the ovaries of younger pupae are so small that many are needed to set up one culture. The ovaries of older pupae are not used because they contain mature eggs, the yolks of which disturb cell growth in the primary culture.

III. Method for Setting Up the Primary Culture

First, the surface of the pupae was sterilized by submersion in 70% ethyl alcohol for at least 5 minutes. The sterilized pupae were then attached to the bottom of the dissecting tray, and the tray was filled with sterilized Ringer–Tyrode's solution, the composition of which is shown in Table I. The dissection was made on the ventral side of the abdomen under Ringer–Tyrode's solution, and the ovaries were pulled out carefully to avoid injury to the digestive system. The ovaries taken out were washed in Ringer–Tyrode's solutions and freed of the adherent fat bodies and tracheae. The isolated ovaries were stored in Ringer–Tyrode's solution placed in the hole of a Maximov slide until enough ovaries for setting up one culture were obtained. Usually the ovaries from five pupae were

TABLE I

Composition[a] of Ringer–Tyrode's Solution (gm/liter)

Chemicals	Dosage
NaCl	7.0
$CaCl_2 \cdot 2H_2O$	0.2
NaH_2PO_4	0.2
KCl	0.2
$MgCl_2 \cdot 6H_2O$	0.1
$NaHCO_3$	0.12
Glucose	8.0

[a] This is the formula for the Ringer–Tyrode's salt solution modified by Carlson (1946). This was originally formulated for the culture of the neuroblast of grasshopper embryos, but is good enough for tissues of various insects. This can be stored in two stock solutions: one is 10 times concentrated $NaHCO_3$; and the other is 10 times concentrated residual components.

pooled to set up one culture. Ringer–Tyrode's solution was then replaced with the culture medium, and the ovaries were cut into small pieces of about 1 mm in length by means of fine needles under a dissecting microscope. The fragmented ovaries suspended in the culture medium were finally brought into a culture vessel with the aid of a Pasteur pipette.

IV. Culture Vessels

Small petri dishes (35 mm diameter) or small T-flasks were used as the culture vessels for the primary culture. Two ml of the culture medium were distributed to each vessel. At the beginning of subculturing, various sizes of culture vessels were used in order to avoid a drastic decrease in the cell density. For the established cell lines, square bottles (75 × 40 × 25 mm) were used with 5 ml of culture medium.

V. Culture Media

The culture medium used for establishing the cell line was MGM-431 (Table II) (Mitsuhashi, 1972, 1973). For the established cell lines, MGM-901 (Table II) was used also. Both media contained 10% of the

TABLE II
Composition of Culture Media (mg/liter)

Ingredients	MGM-431	MGM-901	Ingredients	MGM-431	MGM-901
NaH$_2$PO$_4$ · 2H$_2$O	958	320	DL-Serine	917	917
NaHCO$_3$	292	120	L-Tyrosine	42	42
KCl	1875	200	L-Tryptophan	83	83
MgSO$_4$ · 7H$_2$O	2333	—	L-Threonine	146	146
MgCl$_2$ · 2H$_2$O	1917	100	L-Valine	83	83
CaCl$_2$	833	120			
NaCl	—	7000	Malic acid	558	—
			Succinic acid	50	—
Sucrose	22080	22080	Fumaric acid	46	—
Fructose	417	—	α-Ketoglutaric acid	308	—
Glucose	3333	5000			
			Folic acid	0.16	0.16
L-α-Alanine	263	263	Biotin	0.08	0.08
β-Alanine	167	167	Thiamine hydrochloride	0.16	0.16
L-Arginine hydrochloride	583	583	Riboflavin	0.16	0.16
L-Asparagine	293	293	Calcium pantothenate	0.16	0.16
L-Aspartic acid	293	293	Pyridoxine hydrochloride	0.16	0.16
L-Cystine	21	21	p-Aminobenzoic acid	0.16	0.16
L-Glutamic acid	500	500	Niacin	0.16	0.16
L-Glutamine	500	500	L-Inositol	0.16	0.16
Glycine	542	542	Choline chloride	1.6	1.6
L-Histidine	2083	2083			
L-Isoleucine	42	42	Polyvinylpyrrolidone K90	500	500
L-Leucine	63	63	Fetuin[a]	20	20
L-Lysine hydrochloride	521	521	Fetal bovine serum[b]	100(ml)	100(ml)
L-Methionine	42	42			
L-Proline	292	292	pH	6.5	6.5
L-Phenylalanine	125	125			

[a] Spiro method, Grand Island Biological Co., New York, New York.
[b] Microbiological Associates Inc., Bethesda, Maryland.

fetal bovine serum, and the serum was essential for cell growth. Neither insect hemolymph nor vertebrate sera, such as newborn calf, calf, bovine, horse, ram, or chicken serum, could be used in place of the fetal bovine serum. In the cultivation of the established cell lines, polyvinylpyrrolidone K90, which is a macromolecular substance (MW = 360,000) and was used to enhance the viscosity of the medium, and fetuin, which has been said to be an active substance in calf serum, could be omitted without appreciable effects. Fructose also had little effect on the growth of the established cell lines. Four organic acids, namely, malic, succinic, fumaric, and α-ketoglutaric acid, were not essential for the growth of the established cell lines; however, the growth was impaired if these organic acids were completely removed.

VI. Culture History

In July 1970, 15 cultures were set up. Among them, some were terminated by microbial contamination at the early stages of cultivation. Some other cultures were maintained for considerably longer periods and then deteriorated before or after subculturing. Consequently only two cultures (Px 58 and Px 64) remained alive for more than 1 year, and from them cell lines were established. The culture histories of the two cell lines are shown in Table III.

TABLE III
Culture Histories of Cell Lines

	Px 58		Px 64	
Events	Date	Interval (days)	Date	Interval (days)
Initiation passage	7/ 4/1970	—	7/ 8/1970	—
1st	12/18/1970	178	9/ 3/1970	57
2nd	3/25/1971	97	5/ 7/1971	246
3rd	4/ 6/1971	12	9/27/1971	143
4th	4/15/1971	9	12/18/1971	82
5th	5/13/1971	28	12/25/1971	7
6th	7/10/1971	58	12/31/1971	6
7th	7/16/1971	6	1/ 5/1972	5
8th	7/20/1971	4	1/11/1972	6
9th	7/23/1971	3	1/18/1972	7
10th	7/26/1971	3	1/25/1972	7

VII. Growth in the Primary Culture

The cell migration from explants started immediately after the culture was set up. The cell migration was so active that numerous cells occupied the area around explants within 24 hours. The migrated cells were spherical or spindle shaped (Fig. 1). They suspended themselves in the medium at first and later became attached to glass surfaces. These cells showed a tendency to adhere to scratches made on glass surfaces and often lined up along the scratch (Fig. 2). Most of the explanted fragments of ovaries attached themselves to the glass soon after the culture was set up, whereas some took longer to attach. When the number of migrated cells increased, they formed confluent layers of cells (Fig. 3). The layer of cells was mostly single, but double or multiple layers of cells as well as aggregates of cells were frequently observed nearby explants (Fig. 4). During the first few months of the primary culture, mitoses were scarcely observed. However, it was likely that some cells multiplied by mitoses with long intervals, even at the early stages of the primary culture. The cultures were maintained for several months without noticeable changes by renewing the medium once a week.

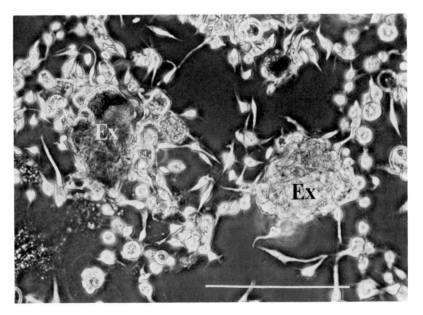

Fig. 1. Cell migration from the explanted fragments of ovaries. Ex, explants. Scale, 200 μm.

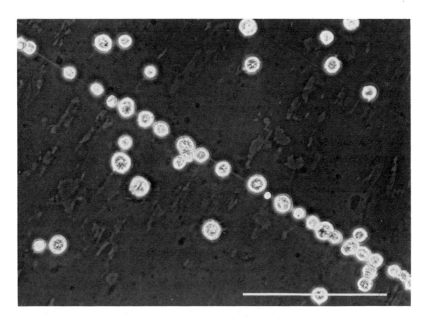

Fig. 2. Cells lined up on a scratch made on the inner surface of a culture vessel. Scale, 200 μm.

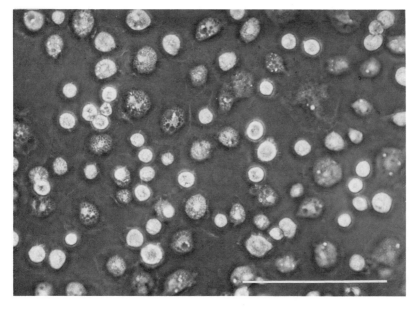

Fig. 3. A confluent monolayer of cells formed in the primary culture of ovarian tissues. Scale 200 μm.

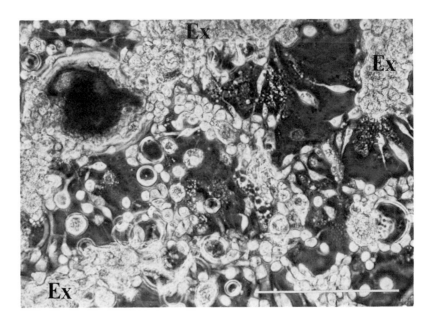

Fig. 4. Multiple layers of cells and aggregates of cells formed in the vicinity of explants. Ex, explants. Scale, 200 μm.

VIII. Subculturing

Several months after the culture was set up, it was noticed that some aggregations of cells were growing to form a center of multiplication. These cells probably derived from the cells that had continued multiplication slowly from the beginning of the culture. The cells forming the multiplication center were mostly spherical and highly reflective, when viewed with bright phase contrast. These cells were usually lightly connected to each other (Fig. 5). Mitoses were frequently observed in these cells. The first subculture was made by flushing the multiplication center with culture medium, and transferring the resultant cell suspension into a new container. The subcultured cells were floating in the medium, but they soon attached themselves to glass. In the first few passages the cells and the original explants were subcultured together. In these cases, the cell migration occurred from the subcultured explants just as in the primary culture. The cell migration also occurred from the cell aggregates, which were transferred to new vessels. It was important to avoid a drastic decrease in cell density when subculturing was made. Otherwise, the subcultured cells soon deteriorated. For this reason, especially in the first

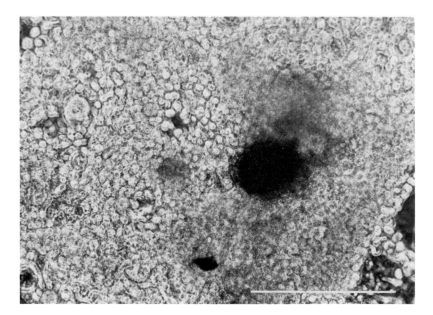

Fig. 5. Aggregation of cells in the multiplication center. These cells were actively multiplying. Scale, 200 μm.

few passages, the cell was transferred gradually from a smaller vessel to a slightly larger vessel by using various sizes of vessels.

The growth of the cell in the subculture was very slow and variable in the first few passages. In Px 58 cells, after the sixth passage, which was about 1 year after the initiation of the culture, cell growth became rapid and constant (Fig. 6). In Px 64 cells, the first subculture could be made earlier than in Px 58 cells, but the cell growth became very slow during the next few passages. The frequent subculture became possible after the fifth passage, which was about 1.5 years after the initiation of the culture. In the early stages of subcultures, the constitution of the cell population was similar to that in the primary culture, but with the advance of the passages, the original explants disappeared, and the cell population came to consist mainly of suspended cells. Thus, the cell population became the form of suspension culture, when the cell lines were established. The Px 58 cell line is more heterogeneous and somewhat larger than the Px 64 cell line (Fig. 7).

After the establishment of the cell lines, subcultures were made by simply shaking the flask and distributing the resulting cell suspension into new flasks with appropriate amounts of fresh medium. The subculturing was repeated with 2- or 3-day intervals.

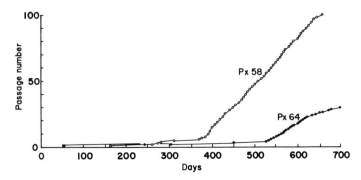

Fig. 6. Records of subculturing, showing changes in multiplication rate of cells.

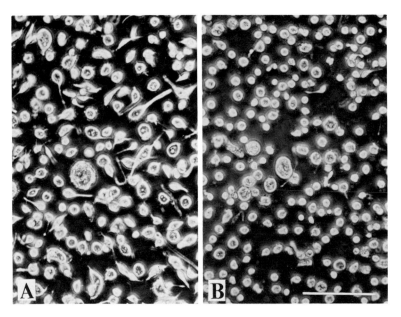

Fig. 7. The mixed population of the established cell lines (A) Px 58 cells; (B) Px 64 cells. (A) and (B) are of the same magnification. Scale, 100 μm.

IX. Storage of the Established Cell Lines

The Px 58 and Px 64 cells could be stored in a refrigerator at 5°C for up to 9 months. When the stored cells were returned to 25°C, the cells took time to recover their original multiplication rate. For instance, the cell kept at 5°C for 6 months required about 1 month to recover.

For long-period preservation of the cell, it was better to employ cyclical treatment, in which the cell was transferred from 5° to 25°C, subcultured at least once, and then returned to 5°C. If this treatment was repeated every 3 months, the cell was kept safely at 5°C for an infinite period.

The storage of the cell at ultra low temperature in the medium containing 10% glycerol was also possible. The frozen cell was kept at —80°C. The frozen cell could recover after several months. The thawing of the frozen cell was done as quickly as possible. Usually the ampule which contains frozen cells was placed directly in hot water until the cell thawed. The glycerol-containing medium was replaced with glycerol-free medium before the cultivation. Otherwise, glycerol deteriorated the cell growth. The cell preserved at —80°C was kept at —20°C for 1 month and then returned to —80°C for further preservation. When thawed and cultured, the cell so treated could recover, although some cells died.

X. Characteristics of the Established Cell Lines

A. MORPHOLOGY

The cells were usually suspended in the medium. The small number of cells, which happened to attach themselves to the glass, could be freed from the glass easily by flushing medium upon them. In the Px 58 cell population, spherical cells of about 20 μm in diameter were predominant (Fig. 8A). This type of the cell had a small nucleus and fine granules. The surface of the cell looked smooth, but when viewed with high magnification, it became evident that these cells had very fine cytoplasmic projections. The fusiform cells were also common (Fig. 8B). They were mostly about 40 μm in length. These cells also contained small nuclei and fine granules. In addition to these two types of cells, there were giant cells with diameters larger than 50 μm (Fig. 8C), tadpole-shaped cells with a long tail (Fig. 8D), and amorphous cells (Fig. 9A) in the mixed cell population. The tadpole-shaped cell often had a long cytoplasmic extension, and sometimes the distal end of the extension swelled. In the extreme case, the cell took a dumbbell shape (Fig. 9B). Slender cytoplasmic extensions sometimes separated from the cell bodies and suspended themselves in the medium. The cell shapes were not rigid cell features, since the fusiform cell or tadpole-shaped cell often rounded, when the culture conditions became unfavorable for the cell. There was a tendency for tadpole-shaped cells to increase in number when the mixed cell population was cultivated in the MGM-901 medium (Table II).

The Px 64 cell also suspended itself in the medium. The most predomi-

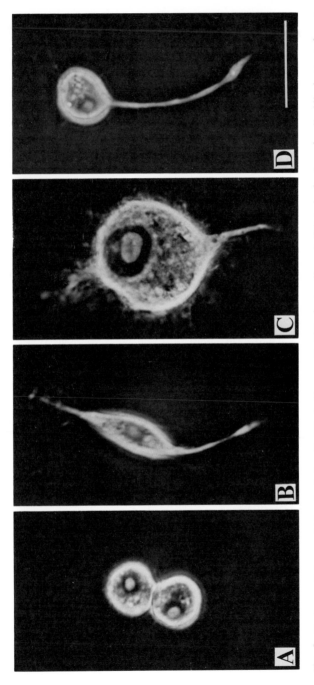

Fig. 8. Various cell types. (A) spherical cell, (B) fusiform cell, (C) giant cell, and (D) tadpole-shaped cell. All photographs are of the same magnification. Scale, 30 μm.

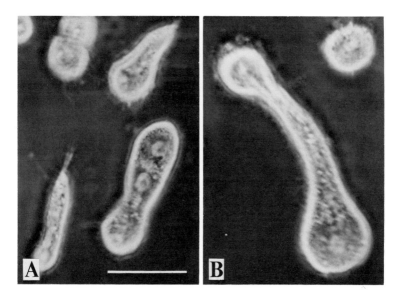

Fig. 9. Amorphous cells (A) and a dumbbell-shaped cell (B). (A) and (B) are of the same magnification. Scale, 30 μm.

nant cells were the spherical cell of about 20 μm in diameter. Other types of cells also could be seen, but not many compared with Px 58 cells. The mixed population of Px 64 cells were, therefore, rather homogeneous in shape and size.

The Px 58 cell was examined with an electron microscope. The cell was prepared in 5% glutaraldehyde in 0.2 M sodium cacodylate buffer and then in 1% osmium tetroxide. The cell pellet was then embedded in Epon 812, and ultrathin sections were cut with a Porter–Blum ultramicrotome. The cut sections were stained with uranyl acetate and lead acetate, and observed with a Hitachi HU-12 electron microscope operating at 75 kV. The common features of the cells were many vacuolelike structures and poor contrast of nuclei (Fig. 10). Neither viruslike particles nor mycoplasmalike bodies were found in or outside the cell. Therefore, the cell may be free of microbial contamination.

B. STAINABILITY

Both the Px 58 and Px 64 cell lines showed low affinity to various dyes when the cells were fixed and stained. The cell was poorly stained with Delafield's hematoxylin after Bouin fixation, with Giemsa after cold methanol or Carnoy fixation, with Feulgen reagent after Carnoy fixation,

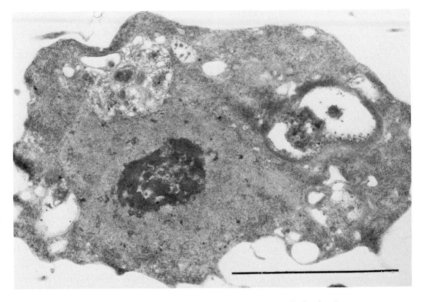

Fig. 10. An electron micrograph of a Px 58 cell. Scale, 5 μm.

with acid fuchsin after Carnoy fixation, and with crystal violet after acetic acid fixation.

C. KARYOTYPE

For the Px 58 cell, its chromosomes were examined. The cell was treated with 1×10^{-6} M colchicine overnight to accumulate metaphase cells. The treated cell was then immersed in the hypotonic solution (one part Ringer–Tyrode's solution to two parts distilled water) for 15 minutes, fixed with Carnoy for 1 hour at 5°C, spread on glass slides, and finally stained with Giemsa. This routine method for preparation of chromosome specimens in vertebrate cells did not work well on the Px 58 cells. All modifications of this method failed to detect chromosomes. The difficulty of obtaining good chromosome specimens in this cell line is probably due to the facts that this cell line has low affinity to dyes and that the cell does not respond to colchicine treatment efficiently. After many trials to stain the cell DNA, acridine orange turned out to be effective. Then the cell was fixed with Carnoy, stained with acridine orange, and observed with a fluorescent microscope. There was great variation in chromosome number of the Px 58 cell. Many cells were diploid ($2n = 50$), but also many polyploid and heteroploid cells were found. In the extreme case,

560 chromosomes were counted in one metaphase plate. Since the Px 58 cell did not respond well to colchicine treatment, and only limited numbers of metaphase plates were obtained, no statistical analysis of the karyotype has been made.

D. GROWTH

The growth curves of the Px 58 cell are shown in Fig. 11. From this figure the population doubling time at 25°C was estimated to be about 60 hours in MGM-431 medium. If the medium was not renewed, the multiplication rate of the cell decreased usually 1 week after the subculture. By renewing the medium, however, it was possible to keep the cell multiplying for 1 week. The saturation density of the cell was estimated to be about 8×10^6 cells per ml. The growth rate of Px 64 cells is somewhat lower than that of Px 58 cells.

The Px 58 cell was usually cultured in the MGM-431 medium, but they could adapt to the MGM-901 medium (Table II), which had simpler composition than MGM-431 medium. The features of this medium were that the composition of the inorganic salts was the same as that in Ringer–Tyrode's salt solution (Carlson, 1946) and that this medium lacked four organic acids, which are contained in MGM-431 medium. About 3 months were required for the adaptation of the cell to the new medium. In MGM-901 medium, the cell grew slowly. The population doubling time in this medium at 25°C was about 6 days (Fig. 11).

E. ISOLATION OF CELL STRAINS

The cells that grew adhering to a glass surface sometimes appeared in both Px 58 and Px 64 cell cultures. From such cultures, the glass-

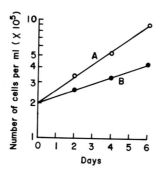

Fig. 11. Growth curves of Px 58 cells at 25°C in MGM-431 medium (A) or MGM-901 medium (B).

attached cells could be isolated. The selection was made at every passage
by simply discarding the suspended cells and subculturing only the cell
attached to the glass surface. The repeated selection resulted in isolation
of a cell strain, which consisted of the cells attached to the glass, but
in such culture some freely floating cells always came out. For keeping
such glass-attached cell populations, it was necessary to repeat selection
whenever the subculture was made. The glass-attached cells usually made
confluent layers of cells (Fig. 12). The cells were mostly fibroblastic in
shape. The layers sometimes piled up on the formed cell sheet. The contact inhibition of growth was not observed in this cell population. The
growth rate of the cells was about the same as the original mixed population. The attachment to glass surface was not firm, and the attached
cells could be separated from the glass by flushing medium onto the cells.
Therefore, the subculture of this cell strain was done without
trypsinization.

F. CHANGE IN pH OF THE MEDIUM

Phenol red is generally incorporated in vertebrate tissue culture media
as a pH indicator. When Px 58 cells or Px 64 cells were cultured in the

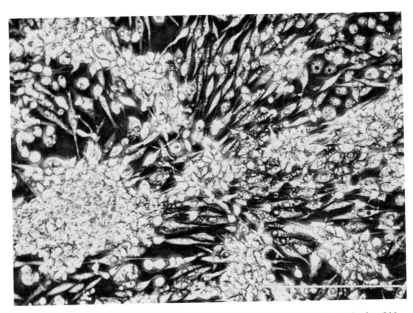

Fig. 12. A strain of Px 58 cells which attached themselves to glass. Scale, 200 μm.

MGM-431 medium containing 0.002% phenol red, the color of the medium did not change markedly even after 8-day cultivation. Then change in pH of the medium was examined. Three cultures in TD-40 flasks, which contained 15 ml of the Px 58 cell suspension (5×10^5 cells per ml), were set up, and an aliquot was sampled from each culture on the fourth and eighth days. On the sampled cell suspension, measurement of pH and the counting of the cell number were carried out (Fig. 13). The cell count was made by means of Thoma's hemocyte meter. It turned out that the pH became only slightly acidic during the 8-day cultivation.

G. REQUIREMENT FOR FETAL BOVINE SERUM

Fetal bovine serum is essential for the growth of Px 58 and Px 64 cells. The MGM-431 medium usually contains 10% fetal bovine serum. Experiments were conducted to know the minimum concentration of fetal bovine serum, which permits the growth of Px 58 cells. Similarly to the previous experiments, three cultures were set up in square bottles with 6 ml of the Px 58 cell suspension (1.5×10^5 cells per ml). The growth of the cell in MGM-431 media containing various amounts of fetal bovine serum was measured. Cell count was made on the fourth and eighth days of the cultivation (Fig. 14). It became evident that the cell could not grow in concentrations below 1% fetal bovine serum. The 2% fetal bovine serum permitted the growth of the cell, but the growth rate was impaired. With MGM-431 medium containing 2% fetal bovine serum, continuous subculturing was difficult. With the medium containing 3% fetal bovine serum,

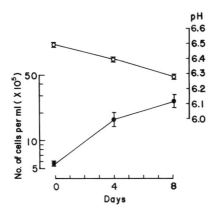

Fig. 13. Change in pH (○) with the increase of cell number (●). Eight-day cultivation of Px 58 cells in MGM-431 medium.

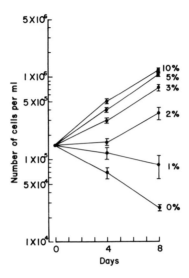

Fig. 14. Growth of Px 58 cells in the media containing various concentrations of fetal bovine serum.

the cell could be subcultured continuously, although the growth rate was considerably lower than in the medium containing 10% serum. In the range of 3–10% of the serum, the cell showed dose-dependent growth.

H. UTILIZATION AND REQUIREMENT OF AMINO ACIDS BY THE CELL LINE

The quantitative changes in free amino acids in the medium were examined. Replicate cultures were set up with the culture tubes containing 5 ml of Px 58 cell suspension (3×10^5 cells per ml). On the fourth and eighth days, three cultures were terminated, and the cell count and the analysis of free amino acids were carried out. For the preparation of the samples for free amino acid analysis, the method described by Hink et al. (1973) was employed. The cell was spun down, and the supernatant was mixed with an equal volume of 70% ethyl alcohol and kept for 3 hours in a refrigerator. Then the sample was centrifuged at 3500 rpm for 30 minutes. The resulting supernatant, which was slightly turbid, was forced to pass through a membrane filter of 0.45 μm in pore size. The sample was finally diluted 25 times with lithium citrate buffer of pH 2.72. The analysis was made by the Na–Li two-column method with Japan Electron Optics Laboratory 6AH automatic amino acid analyzer.

During the 8-day cultivation the cell number increased as follows:

TABLE IV

The Changes in the Free Amino Acids in the Medium during 8-Day Culture of Px 58 Cell Line

	% of amino acid remaining after	
Amino acids	4 days (mean ± SE)	8 days (mean ± SE)
Lysine	115 ± 1.1	98 ± 2.6
Histidine	91 ± 1.5	91 ± 1.7
Tryptophan	110 ± 1.8	97 ± 1.7
Arginine	83 ± 3.0	92 ± 3.4
Serine	98 ± 0.4	98 ± 1.0
Glutamic acid	110 ± 1.8	99 ± 3.8
Glycine	99 ± 2.2	97 ± 2.1
β-Alanine	113 ± 8.5	96 ± 8.3

zeroth day, 3×10^5 cells per ml; fourth day, 7.4×10^5 cells per ml; and eighth day, 17.2×10^5 cells per ml.

Examples of the analytical result are shown in Table IV and Fig. 15. Most amino acids decreased in quantity (Fig. 15), while some of them

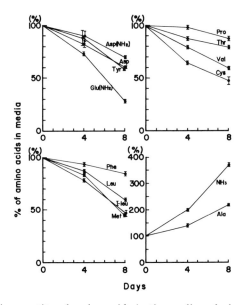

Fig. 15. Changes in quantity of amino acids in the medium during 8-day cultivation of Px 58 cells.

did not show significant changes during the 8-day cultivation (Table IV). α-Alanine was the only amino acid which increased in amount during the cultivation. Ammonia increased in amount also. This was probably due to the degradation of amino acids that were utilized by the cell.

The amino acids, which were not apparently utilized, were thought to be nonessential for the growth of the cell. If these amino acids were removed altogether from the MGM-431 medium, however, the cell immediately ceased its growth. This means that consumption and necessity of amino acids by the cell are not directly related. Then, amino acid requirement by the Px 58 cell was examined by deleting amino acids one by one. The MGM-431 medium contained 10% fetal bovine serum, which contained free amino acids. It is, therefore, impossible to delete a free amino acid completely from the medium. A quantitative analysis of free amino acids in fetal bovine serum, however, showed that the free amino acids in MGM-431 medium, which derived from fetal bovine serum, were almost negligible (Table V). For deletion experiments, replicate tissue cultures were set up with culture tubes containing 2 ml of the cell suspension (1.5×10^5 cells per ml) in the medium lacking one of amino acids. A cell count was made on the eighth day of the cultivation, and the rates of multiplication for the eighth-day culture were compared with that in the complete MGM-431 medium. The result is summarized in Fig. 16.

The growth rate of the cell in the media lacking asparagine, α-alanine, glutamic acid, glycine, and phenylalanine was not significantly different

TABLE V

Composition of Amino Acids and Free Amino Acids (mg/liter MGM-431 Medium)

Amino acids	A.M.[a]	F.B.S.[b]	Amino acids	A.M.[a]	F.B.S.[b]
α-Alanine	263	13.7	Leucine	63	5.7
β-Alanine	167	Trace	Lysine	521	3.3
Arginine	583	Nil	Methionine	42	0.2
Aspartic acid	293	0.9	Phenylalanine	125	3.1
Asparagine	293	Nil	Proline	292	2.9
Cystine	21	Nil	Serine	917	4.7
Glutamic acid	500	14.7	Threonine	146	3.7
Glutamine	500	5.5	Tryptophan	83	Nil
Glycine	542	8.4	Tyrosine	42	2.0
Histidine	2083	0.8	Valine	83	5.5
Isoleucine	42	1.6			

[a] Amino acid mixture, which constitutes MGM-431 medium.

[b] Free amino acids derived from fetal bovine serum added to MGM-431 medium (Microbiological Associates Inc., Bethesda, Maryland).

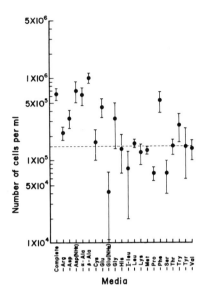

Fig. 16. Number of cells after 8-day cultivation in the media lacking one amino acid.

from that in the complete medium. The cell grew much better in the medium lacking β-alanine than the control. Deletion of other amino acids resulted in the reduction of the growth rate of the cell compared with the control. Among them, the deletion of arginine, aspartic acid, cystine, and tryptophan still permitted limited growth of the cell, while the deletion of glutamine, histidine, isoleucine, leucine, lysine, methionine, proline, serine, threonine, tyrosine, and valine either stopped the cell growth completely or killed the cell.

The long-term cultivation of the Px 58 cell was performed with the media lacking one of amino acids. In the media lacking asparagine, α-alanine, β-alanine, glutamic acid, glycine, and phenylalanine, the cell could be subcultured for an infinite period. These amino acids could be removed altogether without a deleterious effect on cell growth. In the media lacking other amino acids, the cell deteriorated sooner or later. The cell, however, could adapt to the medium lacking serine. In this medium the cell stayed alive for a long time and gradually multiplied; finally, the growth rate of the cell was comparable to that of the control. In the medium lacking glutamine, the cell attached itself to the glass and stayed alive for several months, but apparently did not multiply.

The utilization and requirement of amino acids by the Px 58 cell are summarized as follows. All basic amino acids (lysine, histidine, tryptophan, and arginine) and serine did not change markedly in quantity, but the cell growth was much impaired if one of these amino acids was deleted. Therefore, it is assumed that the presence of these amino acids is important, or these amino acids are essential but seldom utilized.

Glutamic acid, glycine, and β-alanine did not change in quantity, and they are apparently not necessary for the growth of the Px 58 cell. β-Alanine may even be inhibitory to the growth of the cell, because the cell growth was accelerated when it was deleted.

Asparagine and phenylalanine decreased in quantity, but they actually were not necessary for cell growth. The deletion of asparagine even promoted the cell growth to some extent.

Proline, threonine, valine, cystine, aspartic acid, tyrosine, glutamine, leucine, isoleucine, and methionine decreased in quantity, and they were all necessary for cell growth.

α-Alanine increased in quantity and was not necessary for cell growth.

The content of ammonia increased markedly with the cultivation, probably because of the degradation of amino acids.

XI. Discussion

One of the characteristics of the growth pattern in the culture of lepidopteran ovarian tissues is that a long time is required for initiation of active cell multiplication. The cell line from *A. eucalypti* took 10 months before it began explosive multiplication (Grace, 1962). Similarly, the *Bombyx mori* cell line (Grace, 1967) took 9 months, and the *Heliothis zea* (Hink and Ignoffo, 1970) and *Trichoplusia ni* cell lines (Hink, 1970) took 7 months before initiation of vigorous multiplication. The cell of *P. xuthus* described here required about 1 year before initiation of constant multiplication. This does not seem due to the developmental or physiological status of the explant used for the primary culture, because the above-cited cases include the cultivation of ovaries from adult, diapausing pupae, nondiapausing pupae, and larvae, and all took a long time before initiation of active multiplication. In the cultivation of ovarian tissues of dipterous insects, however, the cell usually grew more quickly. In establishing the *Culex quinquefasciatus* cell line (Hsu *et al.*, 1970), for instance, the first subculturing was done 3 weeks after the culture was set up. The long lag period for the cell growth in the primary culture, therefore, may be limited to the lepidopteran ovarian tissue culture. This long lag period, however, may be shortened if the culture medium is improved.

The Px 58 cell line has had more than 300 passages, and the Px 64 cell line already more than 150 passages. In mammalian cell cultures, especially human cell cultures, it is recognized that the cell often is fated to die at around the fiftieth passage. Similar phenomena have also been observed in insect cell cultures (J. Mitsuhashi, unpublished observation). In the culture of the cell of *P. xuthus*, however, there was no such trouble, and the cell continued to multiply at a constant rate.

The exact origin of the cell lines reported here could not be determined. Apparently they did not originate from oocytes, because oocytes came out from the ovariole and degenerated in the primary culture. In general, the cell obtained in the cultivation of lepidopterous ovarian tissues has been considered to originate from the intermediate layer cell (Stanley and Vaughn, 1968). A part of the Px 58 and Px 64 cells were probably derived from the intermediate layer cell also because the shape and behavior of the majority of the cells were similar to those obtained from the ovarian tissues of other lepidopterous insects. Some cells, however, looked like hemocytes too. A possibility that the mixed cell population contains the cell originating from other parts than the intermediate layer, hemocytes for instance, cannot be denied.

The Px 58 cells, as well as the Px 64 cell, could be stored at 5°C for considerably long periods. This may be one of the features of insect cell lines. Rahman *et al.* (1966) stored Grace's *A. eucalypti* cell line at 4°C for 9 months, and Mitsuhashi (1968) stored a *Chilo suppressalis* cell line at 5°C for 4 months.

The mixed population of both cell lines consisted of cells with various shapes. The detailed classification of the cell by its shape, however, seems meaningless, because a cell can change its shape by environmental influence. The transconfiguration of fusiform cells or tadpole-shaped cells into spherical cells was very common when the culture conditions changed unfavorably for the cell.

The statistical analysis of the karyotype of the Px 58 or Px 64 cell lines is difficult, because metaphase plates enough for the analysis can hardly be obtained due to the effectiveness of colchicine. If 1×10^{-6} M colchicine is added to the medium, the cells all rounded within a few hours. This means that the medium changed unfavorably for the cell. If the colchicine is used at a more diluted concentration, which does not induce the morphological change of the cell, little metaphase is obtained. As far as examined, there was a great variation in the karyotype of the cell. The cell was mostly diploid, but there were many polyploid and heteroploid cells. A giant cell was found to contain 560 chromosomes. In the cell line derived from lepidopterous ovarian tissues, extreme polyploidy is not rare. Thomson and Grace (1963) reported a $128n$ cell in the mixed population of the *A. eucalypti* cell line.

The cell lines of *P. xuthus* require fetal bovine serum for their growth. One serum protein, fetuin, cannot replace fetal bovine serum (J. Mitsuhashi, unpublished). There have been many reports that insect cell culture requires fetal bovine serum, or insect cell culture is much promoted by the addition of fetal bovine serum. The isolation of the cell growth promoting substance in fetal bovine serum and clarification of the nature of the substance are important problems for the development of insect cell cultures.

The utilization of amino acids has been studied up to date on *A. eucalypti* cells (Grace and Brzostowski, 1966), *Periplaneta americana* cells (Landureau and Jollès, 1969), and *Carpocapsa pomonella* cells (Hink, 1972; Hink et al., 1973). The requirement of amino acids by the cell by means of deletion experiments, however, has never been done. The amino acids, which were utilized by all the above cell lines including the Px 58 cell line, were aspartic acid, proline, methionine, tyrosine, and cystine. The S-19 medium, which was used for the cultivation of *P. americana* cells, contained cysteine instead of cystine. This cysteine was probably utilized as cystine by the cell, because cysteine is usually quickly oxidized to cystine in solution. The above five amino acids were all essential for the growth of Px 58 cells. Glutamine was the most utilized amino acid by the Px 58 cells. The same is true for other cell lines, except *C. pomonella* cells. Glutamic acid has been reported to be utilized considerably by the cell (Grace and Brzostowski, 1966; Landureau and Jollès, 1969; Hink, 1972). The Px 58 cell, however, neither utilized glutamic acid nor required it for cell growth. Valine and leucine were essential for the growth of the Px 58 cell, and they were utilized by all other cell lines, except *C. pomonella* cells. The cell lines other than *C. pomonella* utilized phenylalanine, but this amino acid seemed to be nonessential for the growth of Px 58 cells. Asparagine and isoleucine were utilized by *A. eucalypti* cells and Px 58 cells. Isoleucine was essential, but asparagine seemed to be nonessential for the growth of Px 58 cells. The basic amino acids were utilized only by *P. americana* cells; they were lysine, histidine, and arginine, but tryptophan was not utilized by any cell line. These four amino acids, however, were found to be essential for the growth of Px 58 cells. Threonine, serine, glycine, and β-alanine were utilized by only one of these cell lines; Threonine and serine were essential, but the latter two seemed to be nonessential for the growth of Px 58 cells. The amount of α-alanine in media increased with cultivation in all cases. It is not certain whether the increase in α-alanine is due to degradation of other amino acids or due to synthesis by the cell from pyruvic acid and the amino radical. This amino acid seemed to be nonessential for the growth of Px 58 cells.

It is impossible to delete a free amino acid completely from the medium containing fetal bovine serum. From the present deletion experiment, therefore, only essential amino acids could be determined. The amino acids, of which incomplete deletion did not cause deterioration of the cell growth, are either nonessential or essential amino acids, of which a small quantity is needed. Such amino acids for Px 58 cells were α-alanine, β-alanine, asparagine, glutamic acid, glycine, and phenylalanine, while other amino acids were essential for the growth of Px 58 cells. In *C. pomonella* cells, it has been reported that the growth rate of the cell did not change for at least two passages, if the concentrations of aspartic acid, glutamic acid, proline, cystine, and methionine were reduced to a great extent.

From the deletion experiment with Px 58 cells, it was recognized that deletion of β-alanine or asparagine promoted the cell growth. These amino acids were not strongly inhibitory to the cell growth, but should be unfavorable to cell growth if present.

The pattern in the utilization of amino acids may be different in different cell lines. *Antheraea eucalypti* cells and Px 58 cells show similar patterns. *Periplaneta americana* cells utilized the most amino acids, while *C. pomonella* cells consumed few amino acids.

From the experiment with Px 58 cells, it became evident that the utilization of amino acids does not immediately reflect the requirement of amino acids by the cell. The deletion experiments, therefore, are necessary for the determination of the essential amino acids.

The Px 58 and Px 64 cells are similar in many respects to the other cell lines derived from lepidopteran ovaries, but they also have many different characteristics. Before supplying these cell lines for various experiments, it is desirable to clarify more characteristics of these cell lines, such as sensitivity to viruses, activities of enzymes, metabolism, and nutritional requirements.

References

Carlson, J. G. (1946). *Biol. Bull.* **90**, 109–121.
Grace, T. D. C. (1962). *Nature (London)* **195**, 788–789.
Grace, T. D. C. (1967). *Nature (London)* **216**, 613.
Grace, T. D. C., and Brzostowski, H. (1966). *J. Insect Physiol.* **12**, 625–633.
Hink, W. F. (1970). *Nature (London)* **226**, 466–467.
Hink, W. F. (1972). *In* "Insect and Mite Nutrition" (J. G. Rodriquez, ed.), pp. 365–374. North-Holland Publ., Amsterdam.
Hink, W. F., and Ignoffo, C. M. (1970). *Exp. Cell Res.* **60**, 307–309.
Hink, W. F., Richardson, B. L., Schenk, D. K., and Ellis, B. J. (1973). *Proc. Int. Colloq. Invertebr. Tissue Cult., 3rd, 1971* pp. 195–208.

Hsu, S. H., Mao, W. H., and Cross, H. J. (1970). *J. Med. Entomol.* **7**, 703–707.

Landureau, J. C., and Jollès, P. (1969). *Exp. Cell. Res.* **54**, 391–398.

Mitsuhashi, J. (1968). *Appl. Entomol. Zool.* **3**, 1–4.

Mitsuhashi, J. (1972). *Appl. Entomol. Zool.* **7**, 39–41.

Mitsuhashi, J. (1973). *Appl. Entomol. Zool.* **8**, 64–72.

Rahman. S. B., Perlman, D., and Pradt, S. S. (1966). *Proc. Soc. Exp. Biol. Med.* **123**, 711.

Stanley, M. S. M., and Vaughn, J. L. (1968). *Ann. Entomol. Soc. Amer.* **61**, 1067–1072.

Thomson, J. A., and Grace, T. D. C. (1963). *Aust. J. Biol. Sci.* **16**, 869–876.

3

Phenotypic Variations of Cell Lines from Cockroach Embryos

TIMOTHY J. KURTTI

I. Introduction

The successful application of an insect cell line to a biological research program depends on the phenotypic characteristics of the line which is used for experimentation. The required characteristics will depend on the study and may vary from microbial susceptibility to hormonal sensitivity. The pattern of growth of the cells, e.g., whether or not they form an attached monolayer, may be important in other studies. It is an advantage to be able to predict which cell type when placed *in vitro* will give rise to a cell line of a given phenotype. However, such procedures are poorly defined in the culture of cells from insect embryos. There are several related problems in the culture of specific cell types from embryos. The small size of embryos encumbers the acquisition of sufficient numbers of these cells for culture. The cell dispersion and separation procedures are frequently ill-defined and incomplete, making it difficult to sort out the cell types. Consequently, all the cells of an embryo are used to inoculate a primary culture. This results in primary cultures that are heterogeneous in cellular composition and in the development of cell lines of

unknown tissue origin. These shortcomings point to an area of needed research in invertebrate tissue culture, namely, the improvement of tissue
dissociation procedures and the sorting out or selecting of specific cell
types from a heterogeneous mixture of cells.

We have developed several cell lines, each with a different phenotype,
from embryos of the German cockroach *Blattella germanica*. The methods included the use of several embryonic stages as primary inocula, and
refinements in cell dispersal and selection procedures. This article reviews
the approach we used to develop these cell lines.

II. Dissociation of Embryos

The procedures for cell dispersal have applications in studies of the
cytodifferentiation (St. Amand and Tipton, 1954; Kurtti and Brooks,
1970; Poodry and Schneiderman, 1971; Seecof *et al.*, 1971), cellular
affinities (Lesseps, 1965; Walters, 1969, 1974), and membrane structures
(Satir and Gilula, 1973) of insect cells. They are also useful in the development of cell lines to disperse cells for primary cultures and subcultures.

A. PROCEDURES AND THEIR EVALUATION

Solutions and procedures have been developed (Kurtti and Brooks, in
press) which give good yields and dispersion of single cells from *B. germanica* embryos. The formulas for the solutions are given in Table I.
The saline solution BG-SSA is used to dissolve the dissociation agent

TABLE I

Saline Solutions for *Blattella germanica* **Cells**

	Concentration (gm/liter)	
Components	BG-SSA	BG-SSB
NaCl	10.30	8.40
KCl	1.46	1.25
NaHCO$_3$	0.36	0.36
NaH$_2$PO$_4$ · H$_2$O	0.21	0.21
NaH$_2$PO$_4$	1.34	1.34
Glucose	3.00	3.00
Fetal bovine serum[a]	—	200 ml

[a] Fetal bovine serum is heat inactivated (56°C for
30 minutes). The saline solutions have a pH of 7.4

(usually trypsin 1-250). Solution BG-SSB contains fetal bovine serum and is used to protect dispersed cells. These solutions have the same osmotic pressure as the hemolymph of *B. germanica*, 410 mOsm per kg of H_2O (Landureau, 1966), and the culture medium. The standard procedures used to dissociate embryos are summarized in Table II; the procedure for the dissociation of fat body tissues of nymphs is given for comparison.

The above methods were adopted as the result of studies in which the effectiveness of various dissociation procedures were evaluated. Conditions were sought which gave the best degree of dissociation with the highest yield of single cells but the least amount of cell damage. Single cells and clumps of cells were generally recovered; each cell or clump was evaluated as a particle. The proportion of particles which were single cells indicated the degree of dissociation. The particle yield was quantitated because increased degrees of dissociation frequently resulted in reduced cell yields. Cell condition was evaluated by phase-contrast microscopy; the indicators of trauma were membrane blebs, hypertrophy, Brownian movement in the cytoplasm, and the loss of refractility. Damaged cells were not included in the counts. Particle yields were expressed as the number recovered per embryo. The collection of this kind of data formed the basis for the suggested procedures and the predicted particle yields which are given in Table II.

Trypsinization was performed as follows. The embryos were minced in a BG-SSA solution of trypsin. The resulting suspension was transferred to a 10-ml erlenmeyer flask and incubated with stirring at room temperature. The total time of trypsinization was usually 20 minutes. During the trypsinization the single cells were regularly separated from the larger fragments and transferred to a solution of BG-SSB; this was most essential when the trypsin concentration was 0.25% or more. At 5- or 10-minute intervals, the large fragments were allowed to settle and the cloudy suspension was transferred to a round bottom culture tube; a fresh solution of trypsin was added to the remaining fragments. The fraction recovered from each period was mixed with an equal part of BG-SSB. At the end of trypsinization, the remaining fragments were transferred to the culture tube and pipetted. The resulting suspension was centrifuged at 275 g for 5 minutes. The supernatant fluid was discarded and the pellet was suspended in culture medium and transferred to a culture flask.

B. FACTORS AFFECTING DISPERSION

The factors that influence the dispersion of animal cells have been reviewed by Rinaldini (1958) and Waymouth (1974). The following factors

TABLE II
Procedures to Disperse Cells of *Blattella germanica*

Tissue	Stage of development	Cell types	Dissociation treatment			Expected particle yields from each embryo $\times 10^{3}$ [b]	
			Trypsin concentration	Length of interval (minutes)[a]	Total time (minutes)	Number of single cells	Total number of particles
Embryonic	Segmented germ band (5 days)	Mixed	0.01%	5	20	4.8 ± 2.2	7.1 ± 2.9
Embryonic	Organogenesis (10 days)	Mixed	0.05%	5	20	15.2 ± 1.7	20.9 ± 5.4
Embryonic	Fully formed embryo (15 days or older)	Mixed	0.25%	5	20	5.7 ± 0.4	8.1 ± 0.9
Fat body	Last-instar female nymph	Fat cells, mycetocytes, and urate cells	0.2%	10	30–60	1.4×10^{6} fat cells per female	9.4×10^{4} mycetocytes per female

[a] Intervals of time after which the single cells are removed from the dissociation mixture. See text (Section II,A) for details.
[b] ± 1 SD.

affect the dissociability of *B. germanica* embryos: the agent of dissociation, the developmental age of the embryo, the length of treatment, and the strength of the dissociation agent. Trypsin is effective in dissociating *B. germanica* embryos; it is more effective than a 0.02% solution of ethylenediamine tetraacetate (EDTA) or alkaline (pH 9.5) saline solution. Trypsin is also effective in dissociating dorsal vessels of postembryonic *Blaberus craniifer*, *Leucophaea maderae*, and *Periplaneta americana* (Vago *et al.*, 1968), and embryos of *B. craniifer* and *P. americana* (Landureau, 1966). The stage of embryogenesis affects dispersion. Embryos in the germ band, histogenesis, and early organogenesis stages are readily dissociated by trypsin; those with well-differentiated organ systems dissociate poorly. Concentrations of 0.01 and 0.05% are usually sufficient to dissociate young embryos (10 days or less); a concentration of 0.25% trypsin often results in reduced yields of single cells. However, a concentration of 0.25% trypsin is needed to dissociate the older embryos. There may also be species differences in the concentration of trypsin needed to dissociate cockroach embryos; Landureau (1966) reported that a 0.5% solution was needed to dissociate *B. craniifer*, whereas a 0.1% solution was sufficient to dissociate *P. americana* and *B. germanica*.

Intercellular membrane structures, e.g., septate junctions, between apposed cells influence tissue dissociability. Proteolytic enzymes and chelators of divalent cations disrupt these structures (Satir and Gilula, 1973). The extent to which these structures affect the dissociability of cockroach embryos needs further study, as well as the factors which influence their disruption.

We feel that the application of the above methods and information has yielded primary cultures in which a greater variety and proportion of the cells developed into cell lines. This was mainly the result of using short exposures to minimal concentrations of trypsin.

III. Cultivation and Selection of Phenotypic Variants of Embryo Cell Lines

A. SELECTION OF EMBRYONIC STAGE FOR CULTURE

Insect tissue culturists often advocate the use of a particular stage of embryogenesis as starter material for cell culture (Hirumi and Maramorosch, 1964; Varma and Pudney, 1967; Landureau, 1966; Echalier and Ohanessian, 1970). This attitude stems from the observation that cells isolated from embryos of certain ages give rise to successful primary cultures, whereas cells from other ages are refractory to culture. Some of the characteristics of cells of this latter category are poor dispersal, poor attachment to the culture flask, excessive fragility, insufficient numbers, and low mitotic potential. Nevertheless, these difficulties should not be used as a basis to preclude the selection of a certain stage when attempting to develop a cell line. A variety of stages have been used to develop cell lines of *Drosophila melanogaster* (Horikawa *et al.*, 1966; Echalier and Ohanessian, 1970; Kakpakov *et al.*, 1969; Schneider, 1972). We have developed cell lines from three embryonic ages of *B. germanica:* germ band, histogenesis, and organogenesis. Clearly, more studies are needed in the areas of preparing embryo cells for culture and in determining if specific cell types derived from certain stages have greater propensities for growth *in vitro*.

B. PRIMARY CULTURES AND THE EVOLUTION OF CELL LINES

The systems used to set up primary cultures and to develop cell lines from cockroaches are described in detail elsewhere (Landureau, 1966, 1968; Marks *et al.*, 1967; Quiot, 1971; Landureau and Grellet, 1972, 1975). The following discussion summarizes the approach used in our laboratory to obtain cell lines from embryos of *B. germanica*. The egg case (ootheca) was surface-sterilized by washing it in a 0.2% aqueous solution of Hyamine 1622 (Rohm and Haas, Philadelphia, Pennsylvania). Landureau (1966) found it necessary to surface-sterilize the embryos of *P. americana* in a solution of sodium hypochlorite; this treatment prevents the use of embryos not protected by a cuticle. For *B. germanica*, surface sterilization of the ootheca was sufficient, and this permitted us to use embryos without a cuticle. The embryos were trypsinized using the procedures outlined in Section II,A. The contents of one ootheca were used to seed a single T-15 culture flask (Kontes Glass Co., Vineland, New Jersey). Several media were tried, and a modification of the S-19 medium (Landureau and Jollès, 1969) was adopted (Table III). This medium is used to maintain all of the *B. germanica* cell lines which were derived in this laboratory. The original medium which was formulated by Landureau (1966) was patterned after cockroach hemolymph; it underwent several modifications (Landureau and Grellet, 1972) and, with appropriate adjustments in osmotic pressure, can be used to culture cells from several species of cockroaches.

Primary cultures were composed of monodispersed cells, cell clusters, embryo fragments, and multicellular spheres (vesicles). Within several weeks, a confluent monolayer of cells was formed, and the cells were sub-

TABLE III

**Basal Medium (Modified S-19) for the Cultivation of Embryonic
Cells of** *Blattella germanica*[a]

Components	mg/liter	Components	mg/liter
Amino acids		Inorganic Salts	
L-Aspartic acid	250	NaCl	8500
L-Glutamic acid	1500	KCl	1050
α-Alanine	120	NaHCO₃	360
L-Arginine HCl	800	H₃PO₃	900
L-Cysteine HCl	260	CaCl₂	490
L-Glutamine	300	MgSO₄ · 7H₂O	1260
L-Glycine	750	MnSO₄ · H₂O	65
L-Histidine	300		
L-Isoleucine	120	Sugar	
L-Leucine	250	Glucose	3000
L-Lysine	160		
L-Methionine	500	Vitamins	
L-Phenylalanine	200	Biotin	0.01
L-Proline	750	Choline HCl	181
L-Serine	80	Folic acid	0.02
L-Threonine	200	Isoinositol	0.05
L-Tryptophan	200	Nicotinamide	0.04
L-Tyrosine	180	Calcium pantothenate	0.5
L-Valine	150	Pyridoxine HCl	0.04
		Riboflavin	0.19
Miscellaneous		Thiamine HCl	0.67
Yeast extract	1500	Cyanocobalamin	0.04
Lactalbumin hydrolysate	1500		

[a] The final medium was made by mixing 1 part of fetal bovine serum (heat inactivated at 56°C for 30 minutes) with 5 parts of the basal medium (pH 7.5).

cultured. Landureau (1968) has advised the use of conditioned medium for the first few subcultures and to feed cultures of low cell density. We have used conditioned medium, but its employment was not essential in the development of cell lines. Subculturing was initiated as early as 6 weeks after isolation, but 10 to 20 months (5 to 10 serial transfers) were required before it was possible to subculture on a regular basis. A long adaptation period has also occurred in the evolution of cell lines from embryos of other species of insects (Chiu and Black, 1967; Landureau, 1968; Echalier and Ohanessian, 1970; Schneider, 1972). The reasons for the length of this period are unknown. There is evidence that in some cases it reflects technique, because the time was reduced as the operator gained familiarity with the culture system (Chiu and Black, 1967; Schneider, 1972).

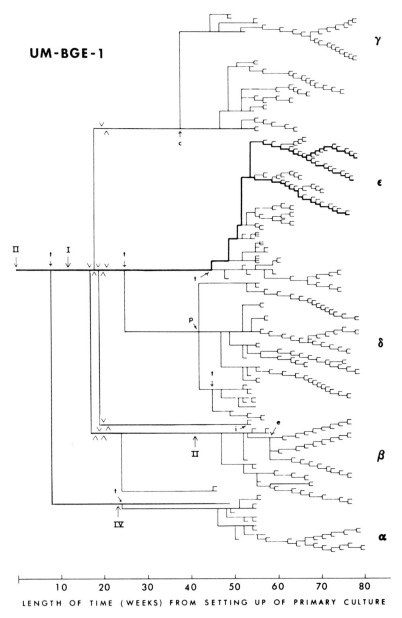

Fig. 1. The evolution of the sublines of UM-BGE-1. Horizontal lines are on the time scale and their lengths are proportional to the length of time the culture flask was maintained. Vertical lines represent the times of subculture. The times at which the sublines were shifted to a modified medium is indicated by arrows with Roman numerals. Large V indicates the times at which conditioned medium was used. Various subculture procedures were introduced at the times indicated by the small arrows

A successful subculture involved a one-to-two dilution of the cells; transfer densities lower than 3×10^5 cells per ml resulted in the loss of the growth momentum. The first subculture removed only a portion of the monolayer, and for this reason the primary cultures were continued and allowed to reestablish a confluent monolayer. The primary culture flasks were usually subcultured several times. The maintenance of several flasks from a single primary culture lead to the divergent development of sublines. Sublines were used to insure against the loss of the line. They were also perpetuated to study what factors affected the length of the adaptation period and the type of cells which predominated. Figure 1 shows the evolution of the sublines of UM-BGE-1. The epsilon subline (bold line) is currently maintained in our laboratory. Several modifications of Landureau's media were compared in the development of this line, as well as several subculture procedures. These two parameters had little influence on the length of the adaptation period since all of the sublines "matured" at approximately the same time (maturation here means the point at which a regular subculture interval could be maintained). The procedures did, however, affect the type of cell which was selected in the sublines. This divergence in cell types was observed in the evolution of several cell lines. For example, line UM-BGE-5 has two sublines (Table IV), one (beta) in which the cells attach firmly to the culture flask and the other (alpha) in which the cells do not.

C. EMBRYO CELL LINES AND THEIR CHARACTERISTICS

The first cell line of *P. americana* was developed by Landureau (1968) with cells isolated from embryos in 1965. After 10 months of culture, islands of cells were noticed which eventually overgrew the culture. A conditioned medium containing lactalbumin hydrolysate was used to increase the success of the first few subcultures. The cells which comprise the EPa line have been described as "fibroblastic." In the spent medium they are easily dispersed and when transferred to fresh medium they reattach to the culture flask. The cells require the amino acids arginine, histidine, leucine, methionine, phenylalanine, tryptophan, tyrosine, and valine (Landureau and Jollès, 1969). The vitamins folic acid, cyanocobalamin, pantothenic acid, choline, and inositol are essential (Landureau, 1969; Landureau and Steinbuch, 1969). Another feature of this line is that the cells release into the medium a factor(s) which destroys microorganisms.

---

with an accompanying lower case letter (t, trypsin; c, collagenase; p, rubber policeman; i, cold, i.e., 4°C; and e, EDTA). The bold line shows the evolution of the epsilon subline which was selected for continuance in this laboratory.

TABLE IV
Cell Lines from Embryos of *Blattella germanica*

Cell line designation	Stage of embryogenesis used as primary explant	Date primary culture was initiated	Number of subcultures (as of February 1975)	Subculture interval	Subculture ratio	Chromosomal makeup	Characteristics of the line
UM-BGE-1	Segmented germ band (5 days)	17 February 1972	97	Weekly	1:2 or 3	Polyploid	Round to spindle-shaped cells; dispersed by pipetting
UM-BGE-2	Histogenesis; dorsal closure	27 January 1972	75	Weekly	1:2 or 3	Diploid	Multipolar cells; dispersed by pipetting
UM-BGE-4	Germ band and histogenesis (5 and 7 days)	12 March 1972	71	Weekly	1:2	Diploid and tetraploid	Vesicles which cannot be dispersed with EDTA or trypsin
UM-BGE-5 (alpha)	Organogenesis; nerve cord and midgut formation (10 days)	27 August 1971	57	Biweekly	1:2	Not determined	Round cells; no attachment to flask
(beta)	Organogenesis; nerve cord and midgut formation (10 days)	27 August 1971	32	Triweekly	1:2	Not determined	Multipolar cells; attached monolayer; dispersed with 20 mM EDTA

The agent is a lytic enzyme, chitinase, which is similar to lysozyme (Landureau and Jollès, 1970). Intracytoplasmic granules, believed to be chitinase, have been observed (Deutsch and Landureau, 1970); the appearance of bacteriolytic activity (against *Micrococcus lysodeikticus*) in the culture medium has been correlated with the disappearance of the granules in the cytoplasm. Two chitinases have been isolated from the metabolized medium and characterized; they have the electrophoretic and chromatographic behavior similar to the chitinases from hemolymph (Bernier *et al.*, 1974). The line has been cloned, and strains positive and negative for chitinase production have been isolated (Landureau and Grellet, 1972). This indicates that the line is heterogeneous in cellular makeup.

The *B. germanica* cell lines developed and currently (February, 1975) maintained in our laboratory are listed in Table IV. These lines are maintained under a program similar to that outlined by Wolf and Quimby (1973) and have a weekly to triweekly subculture interval (based on a one-to-two dilution of the cells).

Terms used to identify types of vertebrate cells in culture are frequently applied to insect cells of similar morphologies. Some of these terms define cells characterized as to derivations, behavior, and biosynthetic capabilities; the vertebrate cells may not be homologous with insect cells of similar morphology. For this reason we elected to use descriptive phrases to identify the various types of cockroach cells, at least until a standardized nomenclature for insect cell types is devised. Four types of cells comprise cockroach cell lines: (a) Monodisperse cells which are round to monopolar, bipolar, or multipolar. These cells grow rounded up but may temporally attach and flatten to the surface of the culture flask, particularly after transfer to fresh medium. They show tendencies to clump and are easily dispersed by pipetting. Two lines of *B. germanica* (UM-BGE-1 and UM-BGE-2) and the EPa line consist of this type of cell. (b) Round to ellipsoidal cells which do not attach to the culture flask, even after transfer to fresh medium. UM-BGE-5 alpha comprises these cells. (c) Cells which attach firmly to the culture flask throughout the interval of time between subcultures. These cells have cytoplasmic extensions which contact one another, forming a partially contiguous monolayer with intercellular spaces. They cannot be dispersed with simple pipetting, so a 20-mM solution of EDTA (in BG-SSA) is used to lift them off of the culture flask. This type of cell is found in the subline UM-BGE-5 beta. (d) Cells which form contiguous sheets attached to the flask or forming hollow spheres (vesicles). The sheets are one cell layer in depth, and the cells are closely apposed to one another. The spheres grow free in the medium and this facilitates their subculture.

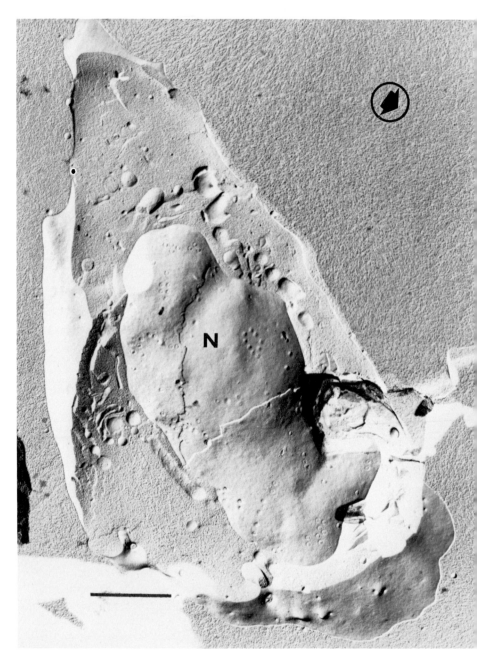

Fig. 2. Freeze-etch replica of a cell of the UM-BGE-1 epsilon subline; N, nucleus. Enclosed arrow indicates the direction of shadowing. Scale, 2 μm. Courtesy of K. R. Tsang.

No means of dispersing the individual cells has been found; solutions of trypsin or EDTA are ineffectual. Line UM-BGE-4 is made up of hollow spheres.

The cells of the lines UM-BGE-1 and UM-BGE-4 have been studied by freeze-fracture (Tsang *et al.*, 1974). A replica of a bipolar cell, commonly found in line UM-BGE-1, is shown in Fig. 2. The nucleus (N) is lobulated, and nuclear pores are present. The cytoplasm contains mitochondria, Golgi apparatus (Fig. 3A), and lipoid inclusions (Fig. 3B). A replica of a cell of the vesicle line, UM-BGE-4, is shown in Fig. 4; the cytoplasm also contains lipoid inclusions (LI). The nuclei of these cells are not as lobulated as those of the UM-BGE-1 line, and clusters of gran-

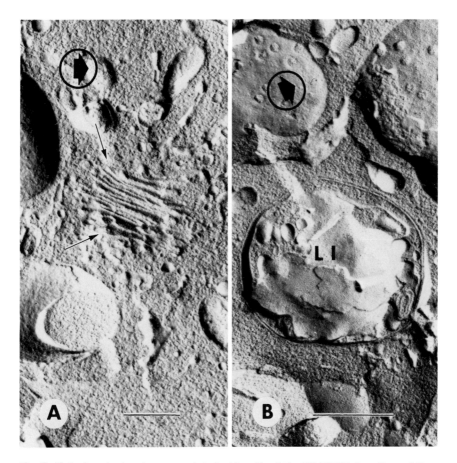

Fig. 3. Cytoplasmic structures associated with cells of the UM-BGE-1 epsilon subline. (A) Golgi apparatus (arrows). Scale, 0.5 μm. (B) Lipoid inclusion, LI. Scale, 1 μm. Enclosed arrows indicate the direction of shadowing. Courtesy of K. R. Tsang.

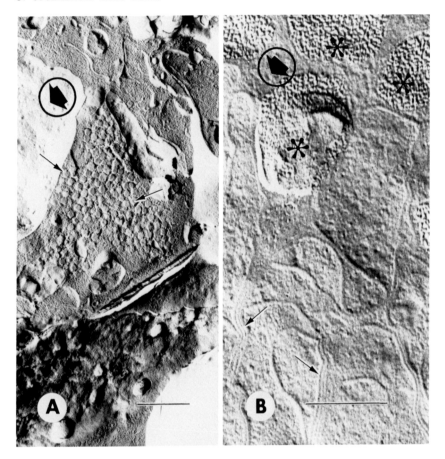

Fig. 5. Structures associated with cells of the UM-BGE-4 line. (A) Cluster of granules (arrows) commonly found in the cytoplasm near the nucleus. Scale, 1 μm. (B) Membrane structures found between apposed cells. Gap junctions (*) and septate junctions (arrows) are both shown. Scale, 0.5 μm. Enclosed arrows indicate the direction of shadowing. Courtesy of K. R. Tsang.

ules are often seen near the nucleus (Fig. 5A). The membranes between apposed cells are rich in gap (*) and septate (arrows) junctions (Fig. 5B).

D. SELECTION OF CELL LINES

The various phenotypes of the *B. germanica* cells described above (Section III,C) can be obtained on the basis of their culture characteris-

---

Fig. 4. Freeze-etch replica of a cell of the UM-BGE-4 line; N, nucleus; LI, lipoid inclusion. Scale, 2 μm. Enclosed arrow indicates the direction of shadowing. Courtesy of K. R. Tsang.

tics. Subculturing inherently fosters the selection of cells that can be dispersed and recovered, while those that remain attached to the surface are lost when the flask is discontinued. We have sought to devise methods of subculture in which the majority of the cells on the monolayer could be dispersed and subsequently recovered. Some of the things we have tried are chilling (4°C), enzymes (trypsin, collagenase, and pronase), and a chelating agent (EDTA). The best agent appears to be a 20-mM solution of EDTA in BG-SSA.

Loosely attached cells were observed in the primary cultures and in the early subcultures. Lines were selected by transferring only those cells that were easily suspended by pipetting. They were recovered from the spent medium by centrifugation and inoculated back into the original flask or into a subculture flask.

Lines forming attached monolayers were developed in a manner similar to that used by Varma and Pudney (1969). The sublines with predominantly attached cells were selected for continuance. The free-floating cells were discarded with the spent medium and, if necessary, the monolayer was washed with a saline solution to flush off the easily dispersed cells. A 20-mM solution of EDTA was used to strip attached cells from the flask for subculture.

Cells that colonize as spheres were selected by harvesting them from primary cultures. Gentle pipetting was used to sever them from the monolayer or fragments of embryos. If the yield of spheres was low, several flasks were used to acquire the numbers needed for a subculture. In the early transfers, patches of attached sheets were observed and these were selected against by transferring only the spheres. Low centrifugal forces (30 g) were used to harvest the spheres from the medium; this aided in the selection against single cells.

The procedure of harvesting cells easily dispersed by pipetting was important in developing the UM-BGE-1, UM-BGE-5 alpha, and UM-BGE-4 lines. Subline UM-BGE-5 beta was obtained by using 20-mM EDTA to harvest the cells attached firmly to the culture flask. The pooling of similar cell types was done to obtain subcultures of greater initial densities; either the cells from several flasks were used to inoculate the same subculture flask or the subculture flask was reinoculated with a new crop of cells from the same donor flask for several subsequent weeks. This manipulation was used to develop the UM-BGE-4 cell line. Richardson and Jensen (1971) also combined cells from primary cultures of similar composition to develop a cell line from embryos of the leafhopper *Colladonus montanus*. Feeder cultures or conditioned medium may also be used to increase the survival of cells in the first few transfers (Landureau, 1968; Echalier and Ohanessian, 1970). The use of different stages of embryogenesis as primary inocula and the development of sublines were perhaps the most important factors in developing *B. germanica* cell lines of different phenotypes and permitted us to acquire a repertoire of lines for further studies. Hink and Ellis (1971) developed two sublines from a single primary culture of embryos of the codling moth *Carpocapsa pomonella;* the two sublines differed in growth rates, chromosome numbers, and cell size.

IV. Conclusion

Substantial progress has been made in the development of cell lines from embryos of cockroaches, in which several different phenotypes can be maintained throughout the subcultures. However, work is still needed, particularly in the area of defining procedures for isolating and developing lines of specific cell types. This will rest on improved tissue dissociation and the cloning or sorting of cells from a heterogenous population.

Specialization does not necessarily abolish the ability of a cell to divide. The ability of mammalian cell lines and strains to express differentiated functions *in vitro* has been well documented (Green and Todaro, 1967). Cell lines from insect embryos are also capable of expressing differentiated functions. Many insect cell lines have been developed (Hink, 1972), but most of them have not been identified as to their tissue of origin. The origins of cell lines can be surmised by comparing them with specific cell types. The cockroach line EPa produces two chitinases which are similar to those found in the hemolymph (Bernier *et al.*, 1974). Electron micrographic evidence also supports the hemocytic origin of this cell line (Deutsch and Landureau, 1970). The pattern of growth and the secretory nature of an *Aedes aegypti* vesicle line indicates that these cells have also retained specialized functions (Peleg and Shahar, 1972), but their identity is unknown. Several clones or lines of *D. melanogaster* cells have isozyme patterns similar to those of the imaginal and brain tissues of third-instar larvae (Debec, 1974). It is our aim to develop a spectrum of cell lines from embryos of *B. germanica,* each from a different tissue and capable of expressing a differentiated phenotype *in vitro.*

Acknowledgment

This review was written while supported by U.S. Public Health Service Research Grant No. AI 09914 from the National Institute of Allergy and Infectious Diseases to Professor Marion A. Brooks. This is Paper No. 9027, Scientific Journal Series, Minnesota Agricultural Experiment Station.

References

Bernier, I., Landureau, J.-C., Grellet, P., and Jollès, P. (1974). *Comp. Biochem. Physiol.* **47**, 41–44.

Chiu, R.-J., and Black, L. M. (1967). *Nature (London)* **215**, 1076–1078.

Debec, A. (1974). *Wilhelm Roux' Arch. Entwicklungsmech. Organismen* **174**, 1–19.

Deutsch, V., and Landureau, J.-C. (1970). *C.R. Acad. Sci.* **270**, 1491–1494.

Echalier, G., and Ohanessian, A. (1970). *In Vitro* **6**, 162–172.

Green, H., and Todaro, G. J. (1967). *Annu. Rev. Microbiol.* **21**, 573–600.

Hink, W. F. (1972). *Advan. Appl. Microbiol.* **15**, 157–214.

Hink, W. F., and Ellis, B. J. (1971). *Curr. Top. Microbiol. Immunol.* **55**, 19–28.

Hirumi, H., and Maramorosch, K. (1964). *Science* **144**, 1465–1467.

Horikawa, M., Ling, L.-N., and Fox, A. S. (1966). *Nature (London)* **210**, 183–185.

Kakpakov, V. T., Gvosdev, V. A., Platova, T. P., and Polukarova, L. C. (1969). *Genetika* **5**, 67–75.

Kurtti, T. J., and Brooks, M. A. (1970). *Exp. Cell Res.* **61**, 407–412.

Kurtti, T. J., and Brooks, M. A. (1975). *In Vitro*, in press.

Landureau, J. C. (1966). *Exp. Cell Res.* **41**, 545–556.

Landureau, J. C. (1968). *Exp. Cell Res.* **50**, 323–337.

Landureau, J. C. (1969). *Exp. Cell Res.* **54**, 399–402.

Landureau, J.-C., and Grellet, P. (1972). *C.R. Acad. Sci.* **274**, 1372–1375.

Landureau, J. C., and Grellet, P. (1975). *J. Insect. Physiol.* **21**, 137–151.

Landureau, J. C., and Jollès, P. (1969). *Exp. Cell Res.* **54**, 391–398.

Landureau, J. C., and Jollès, P. (1970). *Nature (London)* **225**, 968–969.

Landureau, J. C., and Steinbuch, M. (1969). *Experientia* **25**, 1078–1079.

Lesseps, R. J. (1965). *Science* **148**, 502–503.

Marks, E. P., Reinecke, J. P., and Caldwell, J. M. (1967). *In Vitro* **3**, 85–92.

Peleg, J., and Shahar, A. (1972). *Tissue & Cell* **4**, 55–62.

Poodry, C. A., and Schneiderman, H. A. (1971). *Wilhelm Roux' Arch. Entwicklungsmech. Organismen* **168**, 1–9.

Quiot, J. M. (1971). *In* "Invertebrate Tissue Culture" (C. Vago, ed.), Vol. 1, pp. 267–294. Academic Press, New York.

Richardson, J., and Jensen, D. D. (1971). *Ann. Entomol. Soc. Amer.* **64**, 722–729.

Rinaldini, L. M. J. (1958). *Int. Rev. Cytol.* **7**, 587–647.

St. Amand, G. S., and Tipton, S. R. (1954). *Science* **119**, 93–94.

Satir, P., and Gilula, N. B. (1973). *Annu. Rev. Entomol.* **18**, 143–166.

Schneider, I. (1972). *J. Embryol. Exp. Morphol.* **27**, 353–365.

Seecof, R. L., Alléaume, N., Teplitz, R. L., and Gerson, I. (1971). *Exp. Cell Res.* **69**, 161–173.

Tsang, K. R., Kurtti, T. J., and Brooks, M. A. (1974). *In Vitro* **10**, 348–349.

Vago, C., Quiot, J. M., and Luciani, J. (1968). *Proc. Int. Colloq. Invertebr. Tissue Cult., 2nd, 1967* pp. 110–118.

Varma, M. G. R., and Pudney, M. (1967). *Exp. Cell Res.* **45**, 671–675.

Varma, M. G. R., and Pudney, M. (1969). *J. Med. Entomol.* **6**, 432–439.

Walters, D. R. (1969). *Biol. Bull.* **137**, 217–227.

Walters, D. R. (1974). *J. Insect Physiol.* **20**, 49–54.

Waymouth, C. (1974). *In Vitro* **10**, 97–111.

Wolf, K., and Quimby, M. C. (1973). *In Vitro* **8**, 316–321.

4

Snail Tissue Culture:
Current Development and
Applications in Parasitology,
An Introductory Statement

DONALD HEYNEMAN

The study of trematode development *in vitro* and of the varied responses by their molluscan hosts has long been delayed by the lack of a reliable *in vitro* molluscan tissue culture technique. Efforts have therefore been focused on other aspects of parasite development, such as *in vivo* maintenance and cultivation of trematode parasites in abnormal sites (the rabbit eye, body cavity of different hosts, and chick chorioallantois) by Fried and associates, as well as efforts to culture other parasitic worms directly in complex media, various xenic cultures, and with vertebrate tissue cultures. The emphasis in these *in vitro* studies has been on (1) axenizing the culture, (2) simplifying the culture medium to known constituents, and (3) duplicating *in vitro* as nearly as possible the host cell or tissue environment. Considerable success in cultivation of free-living and several parasitic nematodes in simplified media has been achieved, and most stages of the developmental cycle of several cestodes (both among pseudophyllid and cyclophyllid groups) have also been successful in complex culture media. These achievements, accomplished chiefly over the past 15 years, represent significant advances. However, detailed review of these studies is not suitable in a volume focused on invertebrate tissue culture, as most of the *in vitro* studies of parasite development, especially those using helminths, have employed cell-free media. In any case, a well-documented summary covering much of this field is available in the excellent compilation by Taylor and Baker (1968). The specific use of invertebrate tissue culture for *in vitro* study of animal parasites is reviewed *in extenso* in another excellent summary by Chao (1973). These two reviews together provide the most ready access to the majority of publications in English on parasite cultivation.

Yet, even in Chao's comprehensive review of invertebrate tissue culture for parasite studies, the subject of snail tissue and larval trematode culture was apportioned brief consideration, a reflection of the difficulty and relative lack of progress achieved. Nonetheless, molluscan tissue culture gives promise of providing unusual insight into an inordinately complex host–parasite system, the multiple stages of digenetic trematode embryonic forms in the tissues of their snail hosts. Cultivation of these intramolluscan stages of trematodes has been a major objective and remains a crucial stumbling block for workers interested in these important parasites. The ancient association that has evolved between mollusks and trematodes has provided time for the latter to develop a close metabolic dependency on their hosts and for the snail hosts to develop equally complex compensatory metabolic, growth, and defensive responsive mechanisms. Owing to the intimacy and age of this interrelationship, perhaps best considered as a single complex of host and parasite, it has proved to be an extraordinarily difficult task, and their cultivation and normal development *in vitro* so far have resisted all our efforts to view them through the successive stages they undergo, although several important isolated steps have been taken.

Organ, tissue, and cell-line cultures of pulmonate snails and other mollusks have been attempted, but with relatively limited success until recently. In the past 2 years, several collaborating laboratories under NIH contractual support have worked toward these objectives, following separate tracks, but with continuous communication and cross-fertilization of ideas. These include the laboratories of Paul F. Basch, Department of Community and Preventive Medicine, Stanford University School of Medicine, Stanford, California; Christopher J. Bayne, Department of Zoology, Oregon State University, Corvallis, Oregon; Eder L. Hansen, 561 Santa Barbara Road, Berkeley, California (formerly at Clinical Pharmacology Research Institute, 2030 Haste Street, Berkeley, California); and Marietta Voge, Department of Infectious Diseases, UCLA School of Medicine, Los Angeles, California.

A significant advance has been made by Hansen and her co-workers, who have succeeded in developing an embryonic tissue cell line from *Biomphalaria glabrata*, the snail host of *Schistosoma mansoni*. This line has been maintained over 1 year as of March 1975. Of 740 cultures attempted by Hansen's group, only one (no. 638) gave rise to colony foci. New cells appeared in about 13% of those primary cultures that persisted for 14 days or more (about 39% of all initial cultures). But in only one of these, called the Bge line (*B glabrata* embryo) could a successful cell line be established and maintained. The line now appears to be relatively easy to sustain, providing that the exact recommended trypsinization procedures are followed, and the use of the precise serum recommended is utilized, as discussed by Dr. Hansen in Chapter 6. She describes the critical procedures followed in the course of this remarkable achievement: the first successful snail cell line to have been established. This phase of her work was performed during her 1974 studies on *in vitro* development of the sporocysts of *S. mansoni*, one of the NIH-funded investigations on different aspects of snail tissue and schistosome trematode *in vitro* cultivation. In her snail-to-snail sporocyst transplant experiments, a successful production of progeny sporocysts from daughter sporocysts *in vitro*, and the development of daughter sporocysts and maturation of cercariae *in vitro* using the new molluscan cell line, Hansen has made remarkable progress. References to published results will be found in her chapter. The new Bge snail embryo cell line currently is being maintained in several laboratories, e.g., those of Imogene Schneider (Walter Reed Army Institute of Research, Washington, D.C.), Paul Basch, Christopher Bayne, and Marietta Voge (addresses given above).

Development of a molluscan cell line hopefully will prove to be the methodological breakthrough that will lead to new progress in *in vitro* trematode embryological and other studies, similar to the research stimulation provided by the development and then the rapid proliferation of insect cell lines that proved so productive in a number of avenues of inquiry over the past 5 years. The invitational article by Dr. Hansen included in this volume is an appropriate and timely contribution, one that should be followed by new interest in the *in vitro* study of the complex and biologically intriguing early life cycle stages of digenetic trematodes.

A review of molluscan organ culture was also prepared for this volume by Christopher J. Bayne, whose laboratory is similarly engaged in mollusk tissue culture studies. His review provides a general survey of progress to date in mollusk organ culture, emphasizing new work stemming from organ explant studies, an essential link between the intact organism and development of molluscan tissue and cell lines. It may well prove that to carry all of the early stages of digenetic trematode development *in vitro* through the successive precercarial stages, the use of intact mollusk tissues and even tissue complexes or organs may be required, owing to the extreme specialization and interdependency of this host–parasite system. The organ explant approach already has proved its worth as a tool for the study of trematode embryogenesis, and still more successfully as an approach to the new subject of molluscan endocrinology, a topic of special interest to the participants in this symposium. For this reason, Dr. Bayne's chapter 5 should provide a useful comparison with the early stage of insect hormonal studies *in vitro,* which have made such remarkable strides in the past half decade.

References

Chao, J. (1973). *Curr. Top. Comp. Pathobiol.* **2,** 107–145.
Taylor, A. E. R., and Baker, J. R. (1968). "The Cultivation of Parasites *in Vitro.*" Blackwell, Oxford.

5

Culture of Molluscan Organs: A Review

CHRISTOPHER J. BAYNE

I. Introduction

A. COMPARISON WITH CELL CULTURE

In contrast with cell or tissue culturists, organ culturists aim to prevent cellular outgrowth from the primary explant; unless the organ maintains its integrity *in vitro*, a good organ culture is not obtained. In contrast also is the necessary acceptance that indefinite culture is not possible. A consideration of molluscan organ culture is relevant here since some of the objectives of cell culture may be achieved better or sooner through use of organ culture methodology. For example, complete *in vitro* culture of intramolluscan stages of trematode parasites still eludes us, and it may be that their requirements are so subtle that success will necessitate the use of intact organs *in vitro*. These, of course, provide an environment that is more closely similar to the *in vivo* situation, especially with regard to mechanical factors.

B. TYPES OF ORGAN CULTURE

Molluscan organs have been maintained *in vitro* in the course of developing two distinct research approaches: short- and long-term studies. The first includes cardio- and neurophysiological studies, which necessitate maintenance of "normal" organ function for periods of no longer than several hours. Much valuable knowledge of heart and nerve functions has resulted from such studies, though these will not be emphasized here. Instead, I shall focus on organ cultures maintained for days, weeks, or months.

Parasitological and endocrinological investigations almost invariably call upon such techniques, although Coles (1969) used short-term oxygen consumption of organ slices to search for a metabolic hormone in snails.

C. PREVIOUS REVIEWS

Research in this field is still scant, yet there have been several useful reviews. The first (Benex, 1966; Bayne, 1968) dealt with the subject in general terms; later Le Douarin (1970) and Gomot (1972) provided more intensive reviews of work up to 1970. More recently Chao (1973) reviewed such work (as well as cell culture) as it relates to parasitology, and Golding (1974) reviewed papers dealing with neuroendocrine phenomena. Excellent collections of papers on the subject were edited by Lutz (1970) and Thomas (1970). The emphasis here will be on the more recent literature and on future applications in molluscan organ culture. Only brief allusions will be made to earlier work, as in Section I,D.

D. EARLY INVESTIGATIONS

The earliest attempts to maintain molluscan explants in a condition that could be termed "organ culture" were those of Zweibaum (1925) using *Anodonta* freshwater mussel gills; Gatenby (1931, 1933) and Gatenby and Hill (1934) using *Helix* garden snail mantle; Köniček (1933) with *Helix* heart, lung, and foot; and Bevelander and Martin (1949) with *Pintada* marine bivalve mantle tissues. None of these investigators was able to maintain healthy explants for longer than a few days.

E. BACKGROUND TO CURRENT STUDIES

Two approaches are presently followed in molluscan organ culture. The first seeks to maintain the explant in a liquid medium only; the second semisolidifies the nutritive medium with agar. These procedures are modifications of techniques first developed by Thomas (1947, for liquid culture) and Wolff and Haffen (1952, using semisolid medium).

II. Applications of Molluscan Organ Cultures

A. DEVELOPMENTAL BIOLOGY

The value of *in vitro* culture techniques in embryological and regeneration studies is clear, though only Brisson (1968) has applied the technique to the study of molluscan development. Brisson maintained entire embryos of *Achatina marginata* in the nutritive fluid from eggs for up to 34 days, though after 19 days development was no longer normal. Brisson removed embryos from their egg shells to provide better visual observation and photography. Although the procedure was not utilized experimentally, the demonstration that development can be maintained *ex ovo* is an interesting one, apparently related to a critical embryo age after which the constituent cells remain adherent rather than separating and reaggregating randomly, as occurs prior to that time.

Hollande (1968, 1970a) modified the liquid culture method of Thomas (1947) to achieve *in vitro* differentiation of mucous glands of the garden snail *Helix pomatia*. Using Eagle's medium at half strength and adding antibiotics 6 hours after placing the explants in culture in Leighton tubes, she was able to maintain mucous glands up to 33 days, with observable mitoses and without detectable cellular dedifferentiation. This required daily changes of medium (occasionally every 2 days). Cultures were maintained at 4°, 23°, or 37°C. Cell structures retained their normal appearance and secretory activity even at ultrastructural levels (Hollande, 1968). A later paper (Hollande, 1970b) presented results of electron microscopic autoradiography of mucous glands *in vitro* and revealed that the elaboration of merocrine secretions proceeds rapidly. In these short-term experiments, Hollande utilized Earle's saline at half dilution at 22°C. ^{3}H-Leucine appeared in the endoplasmic reticulum, ergastoplasmic cisternae, and the Golgi within 5 to 10 minutes at 22°C, and within 2 hours, labeled secretions were released from the explant. ^{3}H-Galactose, initially bound in the ergastoplasm and *vésicules de transition*, within 10 minutes was transferred to the Golgi. These results illustrate the site and rate of synthesis and the later fate of proteins and glycoproteins in these cells, but, more important, they illustrate the value of *in vitro* culture studies relating to cell biology.

Hollande demonstrated (1970a) that secretory activity of mucus cells depends upon the state of activity or estivation of the snail. Gomot

(1973) then investigated the hormonal basis of this activation (reviewed under Section II,C,1).

B. PARASITOLOGY

Benex (1961, 1964, 1966, 1967a,b, 1969a,b) has provided much of the data on parasitological application of molluscan explants. In her early work, *Biomphalaria glabrata*, the intermediate host of *Schistosoma mansoni*, was the donor organism. It is convenient, when developing methods for *in vitro* organ culture, to be able to monitor viability by gross activity of the explant. Benex therefore selected tentacles, which also were used by Bacila and Medina (1970). Her choice was a good one, since the tentacles move continuously by ciliary gliding, are capable of muscular contractions, frequently are the site of schistosome miracidial penetration, and are easy to remove from the snail. Benex emphasized the importance of avoiding antibiotics toxic to the snail (as did Chernin, 1959), and the need to use media that conform osmotically, ionically, and in pH to the internal medium of the donor. She found that survival could be prolonged from 8 to 10 days in saline and up to 4 weeks when enriched media were used. She also showed that excised tentacles maintained their state of differentiation significantly longer if nerve tissues were present. Dedifferentiation, however, inevitably occurred in explants after about 4 weeks.

Since Benex's intent was to provide an *in vitro* system in which penetration and complete development of schistosome miracidia could occur, she investigated various substances capable of augmenting survival of organ cultures (Benex, 1967a). Using the gills and mantle of *Mytilus edulis*, a marine mussel, she assayed the vitamins B_1, B_2, B_6, B_{12}, C, calcium pantothenate, pantothenic acid, and biotin, as well as Novocain, p-aminobenzoic acid, glutamic acid, p-aminobenzene sulfamide and folic acid, each at 0.01% concentration (Benex, 1967b, 1969a).

With *B. glabrata* tentacles maintained *in vitro*, Benex (1967c) followed *S. mansoni* sporocyst development for 2 weeks, from miracidial penetration and mother sporocyst metamorphosis to a late stage of daughter sporocyst differentiation.

Chernin (1964) had attacked the problem of *in vitro* cultivation of intramolluscan trematode stages several years earlier, using a different approach based upon his heart explant study (Chernin, 1963). He demonstrated the feasibility of obtaining viable and, in some cases, infective cercariae of *S. mansoni* from digestive gland and gonad explants derived from previously infected *B. glabrata*. Better results were achieved using a balanced salt solution with glucose, trehalose, and antibiotics than with the use of a complex medium that contained, in addition, vertebrate serum, antibiotic fluid, embryo extract, lactalbumin hydrolysate, and yeast extract. By removing the tissues from snails 19–20 days postinfection when the snails contained only immature cercarial embryos, Chernin showed that maturation of cercariae could occur in these explanted infected tissues. After 9 days in culture, viable cercariae emerged from the explants. Thus, Benex used snail organs *in vitro* for early growth of schistosome larval stages, and Chernin used such a system for late larval stages.

Although the complete cycle of schistosome intramolluscan development has not yet been successful in the absence of snail organs (Bayne, 1972; Voge and Seidel, 1972; DiConza and Hansen, 1973), important progress has been made in monaxenic culture of schistosome sporocysts (Hansen *et al.*, 1973), in axenic studies on these larvae (Hansen *et al.*, 1974a), and in the production of daughter sporocysts in axenic culture (Hansen *et al.*, 1974b).

Burch and Cuadros (1965) devised and used a complex culture medium that allowed them to maintain gonadal tissues for as long as 60 days. Their paper is instructive in its illustration of the wide tolerances *in vitro* shown by snail explants. Such tolerances have been repeatedly confirmed, most recently by Bayne *et al.* (1975), whose explants have remained viable in their laboratory in excess of 1 year.

It is appropriate to emphasize that numerous disease organisms other than trematodes are important in the pathology of mollusks and other invertebrates (Malek and Cheng, 1974). As man's needs to increase biological control of pest species and to maintain healthy conditions in dense cultures of farmed food species, the need for fundamental research into invertebrate internal defense mechanisms becomes more obvious. This subject is well discussed by Vago (1972), and the potential contribution of organ cultures is emphasized in Chao's paper (1973).

C. REPRODUCTIVE ENDOCRINOLOGY

This constitutes the most productive line of research involving molluscan organ cultures to date. Wolff and Haffen's (1952) paper on the semisolid system of maintaining chick embryo explants *in vitro* proved to be a major stimulus to investigations in this field. A French school of researchers has applied this procedure to problems in endocrinology. Most of their work has been with gastropods, but *Sepia officinalis*, a cephalopod, and *M. edulis*, the common mussel, have also been studied.

Sengel (1961a,b, 1964) first applied Wolff and Haffen's method to molluscan explants, though his work with gill lamellae and intestinal diverticulae of the bivalve *Barnea candida* chiefly illustrated the possibility of maintaining organ integrity and function for limited times *in vitro* (8 days). Wilfried Streiff at Toulouse, later at Caen, and Lucien Gomot and his student André Guyard at Besançon began to exploit more fully the possibilities of this research line, using prosobranch and pulmonate mollusks, respectively.

1. Gastropoda

Streiff's early papers (1963, 1964) described evaluations of various media in terms of survival of prosobranch organs. In these cultures, as in all those using the semisolid system, explants are placed on the surface of a gel prepared by dissolving agar in a liquid medium. The latter consists of a balanced salt solution, a monosaccharide, antibiotics, one or more vertebrate fluids (fetal calf serum, chick embryo extract, amniotic fluid, etc.), and a source of small organic molecules (precursors, vitamins, and amino acids) such as yeast extract, lactalbumin hydrolysate, or egg albumin. The explant derives solute requirements from the gel and gaseous requirements from the moist air into which it protrudes in the petri dish, watch glass, or other culture vessel. A balance must be created with enough nutrients to sustain the explant without the medium becoming too rich. With the medium too rich, cells migrate from the explant and the explant loses its integrity. Of course different tissues do best on different media (Streiff, 1963, 1964), and prolonged culture may require periodic transfer of the explant to a new gel. Frequent transfer is counterproductive in endocrine studies in which hormonal agents released by one explant pass to the target organ mostly by diffusion in the gel and only partly by direct contact.

After determining formulae for prolonged survival of explants from marine prosobranch snails, Streiff and his co-workers (Streiff, 1966, 1967a,b; Streiff *et al.*, 1970; Streiff and Le Breton, 1970a,b) investigated endocrine phenomena pertaining to sex reversal and differentiation of reproductive organs in the protandrous hermaphrodites *Calyptraea sinensis* and *Crepidula fornicata*. Nonspecificity of these male and female hormones was shown by interspecific associations of explants, including explants from the male or female *Littorina littorea*, a species that does not undergo sex reversal.

These workers established that differentiation of the external genital tract (penis and seminal groove) of male prosobranchs is controlled by hormones released by the optic tentacles. Regression of these structures in hermaphrodites is directed by substances released by the pleuropedal complex of the CNS. Explants from the female *Crepidula* will develop a penis in 26 days if associated with the *right* optic tentacle from a male of that species (Streiff and Le Breton, 1970b). Explants from female *Littorina*, a species that normally does not change sex, will similarly develop a penis when associated *in vitro* with right optic tentacles from male-phase *Crepidula*.

The penis, once differentiated, does not require hormone. It will maintain its structure if cultured alone. However, when cultured with pleuropedal complex of a *Littorina* female, the penis totally regresses after 18 days. This effect occurs whether the penis explant is from *Littorina* or *Crepidula*. Le Gall *et al.* (1974) placed the male genitalia of *Crepidula fornicata*, *Buccinum undatum*, and *Viviparus viviparus* with the right ocular tentacles or pleural ganglia and no other tissues. They demonstrated that the factor responsible for growth of the penis is present in the right ocular tentacle of all species tested, while the factor responsible for regression of the penis is in the pleural ganglion of males or females of these species. They found that the penis of *V. viviparus* is not responsive to the regression hormone.

Similar experiments on *Calyptraea sinensis* were undertaken by Streiff (1967b), in which he cultured immature gonads, gonads in male phase, and gonads in female phase with nerve ganglia and hemolymph from various other individuals. He developed complex hypotheses on the endocrine control of sexuality in this species (Fig. 1). Undifferentiated gonia undergo autodifferentiation to oocytes in the absence of any hormone. The deposition of yolk, however, requires a previtellogenic hormone that

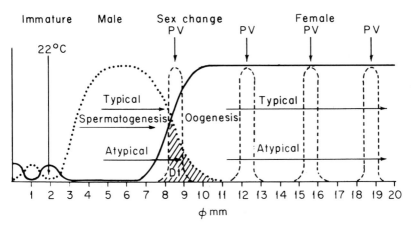

Fig 1. General scheme of the hormonal mechanisms regulating the gonadal cycle of *Calyptraea sinensis*. PV, previtellogenesis. Dotted line, masculinizing cerebral hormone that inhibits ovarian autodifferentiation; dashed line, cerebral hormone that stimulates previtellogenesis; solid line, ovarian autodifferentiation. (With permission, from Streiff, 1967b.)

appears briefly, being released by the female cerebral ganglia. This hormone, present at an effective level only in the hemolymph of previtellogenic females, initiates the formation of nucleoli in cells of the female line. This is the immediate forerunner of vitellogenesis, or yolk formation. Spermatogenesis will occur only if the undifferentiated gonad is associated with the CNS or hemolymph of a male. Lubet and Streiff (1970) discovered subtle differences between *Calyptraea* and *Crepidula* in control of gamete production. *Crepidula*, unlike *Calyptraea*, apparently has no undifferentiated germ cells in the gonad when the animal is in its male phase. This was concluded from failure of male-stage gonads to develop any oocytes when cultured in isolation. The *Crepidula* male cells regressed, but ovarian autodifferentiation, observed under similar circumstances in *Calyptraea*, did not occur. Spermatogenesis requires a hormone produced by the cerebral ganglia of male-phase *Crepidula*. In both genera, the genital tract per se is a "neutral organ," i.e., it maintains its structure and functional capabilities in the absence of hormone. There is no direct hormonal relationship between the gonad and the rest of the genital tract. The responsible hormones interact in an antagonistic balance that leads to the orderly sequence of maturation and functioning seen in normal populations of this animal.

Other investigations of prosobranch endocrinology include those of Choquet (1964, 1965, 1967, 1971) on the intertidal limpet *Patella vulgata*, and Griffond (1973, and personal communication) on the freshwater snail *V. viviparus*. Griffond is continuing these studies, which should yield interesting comparative data from this freshwater prosobranch. She has found that culture of gonad, tentacles, or intestine could not be achieved using Streiff's media, which were developed for marine tissues. Attempts to use Guyard's G_1 and G_2 media, formulated for terrestrial gastropods, were also unsuccessful, highlighting the importance of differences in milliosmolal values in body fluids of marine, terrestrial, and freshwater mollusks. Typical values are 1000 mOsm for marine mollusks (C. J. Bayne, unpublished, 1975), 175–200 mOsm for the terrestrial mollusk *Helix aspersa* (Griffond, 1973), and 80 mOsm for the freshwater mollusk *V. viviparus* (Griffond, 1973). Griffond succeeded in maintaining tissues for 2–4 weeks in solid watch glasses.

Choquet (1964) held his limpet cultures for several days in sterile seawater with 5000 IU penicillin per ml and 0.25 mg per ml streptomycin before the explant operations. Sterile techniques were then used to set gonad fragments, tentacles, ganglia, and ganglia–tentacle complexes into culture on the semisolid gel he employed. Explants were moved to a new gel each 4 days and were kept in the dark at 20°C. Healing was complete in 8–10 days, and survival was satisfactory for more than 3 months, during which time both male and female gonads maintained continuous gamete differentiation. The male cycle is controlled by antagonistic hormones from the cerebral ganglion (which stimulates spermatogenesis) and the optic tentacle (which suppresses the cerebral ganglion effect). The influence of the optic tentacle is dominant after spawning and during wintertime rest, while the cerebral ganglion is dominant during sexual maturation, which culminates in summertime spawning. The change of sex, which occurs after 2 or 3 years of male life, appears to require sustained presence of the suppressing tentacular hormone plus the action of a new vitellogenic hormone produced by the cerebral ganglia. Once the male elements are lost, removal of the tentacular hormone does not lead to their reappearance; the tentacle loses its inhibitory capability.

All other research with organ cultures of gastropods has been concerned with pulmonates. Interest in exploring gametogenesis in *Biomphalaria glabrata* and the relationship of this snail with *Schistosoma mansoni* led Vianey-Liaud and Lancastre (1968) to explore organ culture techniques different from those used by Benex, employing both solid and liquid media in a diphasic system. Tentacles cultured alone dedifferentiated considerably in 20 days, as found earlier by Benex, but adult gonads maintained gametogenesis during 25 days of culture. These authors have not gone on to apply the technique experimentally—a possible sign of problems with the diphasic system.

The pulmonate *Helix aspersa* (the common garden snail of Western Europe, California, and elsewhere) has been the focus of attention for Gomot since the mid-1960's, mostly in collaboration with Guyard, his student. The hermaphroditic gonad of this snail produces eggs and sperm alongside one another in the same acini. The ratio of male-to-female cells drops during the life of the snail. What induces primordial germ cells to develop as spermatozoa or as ova? To answer this question a hormone-free medium was devised (Guyard, 1967; Gomot and Guyard, 1968). On agar gel with a balanced salt solution and glucose, the isolated young gonad survives but maturation of germ cells does not occur. Addition of chick embryo extract promotes oogenesis, while hemolymph from adult *Helix* promotes spermatogenesis. Juvenile gonads cultured with various other organs showed no apparent influence of mature reproductive tract organs on gametogenesis. The cerebral ganglion taken from adult *Helix* during active oogenesis (Spring and summer months) had the effect in culture of favoring the female line of the target gonad. The physiological state of the donor snail clearly is important for the effect observed *in vitro*. For example, ganglia from snails in estivation are ineffective in stimulating gametogenesis. A "feminizing" effect results even when the ganglion is from distantly related snails, such as females of the freshwater prosobranch *V. viviparus*. When ganglia are taken from male *V. viviparus*, the effect is neutral to unfavorable for survival.

Guyard (1969a) summarized previous work on molluscan organ culture and reported that his medium G_2 will povide hormone-free nourishment for *Helix* gonads, allowing them to remain healthy for 50 days; other explants were also well nourished on this medium. The commercial availability of the components make this a convenient formulation. Guyard (1969b) then showed that the primordial gonad will develop into an ovary if grown on G_2 medium in the absence of other tissues. Female autodifferentiation therefore occurs in *Helix* as it does in *Crepidula*, *Viviparus*, *Littorina*, *Buccinum*, and *Calyptraea* (reviewed by Guyard, 1971). In normal snails, a wave of male activity in the gonad is due to (a) inhibition of female autodifferentiation (a premeiotic manifestation), and (b) a mitogenic effect resulting in spermatogonial multiplication. Guyard (1971) therefore suggests the existence of an androgenic–mitogenic factor specific to pulmonates and arising external to the gonad. The cerebral ganglion and optic tentacles are the sites of its manufacture and release.

During oogenic activity in the gonad, the cerebral ganglion of *Helix* produces a factor necessary for vitellogenesis (Guyard, 1971). Readers interested in reproductive control in gastropods are recommended to read Guyard (1971) and Golding (1974) ; Golding also reviews work that has involved ablation of specific ganglia. Nothing is yet known of the chemistry of these factors active in reproductive control.

Reference was made earlier to the demonstration by Hollande (1970a) that mucous-gland secretory activity was dependent upon the state of activity or estivation of the snail. Using the Wolff and Haffen (1952) technique with 50% Medium 199 in place of Tyrode's solution, Gomot (1973) demonstrated that the stimulus of secretory activity is due to a hormone produced by the ovotestis. The effect of this hormone is suppressed by another factor in estivating snails.

Bailey (1973), working in the United Kingdom with *Agriolimax reticulatus* (the Gray Field Slug), utilized organ culture methodology to investigate hormonal needs of various reproductive organs. His system, somewhat more complex than that used by the French workers, led to the suggestion of a masculinizing hormone in the brain–tentacle complex of male-phase slugs. Oocyte maintenance was improved when homologous tissue extracts were added to the culture medium, but this was probably due to "ovarian autodifferentiation" as reported for *H. aspersa* (Guyard, 1969b). Vitellogenesis did not occur *in vitro*, owing to the absence of a required hormone, the absence of an adequately nutritive environment, or to a suppressor effect. Metals involved in manipulation and support of explants in this work were noted by Benex (1966) to be potentially detrimental to explants. Some development of prostate gland cells may have been dependent upon the synergistic interaction of products from the brain–tentacle complex and the ovotestis, but such "development" proved to be abnormal at the ultrastructural level.

Over the past 3 years, research in the author's laboratory has focused on cell culture of the freshwater pulmonate *B. glabrata* (Bayne *et al.*, 1975). But findings in cell culture can also be beneficial in formulation of organ culture techniques. The capability now exists to culture *B. glabrata* cells for over 1 year, during which period mitoses were scarce. Explanted gonads and hearts, when placed on solid surfaces such as glass and plastic, give up large numbers of migrating cells—desirable for cell culture but not for organ culture. The migratory habit can, however, be suppressed by explanting the organs on semisolid gel. The potential, therefore, is presently at hand for prolonged organ culture experiments with freshwater pulmonates. Other species that have been used for explants are *Lymnaea stagnalis* and *Stagnicola emarginata*. Possibly reflecting less sophisticated homeostatic mechanisms in mollusks, explants of these animals are tolerant of wide ranges of pH, osmolality, temperature, and gas phase. They should therefore provide useful material for organ culture research.

2. Bivalvia

Application of organ cultures to problems of reproductive endocrinology of bivalve mollusks has recently begun. Houtteville and Lubet (1974) cultured mantle tissue (containing gonadal elements) of male *M. edulis*, using the procedure developed by Streiff (1966). Explants were cultured alone or with the visceral ganglia or the cerebropleural ganglia from the same individual. The germ cells regressed unless cultured with the cerebropleural ganglia. These ganglia appear to release a factor that induces movement of glycogen and other reserves from connective tissue to germ cells. Antagonizing this, the visceral ganglia enhance accumulation of reserves in storage cells of the mantle and support proliferation of phagocytes, which invade the gonad and clear residual gametes. This is what is seen *in vivo* after the spawning of these bivalves.

3. Cephalopoda

The demonstration by Wells and Wells (1959) that optic glands of *Octopus vulgaris* exert control over maturation of gonads has led to *in vitro* studies of this phenomenon. Durchon and Richard (1967), using the medium of Guyard (1967) and Gomot and Guyard (1968), demonstrated that freeing optic glands of *S. officinalis* (cuttlefish) from cerebral control (by excision and *in vitro* culture) led to their activation and caused them to begin secretory activity. Culture of ovarian fragments with optic glands showed that mitotic activity is increased compared with controls cultured alone. At the same time, vitellogenesis was induced in oocytes with diameters in excess of 0.3 mm; smaller oocytes were unresponsive to the vitellogenic hormone.

III. Future Applications

The culture of intracellular parasites is dependent upon growth of the specific host cells *in vitro*. Similarly, it is probable that successful culture of endoparasites normally located within tissues (rather than in the gut, lungs, etc.) will require culture in that tissue, as indicated in the studies of Chernin and Benex with *S. mansoni*. Recent success in maintaining snail cells *in vitro* (Hansen, 1974; Bayne *et al.*, 1975), as detailed in Chapter 6 by Hansen, represents a significant advance and opportunity for rapid progress in this field.

Hormonal research has great potential in organ culture investigations. The number of possible endocrine and target organs that have been cultured together is still small. Only reproductive control has been examined so far. Our knowledge of the hormonal repertoire will surely be greatly expanded in future work. A histochemical base already exists for studies of water regulation in snails (Highnam and Hill, 1969). The endocrine basis of regeneration is another potentially fertile subject of investigation.

Most important, *in vitro* culture of endocrine organs on semisolid media provides a means of getting at the active molecules, using sensitive preparation and detection methods together with bioassays. The possibility of biochemical analyses, leading to chemical identification and synthesis, is an exciting prospect, since no molluscan hormone has yet been identified, with the possible exception of the bag cell hormone of *Aplysia californicus* by Wilson (1971).

New ground can be broken with organ cultures. Example are the important problems of the site of hemopoiesis in mollusks, and the site of synthesis and release of macromolecules, such as hemoglobin and hemocyanin. Evolutionary relationships at the phylum and class level are in need of data, and organ cultures can contribute to investigations of interspecific similarities among hormones and their actions. Hormones appear to be evolutionarily conservative molecules, and it is conceivable that molluscan organs may be reactive to annelid and arthropod hormones and to turbellarian neural products, though such questions have yet to be experimentally studied. The potential of this approach certainly seems to warrant fuller utilization of molluscan organ cultures.

References

Bacila, M., and Medina, H. (1970). *Ann. Acad. Brasil. Cienc.* 42, 141–146.
Bailey, T. B. (1973). *Neth. J. Zool.* 23, 72–85.
Bayne, C. J. (1968). *Malacol. Rev.* 1, 125–135.
Bayne, C. J. (1972). *Parasitology* 64, 501–509.
Bayne, C. J., Owczarzak, A., and Noonan, W. E. (1975). "Conference on Pathobiology of Invertebrate Vectors of Disease," Vol. 266, pp. 513–527. N.Y. Acad. Sci., New York.
Benex, J. (1961). *C.R. Acad. Sci.* 253, 734–736.
Benex, J. (1964). *C.R. Acad. Sci.* 258, 2193–2196.
Benex, J. (1966). *Ann. Parasitol.* 41, 351–378.
Benex, J. (1967a). *C.R. Acad. Sci., Ser. D* 265, 571–574.
Benex, J. (1967b). *C.R. Acad. Sci, Ser. D* 265, 631–634.
Benex, J. (1967c). *Ann. Parasitol.* 42, 493–524.
Benex, J. (1969a). *C.R. Acad. Sci., Ser. D* 268, 631–634.
Benex, J. (1969b). *C.R. Acad. Sci., Ser. D* 268, 842–844.
Bevelander, G., and Martin, G. (1949). *Anat. Rec.* 105, 614.
Brisson, P. (1968). *Arch. Anat. Microsc. Morphol. Exp.* 57, 345–368.
Burch, J. B., and Cuadros, C. (1965). *Nature (London)* 206, 637–638.
Chao, J. (1973). *Curr. Top. Comp. Pathol.* 2, 107–144.
Chernin, E. (1959). *J. Parasitol.* 45, 268.
Chernin, E. (1963). *J. Parasitol.* 49, 353–364.
Chernin, E. (1964). *J. Parasitol.* 50, 531–545.
Choquet, M. (1964). *C.R. Acad. Sci.* 258, 1089–1091.
Choquet, M. (1965). *C.R. Acad. Sci.* 261, 4521–4524.
Choquet, M. (1967). *C.R. Acad. Sci., Ser. D* 265, 333–335.
Choquet, M. (1971). *Gen. Comp. Endocrinol.* 16, 59–73.
Coles, G. C. (1969). *Comp. Biochem. Physiol.* 29, 373–381.
DiConza, J. J., and Hansen, E. L. (1973). *J. Parasitol.* 59, 211–212.
Durchon, A., and Richard, M. (1967). *C.R. Acad. Sci., Ser. D* 264, 1497–1500.
Gatenby, J. B. (1931). *Nature (London)* 128, 1002–1003.
Gatenby, J. B. (1933). *Arch. Exp. Zellforsch. Besonders Gewebezuecht.* 13, 665.
Gatenby, J. G., and Hill, J. C. (1934). *Quart. J. Microsc. Sci.* 76, 331.
Golding, D. W. (1974). *Biol. Rev. Cambridge Phil. Soc.* 49, 101–224.
Gomot, L. (1972). In "Invertebrate Tissue Culture" (C. Vago, ed.), Vol. 2, pp. 42–136. Academic Press, New York.
Gomot, L. (1973). *Proc. Conf. Eur. Comp. Endocrinol., 7th, 1973* Abstract.
Gomot, L., and Guyard, A. (1968). *Proc. Int. Colloq. Invertebr. Tissue Cult., 2nd, 1967* Abstract.
Griffond, B. (1973). *C.R. Acad. Sci., Ser. D* 276, 2047–2051.
Guyard, A. (1967). *C.R. Acad. Sci., Ser. D* 265, 147–149.
Guyard, A. (1969a). *C.R. Acad. Sci., Ser. D* 268, 162–164.
Guyard, A. (1969b). *C.R. Acad. Sci., Ser. D* 268, 966–969.
Guyard, A. (1971). *Haliotis* 1, 167–183.
Hansen, E. L. (1974). *Int. Res. Commun. Syst.* 2, 1703.

Hansen, E. L., Perez-Mendez, G., Long, S., and Yarwood, E. (1973). *Exp. Parasitol.* **33**, 486–494.

Hansen, E. L., Perez-Mendez, G., and Yarwood, E. (1974a). *Exp. Parasitol.* **36**, 40–44.

Hansen, E. L., Perez-Mendez, G., Yarwood, E., and Buecher, E. J. (1974b). *J. Parasitol.* **60**, 371–372.

Highnam, K. C., and Hill, L. (1969). "The Comparative Endocrinology of the Invertebrates." Amer. Elsevier, New York.

Hollande, E. (1968). *C.R. Acad. Sci., Ser. D* **267**, 1054–1057.

Hollande, E. (1970a). *C.R. Acad. Sci., Ser. D* **270**, 2825–2828.

Hollande, E. (1970b). *C.R. Acad. Sci., Ser. D* **270**, 3111–3114.

Houteville, P., and Lubet, P. (1974). *C.R. Acad. Sci., Ser. D* **278**, 2469–2472.

Köníček, J. (1933). *Arch. Exp. Zellforsch. Besonders Gewebezuecht.* **13**, 709.

Le Douarin, N. (1970). *In* "Invertebrate Organ Cultures" (H. Lutz, ed.), pp. 1–75. Gordon & Breach, New York.

Le Gall, S., Griffond, B., and Streiff, W. (1974). *C.R. Acad. Sci., Ser. D* **278**, 773–776.

Lubet, P., and Streiff, W. (1970). *In* "Invertebrate Organ Cultures" (H. Lutz, ed.), pp. 135–152. Gordon & Breach, New York.

Lutz, H., ed. (1970). "Invertebrate Organ Cultures." Gordon & Breach, New York.

Malek, E. A., and Cheng, T. C., (1974). "Medical and Economic Malacology," pp. 1–398, Academic Press, New York.

Sengel, P. (1961a). *C.R. Acad. Sci.* **252**, 3666–3668.

Sengel, P. (1961b). *Bull. Soc. Zool. Fr.* **86**, 301.

Sengel, P. (1964). *Bull. Soc. Zool. Fr.* **89**, 10–41.

Streiff, W. (1963). *Gen. Comp. Endocrinol.* **3**, 733–734.

Streiff, W. (1964). *Bull. Soc. Zool. Fr.* **89**, 56–59.

Streiff, W. (1966). *Ann. Endocrinol.* **27**, 385–400.

Streiff, W. (1967a). *Ann. Endocrinol.* **28**, 461–472.

Streiff, W. (1967b). *Ann. Endocrinol.* **28**, 641–656.

Streiff, W., and Le Breton, J. (1970a). *C.R. Acad. Sci., Ser. D* **270**, 547–549.

Streiff, W., and Le Breton, J. (1970b). *C.R. Acad. Sci., Ser. D* **270**, 632–634.

Streiff, W., Le Breton, J., and Silberzahn, N. (1970). *Ann. Endocrinol.* **31**, 548–556.

Thomas, J. A. (1947). *C.R. Acad. Sci.* **225**, 148.

Thomas, J. A., ed. (1970). "Organ Culture." Academic Press, New York.

Vago, C. (1972). *Entomophaga* **17**, 111–129.

Vianey-Liaud, M., and Lancastre, F. (1968). *C.R. Acad. Sci., Ser. D* **266**, 1317–1319.

Voge, M., and Seidel, J. S. (1972). *J. Parasitol.* **58**, 699–704.

Wells, M. J., and Wells, J. (1959). *J. Exp. Biol.* **36**, 1–33.

Wilson, D. L. (1971). *J. Gen. Physiol.* **57**, 26–40.

Wolff, E., and Haffen, K. (1952). *Tex. Rep. Biol. Med.* **10**, 463–472.

Zweibaum, J. (1925). *C.R. Soc. Biol.* **93**, 785–787.

6

A Cell Line from Embryos of
Biomphalaria glabrata (Pulmonata):
Establishment and Characteristics

EDER L. HANSEN

I. Introduction

Establishment of a tissue culture from the snail intermediate host of the trematode parasite *Schistosoma mansoni* was undertaken with the ultimate objective of providing a useful culture system to study the growth and development of the parasite. Conditions for culture of organs (Benex, 1966) and of tissue fragments (Chernin, 1964) of the host snail *Biomphalaria glabrata* are reviewed by C. Bayne in Chapter 5. Chernin (1963) has maintained amebocytes in monolayer culture, and cellular outgrowths from explants have been reported by Burch and Cuadros (1965).

Evidence has also accumulated from insect tissue culture (Hink, 1972a; Stanley, 1972; Brooks and Kurtii, 1971) that embryos can provide a valuable source of tissue for initiating a cell line. It was therefore decided to use embryos and modifications of methods used in invertebrate and vertebrate tissue culture to deal with the special characteristics and requirements of mollusk material. Insect tissue culture studies also showed that the medium did not have to reproduce precisely the hemolymph composition (Schneider, 1972), since proteins of a specific hemolymph could be replaced by using vertebrate serum (Jones, 1966).

While this work was in progress, primary cultures from *B. glabrata* were reported by Basch and Diconza (1973) and by Cheng and Arndt (1973). A note on the primary cultures from this laboratory was also published (Hansen and Perez-Mendez, 1973), as was one on establishment of the cell line (Hansen, 1974).

II. Procedures Followed in Development of Cell Line

A. PREPARATION OF PRIMARY CULTURES

Primary cultures were initially prepared by many different methods, as will be revealed below. The sequence described here has given consistently good cultures and was the one used for the primary culture that ultimately gave rise to the mollusk cell line.

The albino mutant of *B. glabrata*, originally developed by Newton (1955), was maintained on lettuce. A supplement of compressed rat diet granules was provided to increase egg production. Recently deposited egg masses were removed from the aquaria, briefly rinsed in 0.2% iodine followed by 0.001% Hyamine-1622 (Rohm and Haas) and incubated at 23°C for 4–5 days in autoclaved aquarium water containing antibiotics (Table IA). The water was changed daily. When the embryos were about 0.35 mm in diameter, at the trochophore to early shell stage of development (Fig. 1), they were teased out of their capsules into buffered saline (Table IB). The saline contained antibiotics at 10-fold the above levels, the saline being decreased to 0.8-fold to compensate for the changed osmolality. The embryos were rinsed three times in 0.4-fold buffered saline containing antibiotics (Table IC). Shells were removed by aspirating the embryos through a fine Pasteur pipette.

The cell suspension was made by fragmenting the embryos in a solution of trypsin-EDTA (Table ID). Fragments were teased and aspirated

TABLE I

Solutions for Preparation of Primary Cultures

Component	Amount
A. Antibiotic solution in autoclaved aquarium water	
Penicillin[a]	100 units/ml
Streptomycin[a]	100 μg/ml
Fungizone[a]	0.25 μg/ml
Chlorotetracycline[b]	5 μg/ml
B. Buffered physiological saline	
NaCl	2.8 gm/liter
KCl	0.15 gm/liter
Na$_2$HPO$_4$	0.07 gm/liter
Glucose	1.0 gm/liter
Trehalose	1.0 gm/liter
(Decreased as required to compensate for osmolality of saline in antibiotic solution)	
C. Antibiotics added to diluted saline	
Nemomycin 531[c]	200 μg/ml
Kanamycin 516[c]	400 μg/ml
Lincocin 560[c]	2000 μg/ml
Polymixin-B 535[c]	2000 units/ml
Gentamicin[d]	500 μg/ml
D. Trypsin-EDTA solution[e]	
Na gluconate	1.5 gm/liter
NaCl	2.8 gm/liter
KCl	0.15 gm/liter
Na$_2$HPO$_4$	0.07 gm/liter
NaHCO$_3$	0.05 gm/liter
Glucose	0.5 gm/liter
Trehalose	0.5 gm/liter
Galactose	0.5 gm/liter
Phenol red	5.0 mg/liter
Trypsin (2× crystallized Sigma T-8253)	2.5 gm/liter
Ethylenediaminetetraacetic acid, tetrasodium salt (MW = 380)	38.0 mg/liter
E. Medium with antibiotics	
Schneider's *Drosophila* medium[f]	25%
Fetal salf serum, inactivated	10%
Penicillin	100 units/ml
Streptomycin	100 μg/ml
Fungizone	0.25 μg/ml
Chlorotetracycline	5.0 μg/ml
Gentamicin	125 μg/ml
Neomycin	50 μg/ml
Polymixin B	500 μg/ml
Lincocin	500 μg/ml
Kanamycin	100 μg/ml

[a] GIBCO-524.
[b] GIBCO-528.
[c] GIBCO catalog number.
[d] Schering.
[e] NaOH to adjust pH to 7.6.
[f] GIBCO-172.

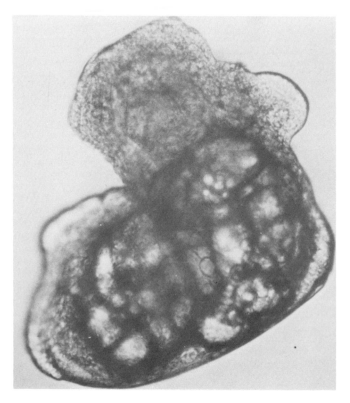

Fig. 1. *Biomphalaria glabrata* embryo, early shell stage used as source of cell suspension, capsule removed.

through a fine pipette and left in the solution for 30 minutes at 37°C. The trypsin solution was then replaced by medium containing high levels of antibiotics (Table IE) and allowed to stand in order to soak out and dilute any remaining trypsin. After 30 minutes the fragments were collected and drained in a 400-mesh sieve (30 μm opening). They were then rinsed free of antibiotics through two successive washes of culture medium. The material was then comminuted by pressing it through the sieve with a Teflon spatula. The sieve was rinsed with medium, the cells and fragments collected, transferred to culture vessels, and allowed to settle for 20 minutes.

These processes required the better part of a day to carry out and were conveniently done in Bureau of Plant Industry watch glasses. The sieves were constructed of mesh fastened with silicone cement to glass rings (1 cm diameter and 5 mm deep). Embryos were handled in groups of 100. The material from all 100 embryos was used to inoculate a plastic T-25 flask while 20–40 embryos were required for each Leighton tube or 5-cm petri dish. Three ml of medium were used in the T-flasks, 1 ml in the petri dishes, and 0.5 ml in the Leighton tubes.

Cultures were incubated at 25°, 27°, or 30°C. The atmosphere was air, with a moist environment being provided when petri dishes were used.

B. TREATMENT OF PRIMARY CULTURE LEADING TO CELL LINE Bge

The culture (628) that finally gave rise to propagable cells was handled in a manner designed to encourage development from any foci of new cells. Sieved embryo material, supplying approximately 1×10^6 cells, was suspended in 0.3 ml of medium M-182 (Table II) and incubated overnight at 25°C in a Leighton tube. Serum-free medium (M-256) was added and then replenished on the following day, one-third of the medium being replaced. On day 5, a medium with 13% fetal calf serum (M-260 in Table II) was added. On day 7 this medium was withdrawn and replaced with the serum-free medium.

Cells grew luxuriantly in the primary culture, which was trypsinized on day 9. The trypsin (0.05% trypsin-0.02 mM EDTA), modified from Table I.D, was applied for 4 minutes at 37°C and then removed. The cells were then transferred to a new Leighton tube. The incubation temperature was changed to 30°C and the culture returned to medium M-260. This was replenished on days 12, 14, and 16. On day 19, a minute colony of small cells was observed among a large number of typical primary culture cells. On day 21 several additional very small colonies were found and the culture was treated, i.e., "reseeded," for 30 seconds with trypsin-EDTA at one-eighth the above concentration. On day 23 there were many colonies of the same small cells. Since reseeding had been successful, it was repeated on day 23 and again on day 26. On day 28, though the small cells still were in the minority, the culture was trypsinized as on day 9 and transferred to a new tube. By day 33 the number of new cells had increased and were well distributed. The culture was again transferred to a new tube. By day 50 the small cells had greatly increased and were subcultured into two Leighton tubes, i.e., the cells were now numerous enough to subdivide at the fourth passage. They continued to grow well and one tube was used on day 54 to inoculate a T-25 flask, and the other tube was transferred into two Leighton tubes. In the subculture made 3 days later a split of 1:3 was made. By day 62, 12 new cultures had been made in Leighton tubes and T-flasks.

The cell line was designated "Bge" (*B. glabrata* embryo) to denote

TABLE II
Media for Initiating and Maintaining Bge Cell Line[a,b]

Component[c]	Medium		
	M-182	M-260[d]	S-301[e]
Complex mixtures			
Fetal calf serum[f] (in %)	13[g]	13	13
Bactopeptone[h] (in mg/liter)	4500	—	—
Lactalbumin hydrolysate[i] (in mg/liter)	—	4500	4500
Yeast hydrolysate (in mg/liter)	—	—	440
Salts and buffers (in mg/liter)			
NaCl	2800	1792	462
KCl	150	96	352
$CaCl_2$	530	339	132
$MgSO_4 \cdot 7H_2O$	450	288	814
Na_2HPO_4	70	45	154
KH_2PO_4	—	—	99
$NaHCO_3$	50	32	88
Amino acids and related compounds (in mg/liter)			
L-Alanine	6.0	3.8	—
L-Arginine	4.2	2.7	88
L-Asparagine	2.0	1.3	—
L-Aspartic acid	6.0	3.8	88
β-Alanine	—	—	110
L-Cysteine	6.0	3.8	13
L-Cystine	2.4	1.5	22
L-Glutamic acid	10.0	6.4	176
L-Glutamine	20.0	20.0	396
Glycine	6.0	3.8	55
L-Histidine	4.0	2.6	88
L-Isoleucine	5.2	3.3	33
L-Leucine	5.2	3.3	33
L-Lysine	5.8	3.7	363
L-Methionine	1.5	1.0	176
L-Phenylalanine	3.5	2.1	33
L-Proline	6.0	3.8	374
L-Serine	12.0	7.7	55
L-Threonine	4.8	3.1	77
L-Tryptophan	0.8	0.5	22
L-Tyrosine	3.6	2.3	110
L-Valine	4.7	3.0	66
Sugars (in mg/liter)			
Galactose	2000	1280	1300
Glucose	860	550	440
Trehalose	860	550	440

TABLE II (*Continued*)

Component[c]	Medium		
	M-182	M-260[d]	S-301[e]
Organic acids (in mg/liter)			
Fumaric acid	—	—	22
α-Ketoglutaric acid	—	—	44
Malic acid	—	—	22
Succinic acid	—	—	22
Vitamins (in μg/liter)			
Biotin	38	24	—
Calcium folinate	38	24	—
D-Calcium pantothenate	75	48	—
Cyanocobalamine	38	24	—
Folic acid	75	48	—
N-Acetylglucosamine	150	96	—
Niacin	75	48	—
Nicotinamide	75	48	—
Pantethine	38	24	—
p-Aminobenzoic acid	75	48	—
Pyridoxal HCl	38	24	—
Pyridoxamine HCl	38	24	—
Pyridoxine HCl	75	48	—
Riboflavin	75	48	—
Thiamine HCl	75	48	—
Thioctic acid	38	24	—
Nucleic acid precursors (in μg/liter)			
Adenosine-3′(2′)phosphoric acid	—	347	—
Cytidine-3′(2′)phosphoric acid	—	323	—
Guanosine-3′(2′)phosphoric acid	—	363	—
Uridine-3′(2′)phosphoric acid	—	324	—
Thymine	—	126	—
Osmolality (in mOsm)	185	154	155

[a] pH is 7.0–7.2. Phenol red can be added at 5 mg/liter.

[b] Gentamicin (Schafer *et al.*, 1972) can be added at 100 mg/liter.

[c] Components in mg/liter, μg/liter, or % as indicated.

[d] In M-260 the base of salts, amino acids, and vitamins is decreased to 64% to lower the osmolality from that of M-182.

[e] Lactalbumin hydrolysate and galactose added to Schneider's *Drosophilia* medium, GIBCO-172, diluted to 22%. A variant medium contained 30% Schneider's and 5% fetal calf serum (Hansen, 1974).

[f] Fetal calf serum heated to 56°C for 30 minutes, stored frozen in small vials.

[g] In M-256 the fetal calf serum was omitted and 1 g/liter bovine serum albumen, crystallized (Sigma A-4378) added. Other components are as in M-182, except for nucleic acid precursors, which are as in M-260. M-182 is a modification of S-130 (Hansen and Perez-Mendez, 1973).

[h] Difco 0118.

[i] Edamin-Type S. Humco-Sheffield Co., Norwich, New York 13815. Five percent solution adjusted with sodium hydroxide to pH 7.0. Lactalbumin hydrolysate, GIBCO-164, can be used.

its source. Subcultures were made after that according to the rate of growth and density of the cells, as described in Section II,C. At the seventh passage, medium S-301 (Table II) was introduced on a trial basis. The cells responded without an adaptation period and S-301 was then employed as the standard medium. In the tenth passage the cultures were tested for mycoplasma in both broth and agar (GIBCO-804) incubated aerobically and anaerobically. No organisms were detected. However, as an additional precaution, four successive subcultures were treated with the antimycoplasmal antibiotics, chlorotetracycline (GIBCO-18007) at 100 μg/ml; gentamicin (Schering) at 500 μg/ml; kanamycin (GIBCO-516) at 100 μg/ml; and Tylocine (GIBCO-522) at 300 μg/ml (Barile, 1973). Subsequently, cultures were carried without antibiotics.

C. SUBCULTURE OF CELL LINE

Subcultures were made by gentle trypsinization with splits of 1:2 to 1:4 at intervals of 3–8 days at 27° and 30°C, or 4–23 days at 25°C. The tube was first tipped and the medium drained and removed. Then 0.5 ml of a diluted solution of trypsin-EDTA (one-fifth to one-tenth the concentration listed in Table ID), warmed to 37°C, was gently flooded over the cells and promptly removed. This was repeated and the container placed upright, allowing the cells to drain for 50 seconds. The drained fluid was removed, rapidly replaced with 2 ml of medium, and the cells loosened by flushing, using a bent-tip Pasteur pipette. The suspension was gently aspirated through a fine-tip Pasteur pipette to disperse the cells. Aggregates were allowed to settle for 1 minute and the supernatant was taken up and distributed to the subculture vessels. Medium was added to the new vessels and to the residual cells in the parent culture. Frequently, as many as four additional subcultures were made at intervals of 2–5 days from the same parent culture. These later subcultures were usually successful, as was particularly noteworthy in the development of the Bge-5 strain (see below).

Departure from this protocol of trypsinization resulted in poorly attached or lysed cells. It proved essential that the trypsin be no more concentrated than 0.05% and that it be drained off the cells and removed before the culture medium is added. The only variation of this procedure that proved usable was the use of a more dilute trypsin, i.e., trypsin-EDTA at one-fourth to one-eighth the above concentration. Subcultures could also be made with EDTA alone, 1 ml of a 0.5 mM solution in buffered saline (pH 7.6), placed on the cells for 5 minutes after a preliminary rinse. In both cases, medium was added and the cells suspended and distributed. Cells could also be suspended mechanically by flushing, but only a few cells attached in the subcultures. Commercial trypsin solution (GIBCO-505) diluted to 0.05% has also been used, the osmolality restored with 12.8% Schneider's medium.

Inoculations were made with 5×10^4 to 2×10^6 cells. Cultures could, however, be started with as few as 1.4×10^4 cells. Cell numbers for growth curves were determined by daily hemocytometer counts after suspending the cells with 0.25% trypsin-0.1 mM EDTA (Table ID), although clumps of cells could not always be dispersed. Viability was tested by trypan blue exclusion.

D. CHROMOSOME PREPARATIONS

Colcemid (GIBCO-521L) at 0.15 μg was added to 1- or 3-day-old cultures. Cells were grown on coverslips in petri dishes containing 1.5 ml of medium. After 4 or 18 hours the cells were swollen by adding two volumes of 0.125% sodium citrate solution. After 40 minutes the fluid was removed by careful decantation and the cells fixed for 10 minutes by adding 2 ml of fixative (alcohol–acetic acid, 3:1). They were air-dried and stained with aceto-orcein. Other cells, not swollen or treated with colcemid, were also fixed and stained. Snail embryo fragments were cultured for 10 hours in colcemid and squashed in aceto-orcein. Karyotypic analyses were not carried out, however, because of the difficulty in spreading the chromosomes, their small size, and the lack of distinguishable centromeres.

E. MEDIA

Media were designed to maintain a dense, confluent layer of clear adhering cells in the primary culture. The composition of the particular media used for initiating and maintaining the cell line is shown in Table II. Media M-182, 256, and 260 were designed to parallel mollusk hemolymph amino acid composition (Gilbertson *et al.*, 1967), salts (Chernin, 1964), and osmolality (Lee and Cheng, 1972). Medium S-301 is based on Schneider's *Drosophila* medium (Schneider, 1964, 1966) diluted to 22% to adjust for the required osmolality.

Concentrated stock solutions of groups of components were filtered through Millipore membranes, 0.2 or 0.3 μm porosity, and stored at 4°C or frozen. The complete medium was usually assembled aseptically the day before use and checked overnight for accidental contamination. A batch was usually exhausted within a week, though prepared media can be stored satisfactorily at 4°C for at least 41 days. The pH was monitored by phenol red and measured with a Beckman combination electrode.

Osmolality was measured with osmometers borrowed, as needed, from the University of California at Berkeley. The effect of minor changes was calculated from the proportional contribution of the ingredients and confirmed later by measurement.

F. CHARACTERISTICS OF PRIMARY CULTURES

Using the method described above, as well as others, a total of 740 primary cultures were established over a 2-year period. The general appearance of these cultures was one of dense fibroblastlike cells together with small irregular cells and occasional epithelioid patches (Fig. 2A). The cells usually started to attach within 3 hours and were well grown by 10 days. They then slowly became granular and decreased in numbers. Some cultures persisted for 2 months or more, particularly if small tissue fragments had been included. Before the sieve technique was developed, some difficulty was encountered due to reaggregation, formation of cilia, and swimming away of the aggregates from the surface of attachment.

A wide variety of media was used, due in part to tolerance of the cells to a considerable range of media compositions, which made it difficult to decide upon an optimum formulation. A method was eventually developed to begin with a well-established primary, completely replace the medium with the new one being tested, and then observe and carefully compare cell morphology under the two conditions. The test was concluded by passaging these cells in the new medium and observing again. In this way the optimum pH and osmolality were determined and different batches of serum tested. These batches varied widely in their effect. One important step, therefore, was to test and discard batches showing toxic or inhibitory effects. We also noted that cell granulation was decreased when the serum content of the medium was lowered.

In addition to the primary cultures designed for prolonged maintenance, 35 cultures were passaged with trypsin for selection of cells and the possibility of stimulating cell division. Development of a suitable trypsin solution proved to be an essential step in the development of a selection method leading to the cell line. Maximum attachment of cells occurred when the trypsin diluent (at the correct osmolality) contained carbohydrates and sodium gluconate (Farris, 1968). Furthermore, by using calcium and magnesium-free diluent with EDTA, a lower effective level of trypsin could be used. The solution was compounded in two portions to permit variation of the level of trypsin-EDTA in the diluent (Table ID).

Though primary cultures could be easily transferred and would readily reestablish themselves in a cell layer, it was soon apparent that little division was taking place. Few metaphases were detected even after col-

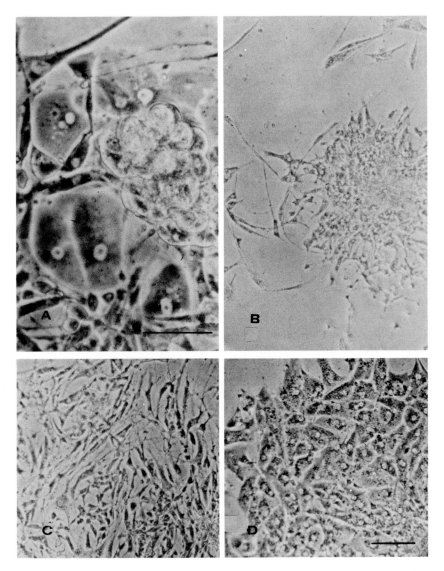

Fig. 2. Primary cultures from *B. glabrata* embryo tissue. (A) Culture 522 at 12 days showing several cell types. Two colonies of smaller cells later developed in this flask; scale, 50 μm. (B) Culture 617 at 20 days showing colony and residual cells of primary culture. (C) and (D) Culture 499 at 81 days showing two morphologically different colonies in T-flask; scale, 100 μm for (B), (C), and (D).

cemid treatment of the cultures. However, exposure to tritiated thymidine showed labeling in occasional small cells, comprising less than 0.1% of the cell layer (Fig. 3). This was evidence that the medium was adequate to support DNA replication in some cells and provided incentive to continue the work.

In February 1974, a discovery was made of colonies of new cells. The first such colony was found in the passage of a 6-day culture (#509). It started as a small group of tightly packed cells distinct from the adjacent larger cells of the primary culture. More colonies were then found in subsequent cultures. In all, 27 cultures developed colonies. They were

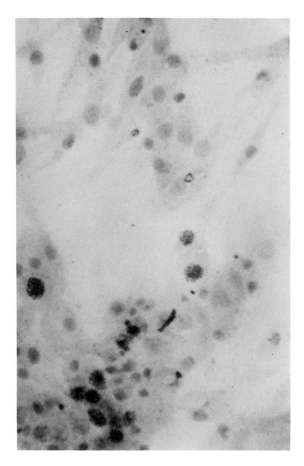

Fig. 3. Primary culture from *B. glabrata* embryo tissue. Two-day-old culture after third passage. Tritiated thymidine for 45 minutes, 50 μCi/ml; fixed, stained with aceto-orcein. Fewer than 1/1000 cells labeled. Photograph of 10 labeled nuclei in best area of slide.

observed in about one-third of all primaries that persisted 14 days or longer. The incidence (8 of 13) was greatest in those primaries that had been treated with a low serum medium. The foci of new cells appeared when the larger cells in the primary were dying (Fig. 2B). They were noted most frequently on the twenty-third to twenty-fourth day, but the time of first appearance of the new colonies varied widely between 6 and 36 days.

Colonies were observed daily, measured, and the number of cells counted. Several cell types were found, usually occurring singly in any one culture. However, on three occasions, two types of colonies occurred together in a single culture (Fig. 2C,D). Colonies were maintained for as long as 11 weeks, with repeated replenishment of the medium, but they failed to continue to develop and slowly died and disappeared. Several attempts were made to passage the colonies but these attempts were also unsuccessful, either because the cells were too old or too few to grow after their introduction into a new vessel. These attempts, however, underlay the decision to introduce trypsinization early. This proved to be the key to the sequence of steps used in deriving the Bge cell line (see Section II,B).

The Bge line originated in culture 628. Other primary cultures initiated in a similar manner were also observed closely for potential new cell lines. For example, culture 627 was handled at the same time as 628. Although a good transfer of the primary culture was made, no colonies developed and the culture was abandoned on day 23. In other primary cultures started subsequently, temperature and medium were varied at different steps; hyaluronidase was used in two cultures instead of trypsin. Although the primary cells were successfully transferred in 14 of these cultures, the cell layers that formed were sparse and in only two cultures did colonies of small cells form. These were lost in subsequent handling.

G. CHARACTERISTICS OF THE Bge CELL LINE

The successful cell line was maintained at three temperatures: 25°, 27°, and 30°C. The number of serial subcultures from April to December 1974, reached 52 at 30°C, 45 at 27°C, and 29 at 25°C. Several series were subcultured at each temperature. The cells grew in large scattered patches, sometimes dense and centrally heaped. The Bge cells in second and third passage and subculture 46 are shown in Fig. 4. The cells are clear and small (20 μm diameter when rounded) with relatively large nuclei (9 μm diameter). When attached, they are fibroblastlike or irregular in shape, with two to five protoplasmic extensions that sometimes are quite long (Fig. 4D). Occasionally, large cells occur (Fig. 5A).

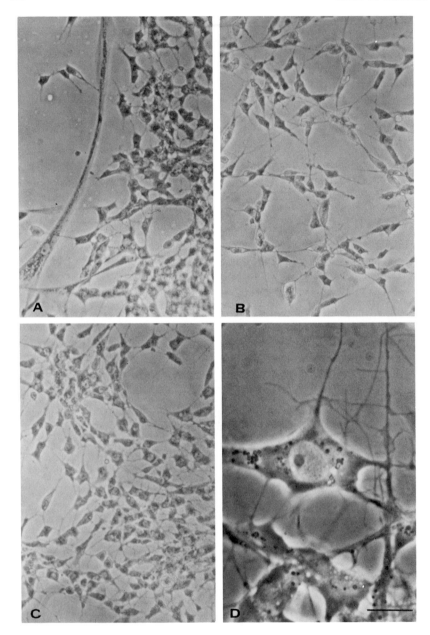

Fig. 4. Early passages of *Biomphalaria glabrata* cell line; scale, 50 μm for (A), (B), and (C). (A) Second passage showing large cells carried over from primary. (B) Third passage showing morphology of small cells that persisted in the cell line. (C) Subculture 46, 4-day culture at 30°C. (D) Cell processes; scale, 10 μm.

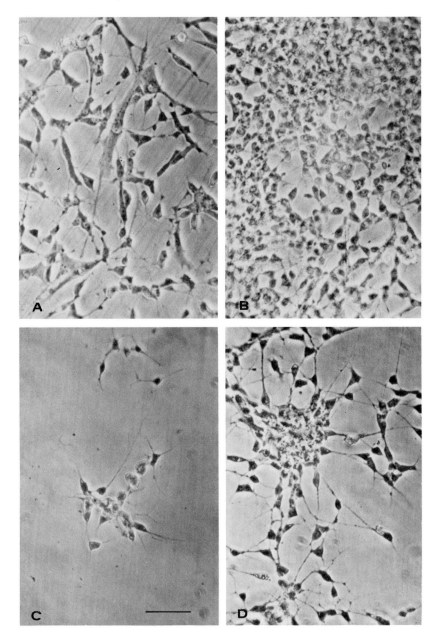

Fig. 5. (A) *Biomphalaria glabrata* cell line, subculture 36, showing large and small cell types. (B) Strain Bge-5, subculture 38, 6-day culture at 27°C. (C) and (D) Strain Bge-5, subculture 41. Appearance and migration of 130 peripheral cells and formation of dense central area. (C) Culture at 24 hours. (D) 12 days later. Scale, 50 μm.

There was at least a 4-fold increase in cell numbers at each subculture, and populations reached 0.5 to 1.2 $\times$ 10^7 cells per T-25 flask, not including cells in clumps that could not be dispersed for counting. Cell viability was 93% or higher. However, there was considerable cell loss at subculture, as shown by the attachment after 17 hours of only 54–74% of the cells inoculated.

Population doubling time of 18 hours at 27°C was determined from daily counts made during passages 44–46, using an inoculum of 5 $\times$ 10^5 cells. With subcultures from parent cultures older than 3 days, the rate of cell growth was slower and doubling time increased to 74 hours from 7-day cultures and 100 hours from 12-day cultures. In such cases the subcultures continued to improve over 16–20 days incubation.

In petri dish cultures there was still greater cell loss. By 3 hours only 0.3% of the cells had attached. This increased to 25% at 18 hours. Plating efficiency tests and attempts at cloning showed that inocula of less than 1 $\times$ 10^4 cells could not be used. Population doubling times in petri dish cultures averaged 46 hours (range, 32–60).

In the thirtieth passage, improved growth was noted in one of the series of cultures (series 5) incubated at 27°C. This characteristic has been retained in subsequent subcultures and the series has been designated the Bge-5 strain. The strain originated in a flask that had been incubated 25 days at 27°C. Two successive subcultures, both of which had yielded clumpy growth, had already been taken from the flask on days 4 and 15. It appears that the successive treatments with trypsin had selected cells that were well attached. Checks were made at that time to be sure that the change in cell growth was not the result of an inadvertent change in the medium composition. The strain has since been subcultured 18 times (through December 1974). A typical culture is shown in Fig. 5B. The strain is marked by a decreased tendency to form clumps. There was less cell loss at subculture, and at 17 hours the population had reached 110% of the inoculum. The doubling time in passage 42, was 20 hours at 27°C, 18 hours at 30°C, and 67 hours at 25°C. Doubling time was also determined by direct observation of cell growth. In cell counts of the area shown in Figs. 5C and 5D, the doubling time was 28 hours.

The mitotic index was calculated on 500 cells from each of four preparations. One- and 3-day cultures from T-47 of the original line and T-45 of the Bge-5 strain were used. The index, which did not appear to be affected by the cell strain, averaged 1.32% and the range was 0.98–1.58%. After colcemid arrest the number of dividing cells increased to 2.5% with a 4-hour exposure to colcemid, and to 3.17% with an 18-hour exposure.

The stability of the chromosomes was continually monitored in the

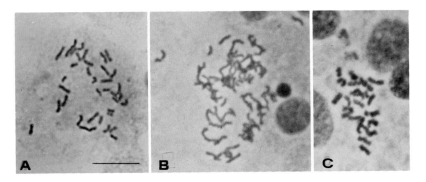

Fig. 6. Chromosomes in *B. glabrata*. (A) and (B) Cell line; colcemid 6 hours, hypotonic swelling 30 minutes, aceto-orcein stain. (A) Subculture 31, $2n = 36$. (B) Polyploid nucleus in subculture 35. (C) Embryo; colcemid 18 hours, aceto-orecin squash; Scale, 10 μm.

original line and in the Bge-5 strain. In the latest subcultures of the original line (T-42 to T-47) 2500 metaphases were examined. The diploid number of chromosomes, $2n = 36$ (Burch, 1960), was retained (Fig. 6A). The chromosomes were similar to those in squash preparations of snail embryos (Fig. 6C). Chromosomes of the Bge-5 strain were checked from T-38 through T-45. Among 5750 metaphases examined, there were five that appeared to be heteroploid (Fig. 6B).

In addition to the above cultures in medium S-301, other series were carried in the initiating medium M-260. In this medium the cells were somewhat larger but populations were lower. The series was abandoned in T-28. Other cultures were tested with a variety of conditions and media. Cultures were not harmed by several hours at 31°C. At 37°C, 20% of the cells of the inoculum attached but there was no increase in cell numbers. With low serum levels (0.5–2%) there was little cell increase but cultures persisted for 3 months with a single medium replacement. Within a pH range of 6.8–7.5, the range of 7.1–7.3 proved to be optimum, the medium then increasing in alkalinity to 7.4 during the incubation period. Cultures to which medium S-130 was added showed the contractility that had been observed in primary cultures (Hansen and Perez-Mendez, 1973).

Occasionally subcultures attached poorly with the cells forming small dense clumps. These were abandoned in favor of cultures proliferating as monolayers. Since the successful series were of similar age, failure of a series was not directly attributable to age. Clumping occurred at all three temperatures tested and was not eliminated by decreasing the serum level to 2%, decreasing the osmolality, or including 0.15% Methocel (15 cps) in the medium. Clumping in diploid lines from *D. melanogaster* embryos has been noted by Echalier and Ohanessian (1970). Attempts to eliminate clumping in our subcultures by changing the dispersing solution or the trypsin level, or by increasing or omitting EDTA were not successful. Finally, however, the appearance of the Bge-5 strain at T-30 circumvented this problem.

H. PROLONGED MAINTENANCE OF THE CELL LINE

With occasional replenishment of the medium, cultures retained good morphology for 6 weeks at 27°C and for 8 weeks at 25°C. Cell viability at subculture was 75%. At intervals, cultures from higher temperatures were stored at 17° or 20°C. Subcultures were made after 13 weeks at 20°C and after 18 weeks at 17°C; cell viability in the latter culture was 61%. Frozen storage is being investigated at other laboratories.

III. Discussion of Factors and Problems in Development of Bge Cell Line

A. SELECTION OF TISSUE SOURCE

Development of a cell line from *B. glabrata* presented three major problems: selection of a tissue source, sterilization of this material, and design of a suitable medium. Embryos were selected for the tissue source since these had been successfully used in starting several insect lines. It was then established that any embryo beyond the blastula stage could be used. A size was chosen that was as large as possible without having an excessive amount of the tough foot tissue and shell, which would interfere with the preparation of cell suspensions. The resulting primary cultures contained many types of cells. The small cell that eventually gave rise to the propagable line had already been noted in a few of the preceding cultures. Its rather low incidence could be due to the considerable loss of unattached cells that occurs at the start of primary cultures. The variety of cells in the primary snail cultures resembles those described for *Drosophila* cultures (Shields and Sang, 1970). However, the small cells of the Bge line appear to be distinctive. With further work it may be possible to identify their source in the developing embryo (Camey and Verdonk, 1970; Kumé and Dan, 1968).

In the early phases of the present work, an increase in the probability of obtaining propagable cells was attempted by decontrolling embryonic development. Early embryos were irradiated or treated with lithium chloride (Raven, 1952). Growth was changed and deformed embryos developed, but cell cultures derived from these embryos did not give rise to dividing cells.

B. STERILIZATION

Sterilization was particularly difficult because of the permeability of the egg capsules and their surrounding envelope (Beadle, 1969). High levels of antibiotics inhibited embryo development, possibly owing to salts included in commercial antibiotic solutions. The use of axenic snails was considered but discarded, owing to the low rate of egg laying in such colonies. Stimulation with vitamin E (Vieira, 1967), which might be effective, has not been tested. Therefore, ordinary aquaria were used and egg laying stimulated by addition of compressed rat food to the standard lettuce diet, as previously noted.

Egg masses were rendered aseptic by repeated washing. The brief treatments with iodine and Hyamine eliminated amebas and micrometazoa without injury to developing embryos. A wide range of antibiotics was used for treating egg capsules and intact embryos. Specific sensitivity tests were not conducted because of the continually changing nature of the microflora in the aquaria. However, after the embryos had been fragmented, the tissues were no longer exposed to antibiotics, thus avoiding possible inhibition of cell division. The adopted schedule proved satisfactory. When attempts were made to abbreviate it, contaminants were encountered.

C. CELL SEPARATION

Many methods of disintegrating the tissues were tried (Paul, 1970). They all resulted in good primary cultures. Individual cells attached readily; fragments attached and became surrounded with a halo of fibroblastlike cells. While such cultures could be maintained for months, there was little cell division. In some cases, large vesicles developed in the fragments. Attempts to liberate cells from these vesicles, as had been reported by Schneider (1972) with *Drosophila* explants, were not successful.

Snail embryo tissue proved to be unusual in that it was resistant to mechanical teasing and grinding. Cell fragments reaggregated and continued to develop as ciliated and differentiated organs. A similar reaggregation and differentiation had been noted by Farris (1968) and was recently reported by Basch and Diconza (1974). Pressing the trypsin-EDTA treated tissue through the fine sieve resulted in a suspension of single cells or small cell groups that were unciliated and did not reaggregate.

D. SELECTION AND TESTING OF MEDIA

The composition of media for mollusk cell cultures has been summarized and discussed by Flandre (1971). Our approach to formulation of a suitable medium was to design a semisynthetic medium based on available, albeit incomplete, information on hemolymph composition.

The most important characteristic of B. glabrata hemolymph is its low osmolality. The carbohydrate galactose seems to have a special role in snail metabolism and occurs in the unique polymer, galactogen (Goudsmit, 1972). Analysis of hemolymph (C. L. Liu and E. L. Hansen, unpublished) showed a concentration of galactose of 0.28 mg/100 ml. Galactose was therefore included in all our formulations. Trehalose, a component of snail tissue, was also included, although we could not detect it in hemolymph at the lowest level of our test method, 2 mg%.

A complete range of amino acids was included. Those reported in hemolymph are included in M-182 at about a 3-fold level. Salt composition followed that of Chernin (1963). Vitamins and other minor components were included, following the composition of F-12, a completely synthetic medium for vertebrate cells (Ham, 1974). Commercially available media were also tried, including RPMI-1640 (GIBCO) and a serum-free medium HI-WO5/BA 2000 (International Scientific Industries). These had to be diluted to about one-third to adjust to the required osmolality. A salt-free concentrate of CMRL-1969 (Healy et al., 1971) was generously provided by G. M. Healy (Connaught Laboratories, Ontario, Canada).

Despite the differences in composition of these media, primary cultures were readily established as long as the osmolality was within the 108–266 mOsm range and the pH was maintained within a 6.9–7.3 range. Still other media were used by Basch and Diconza (1973, 1974). It was similarly noted by Schneider (1972) that certain insect tissue cultures can be successfully carried in a variety of media.

In our attempt to stimulate or select propagable cells, most of the ingredients that had proved useful in other cultures were tried. These included chick embryo extract, chick plasma, egg ultrafiltrate, soy peptone, hemoglobin, vitamin C, high levels of vitamins, taurine, adenosine, and hypoxanthine. A gas phase of increased CO_2-decreased oxygen was tested. In all, 318 variants were made and tested. Although colonies of new cells occurred in several different media, their appearance was not prompted by a particular formulation. The medium in which the cell line originated and was subsequently subcultured was among the simpler formulations employed. Compounds that had stimulated mitosis in other systems, such as zinc (Williams and Loeb, 1973) and phytohemagglutin, elicited no response in these mollusk primary cultures.

Once the cell line appeared, it became possible to test the effect of components on cell growth. These were tested as variants of S-301. Addition of galactose improved cell growth as did the addition of lactalbumin hydrolysate. Lactalbumin hydrolysate is available in several preparations, of which Edamin-S proved to be the best. Two lots were tested. The value of a lactalbumin hydrolysate as a peptone source was discussed by Echalier and Ohanessian (1970). Bactopeptone was used as a satisfactory variant of S-301.

E. SERUM: TYPES TESTED AND ROLE IN THE CELL CULTURE

Synthetic media are conventionally supplemented with homologous serum. *Helix* hemolymph was included in the medium for culture of foot explants (Flandre and Vago, 1963; Hink, 1972b). However, there is increasing evidence from insect tissue culture that the specific proteins of hemolymph could be substituted by vertebrate serum, fetal calf serum being the most widely used. Sera tested in this study include those from goats, horses, sheep, and man. All except fetal calf serum (FCS) were toxic, the cells developing inclusions or becoming detached. The FCS itself was toxic unless "inactivated," i.e., heated at 56°C for 30 minutes. There was considerable variation between serum lots of FCS. It should be particularly noted that serum was toxic in these cultures if it was noticeably red from hemolysis or contained floating lipid material after thawing.

Even among otherwise normal-appearing serum lots, differences in cell response were noted. It appears likely that all serum lots are somewhat toxic and that the differences observed are a matter of degree. The best method to test serum lots employed in this work was to establish a culture in a good medium, replace it with the test medium, and later transfer the culture with the test medium. Several serum lots were screened in this manner, resulting in selection of three that provided an adequate reserve supply. Even with the selected serum there was evidence of toxicity to the cells as seen by formation of granules. Serum was necessary and attachment would not occur without it. The toxicity, however, was greatly reduced by replacement with a serum-free medium once the cell suspension had attached. The small foci of new cells occurred most frequently in primaries that had been subjected to this change of medium.

The rationale for the series of media used in starting the line was that the cells should be exposed first to a medium resembling hemolymph as closely as possible. Then serum-free medium was put on at intervals to decrease possible toxic or inhibitory effects from the serum. Finally, once the cell line had shown vigorous growth, a medium based on Schneider's medium was used. Schneider's medium is commercially available and should improve the ease of handling and general utility of the Bge cell line.

F. DEVELOPMENTS LEADING TO THE CELL LINE

The original expectation had been that appearance of foci of dividing cells would lead directly to a cell line. The new colonies in the snail culture differed from those initiating insect cell lines (e.g., Mitsuhashi, 1973) in that the mollusk dividing cells appeared early in the primary cultures but did not immediately increase to numbers sufficient for subculturing. Rather, colony growth was limited, and after a time the colony consisted of a dense center with a radiating periphery that ceased to extend. It was not possible to stain the colonies at this point in order to check for dividing cells, but their appearance suggested that dividing cells originated in the center and division ceased as this region became crowded. Colonies could be lifted with trypsin and occasionally would remain intact and reattach. It appeared that these cells might be too old to respond to the stimulation of trypsinization, so the procedure of *early* trypsinization was introduced. This had the effect of reducing the confluent layer of nonproliferating cells and of removing degenerating cells. As cells of the new colony appeared, they were partially redistributed by the reseeding process, so there was little or no possibility of contact inhibition. Their continued growth seemed to be stimulated by treatment with dilute trypsin. The trypsinization effective for reseeding the small cells was much less than that used by Pudney and Varma (1971) in starting an *Anopheles stephensi* line, and most of the snail cells remained attached. More extensive trypsinization resulted in complete loss of the colony cells.

The essential requirement for starting the cell line seems to have been the occurrence in the culture of a particular cell type under conditions in which these cells could continue to divide. These conditions apparently were provided by a series of media changes and trypsin treatments. The cell line did not simply emerge from long-term maintenance of a primary culture, but originated from a series of planned moves designed to liberate dividing cells from the restraining influence of nonproliferating cells and to stimulate them to continue dividing.

The sudden increase in cell division that occurred at the start of the cell line after an incubation period of 50 days suggests that a cultural alteration had taken place (Pontén, 1971). This alteration was not accompanied by a change in chromosome number, because the cells remained essentially diploid. The emergence of strain Bge-5 suggests that further spontaneous changes followed by selection may well occur.

The Bge line has maintained good cell growth through 52 subcultures. Additional cultures are needed to determine whether the line can be continued indefinitely or whether it will prove to be finite (Hayflick and Moorhead, 1961). Established near-diploid lines for mouse tissue have been described (e.g., Farber and Liskay, 1974). Any onset of change in the Bge cell line can be studied through maintenance of separate series that show different growth rates at various temperatures.

G. MAINTENANCE

Continuation of the cell line in a vigorous condition is relatively simple as long as several conditions are met. Four university students were briefly trained and, after a few false starts, succeeded in routinely making successful transfer and maintenance of the line. Quality of the serum lot for the medium is especially important. Transfers should be made from young vigorous cultures. It is particularly important for this cell line to use *dilute* trypsin and to remove it as completely as possible. Overtrypsinization has a deleterious effect on the cells. Incubation should be in the 27°–30°C range.

The cell line described here has been distributed to other laboratories as well-attached young cultures. Some difficulties were initially encountered, owing to the problems mentioned above, but subsequent successful maintenance is currently underway. It is hoped that the detailed consideration of the Bge cell line in this presentation will permit other investigators to avoid some of the difficulties that we have encountered and will speed up applications of this potentially valuable resource in host–parasite interaction studies.

IV. Summary

A cell line has been established from embryos of the pulmonate snail *Biomphalaria glabrata*, an important trematode intermediate host. The "Bge" cell line (for *B. glabrata* embryo) was initiated by treating a colony of new cells within a subcultured primary culture several times with highly dilute trypsin and periods of serum-free medium. The line has been maintained through 52 subcultures as of December 1974. It is incubated at temperatures of 25°–30°C and can be held at 17°C for at least 3 months. The population doubling time at 27°C is 18–20 hours. A suitable medium consists of Schneider's *Drosophila* medium diluted to 22% with addition of galactose, lactalbumin hydrolysate, and inactivated fetal calf serum. The cells are essentially diploid ($2n = 36$). A particularly viable strain, Bge-5, was derived from the thirtieth subculture.

The method of liberating and encouraging dividing cells from young primaries differs from techniques usually employed and should be of value in initiating cell cultures from other invertebrate tissue sources— particularly molluscan cells, which have been particularly difficult to culture.

Acknowledgments

The author is indebted to Dr. K. Walen for chromosome studies of snail embryos and for tritiated thymidine treatment of primary cultures. She and Dr. J. R. Allen participated in establishing the primary cultures. Special thanks are due to Marguerite Woodmansee for her careful work and perceptive observations. Thanks are also due to Dr. C. Blank, G. Perez-Mendez, and R. Yescott for design and preparation of the media. The advice and encouragement from Dr. I. Schneider are greatly appreciated. It is a pleasure also to acknowledge the valuable conferences with Drs. M. Voge, C. Bayne, P. Basch, and E. Chernin and their present collaboration in carrying the cell line. The line is also being maintained at the Hooper Foundation, University of California at San Francisco, with the assistance of Dr. K. H. Jeong.

Supported by NIAID Grant A1-07359 and Contract N01-A1-22525 from USPHS for work at the former Clinical Pharmacology Research Institute, Berkeley, California.

References

Barile, M. F. (1973). *In* "Tissue Culture: Methods and Applications" (P. F. Kruse and M. K. Patterson, Jr., eds.), pp. 729–735. Academic Press, New York.

Basch, P. F., and Diconza, J. J. (1973). *Amer. J. Trop. Med. Hyg.* **22**, 805–813.

Basch, P. F., and Diconza, J. J. (1974). *J. Invertebr. Pathol.* **24**, 125–126.

Beadle, L. C. (1969). *J. Exp. Biol.* **50**, 473–479.

Benex, J. (1966). *Ann. Parasitol.* **41**, 351–378.

Brooks, M. A., and Kurtii, T. J. (1971). *Annu. Rev. Entomol.* **16**, 27–52.

Burch, J. B. (1960). *Z. Tropenmed. Parasitol.* **11**, 449–452.

Burch, J. B., and Cuadros, C. (1965). *Nature (London)* **206**, 637–638.

Camey, T., and Verdonk, N. H. (1970). *Neth. J. Zool.* **20**, 93–121.

Cheng, T. C., and Arndt, R. J. (1973). *J. Invertebr. Pathol.* **22**, 308–310.

Chernin, E. (1963). *J. Parasitol.* **49**, 353–364.

Chernin, E. (1964). *J. Parasitol.* **50**, 531–545.

Echalier, G., and Ohanessian, A. (1970). *In Vitro* **6**, 162–172.

Farber, R. A., and Liskay, R. M. (1974). *Cytogenet. Cell Genet.* **13**, 384–396.

Farris, V. K. (1968). *Science* **160**, 1245–1246.

Flandre, O. (1971). *In* "Invertebrate Tissue Culture" (C. Vago, ed.), Vol. 1, pp. 361–383. Academic Press, New York.

Flandre, O., and Vago, C. (1963). *Ann. Epiphyt.* **14**, 161–171.

Gilbertson, D. E., Etges, F. J., and Ogle, J. D. (1967). *J. Parasitol.* **53**, 565–568.

Goudsmit, E. M. (1972). *In* "Chemical Zoology" (M. Florkin and B. T. Scheer, eds.), Vol. 7, pp. 219–243. Academic Press, New York.

Ham, R. (1974). *In Vitro* **10**, 119–123.

Hansen, E. L. (1974). *Int. Res. Commun. Syst.* **2**, 1703.

Hansen, E. L., and Perez-Mendez, G. (1973). *Int. Res. Commun. Syst.* (73–11) 1-0-12.

Hayflick, L., and Moorhead, P. S. (1961). *Exp. Cell Res.* **25**, 585–621.

Healy, G. M., Teleki, S. V., Seefried, A. V., Walton, M. J., and Macmorine, H. G. (1971). *Appl. Microbiol.* **21**, 1–5.

Hink, W. F. (1972a). *Advan. Appl. Microbiol.* **15**, 157–214.

Hink, W. F. (1972b). *In* "Invertebrate Tissue Culture" (C. Vago, ed.), Vol. 2, p. 380. Academic Press, New York.

Jones, B. M. (1966). *In* "Cells and Tissues in Culture" (E. N. Willmer, ed.), Vol. 3, pp. 397–457. Academic Press, New York.

Kumé, M., and Dan, K. (1968). "Mollusca," Chap. 2, pp. 485–525. Nolit Publishers, Belgrade, Yugoslavia. (translated from Japanese by J. C. Dan, TT-67-58050. National Library of Medicine, U.S. Pub. Health. Serv., Washington, D.C.).

Lee, F. O., and Cheng, T. C. (1972). *Exp. Parasitol.* **31**, 203–216.

Mitsuhasi, J. (1973). *Ann. Entomol. Zool.* **8**, 64–72.

Newton, W. L. (1955). *J. Parasitol.* **41**, 526–528.

Paul, J. (1970). "Cell and Tissue Culture," 4th ed. Williams & Wilkins, Baltimore, Maryland.

Pontén, J. (1971). "Spontaneous and Virus-Induced Transformation in Cell Culture." Springer-Verlag, Berlin and New York.

Pudney, M., and Varma, M. G. R. (1971). *Exp. Parasitol.* **29**, 7–12.

Raven, P. (1952). *J. Exp. Zool.* **121**, 1–78.

Schafer, T. W., Pascale, A., Shimonaski, G., and Came, P. E. (1972). *Appl. Microbiol.* **23**, 565–570.

Schneider, I. (1964). *J. Exp. Zool.* **156**, 91–104.

Schneider, I. (1966). *J. Embryol. Exp. Morphol.* **15**, 271–279.

Schneider, I. (1972). *J. Embryol. Exp. Morphol.* **22**, 353–365.

Shields, G., and Sang, J. H. (1970). *J. Embryol. Exp. Morphol.* **23**, 53–69.

Stanley, M. S. M. (1972). *In* "Growth, Nutrition, and Metabolism of Cells in Culture" (G. H. Rothblat and V. J. Cristofalo, eds.), Vol. 2, pp. 327–370. Academic Press, New York.

Vieira, E. C. (1967). *Amer. J. Trop. Med. Hyg.* **16**, 792–796.

Williams, R. O., and Loeb, L. A. (1973). *J. Cell Biol.* **58**, 594–601.

Insect Hormones

7

Biosynthesis of α-Ecdysone by Prothoracic Glands *in Vitro*

HARUO CHINO

I. Introduction

Since the classic experiments by Fukuda (1940) on the commercial silkworm *Bombyx mori,* it has generally been believed that the hormone responsible for the molting process in all insects is a product of the prothoracic glands or analogous organs. This crystalline molting hormone was first isolated from the pupae of *Bombyx* by Butenandt and Karlson (1954) and named ecdysone. Since it has been shown that this compound elicits all the effects of active prothoracic glands, ecdysone has naturally been assumed to be their product. However, there has been no direct evidence to establish this. In fact, when Butenandt and Karlson (1954) isolated ecdysone from *Bombyx* pupae, they extracted it from the whole body of animals but not from the prothoracic glands themselves. In addition, Nakanishi *et al.* (1972) have found ^{3}H-ecdysone in the extract from isolated abdomens of *Bombyx* larvae, which were free of prothoracic glands and had been injected with ^{3}H-cholesterol, a precursor of ecdysone (Karlson and Hoffmeister, 1963). Locke (1969) has suggested from his electron microscopic observations of *Calpodes ethlius* that the oenocytes (subspiracular glands) may be responsible for the secretion of ecdysone. This suggestion was also supported by Bonner-Weir (1970), who conducted ligation experiments on the same species. All these endocrinological studies, together with the findings that considerable amounts of ecdysones or ecdysonelike substances exist in a large variety of plants (Nakanishi *et al.*, 1966; Takemoto *et al.*, 1967), have compounded the fact that we have little knowledge of the insect molting hormone with respect to the elementary question of whether the prothoracic gland is really the site of ecdysone synthesis.

We believe that the crucial experiment to answer this elementary question should be done by means of organ culture of the isolated prothoracic glands. In fact, several investigators have been utilizing this technique to resolve this question. Willig *et al.* (1971) have reported that the cultured ring gland complexes of *Calliphora erythrocephala* (the organ analogous to the prothoracic gland) produce an inactive hormone precursor, presumably an α-glucoside of ecdysone, which may be converted to active ecdysone *in vivo*. Kambysellis and Williams (1971) have observed that meiosis and spermatogenesis of cultured germinal cysts of *Samia cynthia* occur only in the presence of ecdysones or active prothoracic glands, suggesting that prothoracic glands secrete ecdysones or related compounds. A similar report by Agui *et al.* (1972) demonstrated that the larval integument of *Chilo suppressalis* undergoes molting when the tissues are cultured together with active prothoracic glands or when cultured in a medium in which active prothoracic glands have been precultured.

These studies seem to favor the theory that the prothoracic gland is the site of ecdysone synthesis. However, not only the chemical data presented are insufficient, but also the amount of hormone produced in such culture systems has been too small to provide definite proof that this gland is really responsible for ecdysone synthesis.

II. An Improvement of the Culture System

A. CULTURE MEDIUM

In most of the above investigations, synthetic culture media such as Grace's and Mitsuhashi's have been employed. These synthetic media do not contain cholesterol. Since cholesterol is known to be a precursor of ecdysone (Karlson and Hoffmeister, 1963), the prothoracic glands may require cholesterol when synthesizing ecdysone. In this connection, one should bear in mind that, unlike mammalian tissues, insect tissues are unable to synthesize sterols from small molecules such as acetate, and that insects, therefore, must obtain sterols from their food source. Since insects have an open circulatory system, cholesterol may be directly obtained from cholesterol in the hemolymph for *in vivo* synthesis of ecdysone.

This author has isolated and purified two major lipoproteins from the pupal hemolymph of the silkworm *Philosamia cynthia* while studying lipid transport in insects (Chino *et al.*, 1969; see also review of Gilbert and Chino, 1974). These two lipoproteins (named lipoprotein I and lipoprotein II) are globular in shape (see Fig 1), and the molecular weights are 700,000 and 500,000, respectively. Both lipoproteins contain a considerable amount of cholesterol in addition to a diglyceride which is a major neutral lipid in both lipoproteins. In addition, Chino and Gilbert (1971) have found that these lipoproteins, particularly lipoprotein I, have the physiological function of taking cholesterol from the midgut and carrying it to the site of utilization. We have also shown that cholesterol does not exist in hemolymph in a free form, but that all cholesterol present in hemolymph is associated with either lipoprotein I or II. These studies have naturally led us to attempt culturing prothoracic glands in insect hemolymph itself, since it is assumed that hemolymph would be physiologically the most satisfactory medium and that the cholesterol associated with the lipoproteins mentioned above may be utilized as the substrate of ecdysone (Chino *et al.*, 1974).

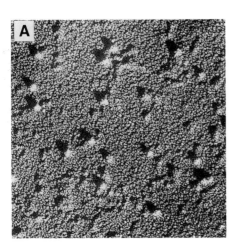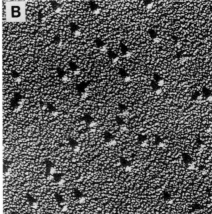

Fig. 1. Electron micrographs of hemolymph lipoproteins by shadowing method. (A) lipoprotein I; (B) lipoprotein II (×125,000).

Therefore, we have used hemolymph from diapausing (dormant) pupae of the *Cynthia* silkworm, after removing the hemocytes by centrifugation. This hemolymph is uniform in its chemical nature throughout the diapausing period. However, an expected technical difficulty arose when the hemolymph darkened soon after exposure to air, because of a tyrosine–tyrosinase reaction. It may theoretically be possible to use some chemicals such as phenylthiourea to prevent such darkening, but these reagents are poisonous, or at least physiologically deleterious, to the tissues. Therefore, we subjected the hemolymph to a Sephadex column to separate the protein fraction from the low molecular substances, including tyrosine. For each separation, 2–3 ml of hemolymph were applied to a Sephadex column (1 × 15 cm) equilibrated with saline solution (2.98 gm KCl, 3.04 gm $MgCl_2 \cdot 6H_2O$, and 3.70 gm $MgSO_4 \cdot 7H_2O$ in 1000 ml redistilled water). During the chromatographic process, yellow pigments (carotenoids) associated with both lipoproteins I and II acted as a marker for following the migration of the protein fraction (Chino *et al.*, 1969). The collected protein fraction was diluted with an equal volume of Wyatt's culture medium (Wyatt, 1956) omitting tyrosine, filtered through a Millipore filter, and then used as the culture medium. This medium, referred to as "hemolymph medium," contained about 40% of the original hemolymph based on protein concentration, and never darkened even after prolonged exposure to air.

B. SUPPLY OF EXCESS OXYGEN TO PROTHORACIC GLANDS

Ishizaki (personal communication) has suggested that there is a close subcellular similarity between prothoracic glands and the mammalian adrenal cortex. In the latter tissue, it is known that a number of steroidal hormones are synthesized by hydroxylation of cholesterol in the presence of a "mixed function oxygenase system." Although the enzymatic pathway of ecdysone synthesis from cholesterol is still unknown, it may be assumed that there is a similar pathway for the synthesis of ecdysone, i.e., the hydroxylation of cholesterol by a mixed function oxygenase system. In this connection, it should be noted that the adrenal cortex has a well-developed aortic system, while prothoracic glands are connected with a highly developed tracheal system. These anatomical facts strongly suggest that both the adrenal cortex and the prothoracic glands require an ample supply of oxygen to function adequately. Thus, we continuously supplied an excess of oxygen (0.5 unit partial pressure) to the prothoracic gland culture.

III. Synthesis of Molting Hormone

A. CULTURE OF PROTHORACIC GLAND AND BIOASSAY OF THE MOLTING HORMONE

The larvae of *B. mori*, 1 day after spinning, were immersed for 2 minutes in 0.05% mercuric chloride in 70% ethanol, followed by two rinses with sterile distilled water. The prothoracic glands were carefully dissected from the sterilized larvae and placed in a small petri dish filled with insect Ringer's solution, where the contaminating fragments were removed. A pair of glands was then transferred into a drop (0.04–0.06 ml) of the culture medium in another sterile petri dish, which was then inverted and placed in a small incubator (30 × 30 × 30 cm). A mixture of 60% air and 40% oxygen (equivalent to 0.5 unit oxygen partial pressure) was continuously supplied to the incubator at the flow rate of 10 liters per hour. In one series, more than 20 cultures were maintained simultaneously, and the culture period was usually 3–5 days at 25°C.

After cultivation, the hormone was extracted from the glands or the culture media according to the method of King and Siddall (1969), using organic solvents such as methanol and butanol. The hormone extract was then bioassayed with the use of *Sarcophaga* test abdomen to determine the hormonal activity. In this bioassay technique, developed by Ohtaki *et al.* (1968), the activity was calculated from the standard curve and puparium index for β-ecdysone and expressed in nanograms of β-ecdysone equivalent.

B. PRODUCTION OF MOLTING HORMONE

The improved culture system, in which the dissected prothoracic glands were cultured in hemolymph medium under a continuous supply of excess oxygen, led to the first successful result. A typical set of data shown in Table I reveals that, before cultivation, the amount of hormonal activity detected is extremely low in the glands and completely lacking in the culture medium, whereas, after cultivation, considerable activity is found only in the medium and none found in the glands themselves. These results indicate that as soon as the hormone is synthesized by the glands, it is secreted into the medium, which may explain why other investigators have failed to detect an appreciable amount of molting hormone in extracts from the prothoracic gland itself.

Table II shows the efficiencies of various culture media; in comparison

TABLE I

Amounts of Molting Hormone Found in Glands and Media before and after Cultivation

Medium or organ	Gland cultures (No.)	Amount of hormone[a]
Before cultivation		
Prothoracic gland	40	2.5
Hemolymph medium (1 ml)		0.0
After cultivation		
Prothoracic gland	25	0.0
Hemolymph medium	25	120.0

[a] Expressed in nanograms of β-ecdysone equivalent per pair of glands. Culturing conditions are described in the text. From Chino et al. (1974); Science **183**, 529. Copyright 1974 by the American Association for the Advancement of Science.

TABLE II

Effect of Culture Medium on the Synthesis of Molting Hormone by Prothoracic Glands

Medium	Gland (No.)	Amount of hormone[a]
Grace's	14	4.8
Wyatt's	24	17.7
Hemolymph	25	120.0
Lipoprotein I	21	148.0
Lipoprotein II	16	108.0

[a] The amount of hormone found in culture medium after cultivation is expressed in nanograms of β-ecdysone equivalent per pair of glands. The amount of proteins contained in one culture medium was 1.3 mg for hemolymph, 0.54 mg for lipoprotein I, and 0.50 mg for lipoprotein II. From Chino et al. (1974); Science **183**, 529. Copyright 1974 by the American Association for the Advancement of Science.

with glands cultured in hemolymph medium, those in synthetic culture media produce only a small amount of hormone. It should be noted here that the amount of hormone shown in Table II represents the amount produced when the glands were cultured under an excess oxygen supply. When the glands were cultured in air, as described later, the amount of hormone produced was reduced to about one-fourth the values shown in Table II. Therefore, one can understand the poor yield of hormone (usually less than 5 ng per pair of glands) when the glands are cultured under such conditions. The reason why hemolymph medium, as opposed to synthetic media, is more effective in supporting hormone synthesis, as pointed out in Section II, may be partly explained by the assumption that the cholesterol associated with the two major hemolymph lipoproteins (lipoproteins I and II) may be utilized as a substrate for ecdysone synthesis. To test this assumption, lipoproteins were added to the culture medium. Lipoproteins I and II were isolated and purified from pupal hemolymph of the *Cynthia* after the method of Chino *et al.* (1969). The final lipoprotein fractions obtained were dialyzed against saline (2.98 gm KCl, 3.04 gm $MgCl_2 \cdot 6H_2O$, and 3.70 gm $MgSO_4 \cdot 7H_2O$ in 1000 ml of distilled water) for several hours, diluted with an equal volume of Wyatt's culture medium, filtered through a Millipore filter, and then used as culture medium. The results are given in Table II. It is seen that both lipoproteins I and II, as well as the hemolymph medium, are satisfactory as culture media and appear to be more efficient than hemolymph per milligram of protein.

As seen in Table III, a continuous supply of excess oxygen was also beneficial. Hormonal production by glands cultured under excess oxygen was almost four times greater than glands cultured in air.

We next examined the duration of *in vitro* hormone synthesis in our culture system. As shown in Table IV, the amount of hormone found in the culture medium reached its maximum on the third day of cultivation. After 3 days, hormone production ceased.

TABLE III

The Effect of Excess Oxygen Supply on *in Vitro* Synthesis of Molting Hormone by Prothoracic Glands

Partial pressure of oxygen (atm)	Gland cultures (No.)	Amount of hormone[a]
0.2	15	25
0.5	25	97

[a] As in Table II.

TABLE IV

Duration of *in Vitro* Synthesis of Molting Hormone by
Prothoracic Glands

Cultivation period (days)	Gland cultures (No.)	Amount of hormone[a]
1	25	38
3	29	96
5	29	81

[a] As in Table II.

C. CHANGE OF PROTHORACIC GLAND ACTIVITY DURING THE LAST INSTAR

Since it has been suggested by Fukuda (1944) and Shaaya and Karlson
(1965) that prothoracic glands of *Bombyx* larvae are most active in hormone secretion 1 day after spinning, all of the aforementioned experiments were done with *Bombyx* larvae at this stage. Thus we investigated
hormonal synthesis during the last instar. As seen in Fig. 2, hormonal

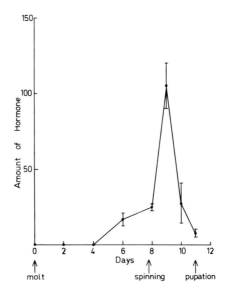

Fig. 2. Changes in the activity of prothoracic glands to synthesize molting hormone
during the last-instar larval stage. Ordinates represent the amount of hormone produced by a pair of glands expressed in nanograms of β-ecdysone equivalent.

synthesis was first detected on the sixth day. There was a marked increase in activity at the time of spinning. After reaching the maximum level 1 day after spinning, there was a sharp reduction of hormonal synthesis, reaching its lowest level at pupation time. These results are in agreement with the ecdysone titers reported by Shaaya and Karlson (1965).

IV. Chemical Nature of the Hormone Produced by the Cultured Prothoracic Gland

A. THIN-LAYER CHROMATOGRAPHY

Since it is well known that two different ecdysones, α-ecdysone and β-ecdysone, naturally exist in insect tissues (see Fig. 3 for the chemical structure), it seemed important to examine the chemical structure of the molting hormone produced *in vitro*. For this test, we first used thin-layer chromatography (TLC). The hormonal fraction was extracted from the culture medium after cultivating 40 pairs of glands. One-half the hormonal fraction was directly bioassayed to determine hormonal activity, and the other half was applied to a TLC plate (commercially prepared silica gel plate, Merck). Pure β-ecdysone was run at the same time as a marker. After the plate was developed, the spots were eluted with tetrahydrofuran from several regions, including the solvent front and the origin, and was then bioassayed. The results given in Table V clearly reveal that all the hormonal activity is located in a distinct region corresponding to β-ecdysone ($R_f = 0.53$). However, since it is well known that α- and β-ecdysone show close, sometimes overlapping, R_f's on TLC, these results were inconclusive.

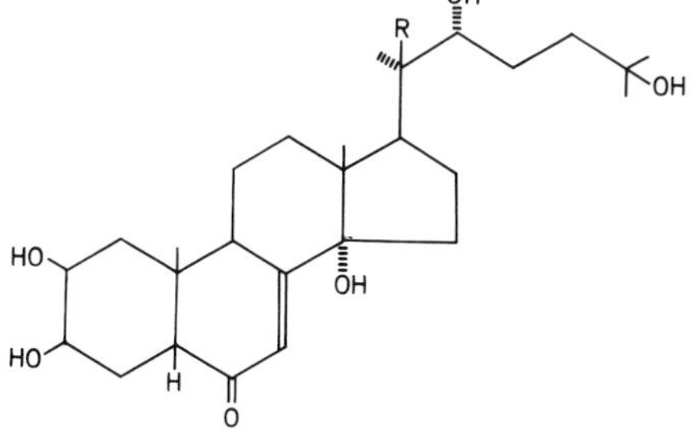

Fig. 3. Structure of ecdysones. R, H for α-ecdysone; R, OH for β-ecdysone.

TABLE V

Separation of the *in Vitro* **Produced Hormonal Fraction on Thin-Layer Chromatography**[a]

Region tested (Rf)	Hormonal activity[b] (%)
0.00–0.10 (mean = 0.05)	0
0.10–0.30 (mean = 0.20)	0
0.46–0.60 (mean = 0.53)	100
0.72–0.86 (mean = 0.79)	0
0.86–1.00 (mean = 0.93)	0

[a] The fraction (equivalent to 20 gland cultures) containing 2150 ng of the hormone as β-ecdysone equivalent was applied to a TLC plate which was developed with a mixture of chloroform and methanol (2:1). The spot of standard β-ecdysone was located at an R_f between 0.50 and 0.55 (mean = 0.53).

[b] The ratio of the hormonal activity found in each region to the total activity of the hormone originally applied to the TLC plate.

B. LIQUID CHROMATOGRAPHY

Further evidence of the chemical nature of the hormone was obtained from high-pressure liquid chromatography. An aliquot of hormonal fraction extracted from the culture medium equivalent to 200 pairs of glands was first applied to TLC. A portion of the fraction eluted from the region corresponding to ecdysone on TLC was subjected to bioassay, and the remaining portion was submitted to liquid chromatography. All fractions detected were collected individually and bioassayed. A typical chromatogram is shown in Fig. 4A. This chromatogram clearly demonstrates the presence of eight or more fraction peaks, indicating that the extract from the TLC plate still contains many different compounds. As seen in Fig. 4, fraction 8 is identical with α-ecdysone, but not with β-ecdysone with respect to retention time. Furthermore, fraction 8, when chromatographed with standard α-ecdysone, gave a distinct single peak. On the other hand, the results of the bioassay clearly revealed that eight fractions (first to seventh and ninth) had no hormonal activity at all, and that only fraction 8 had such activity, which was exactly equal to the total activity of the material applied originally to the chromatograph.

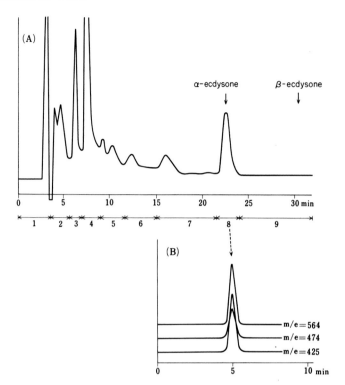

Fig. 4. (A) Liquid chromatogram of the hormonal fraction eluted from a TLC plate. High pressure liquid chromatography: instrument, Dupont 830 liquid chromatograph; column, Zorbax SIL (25 cm × 2.1 mm i.d.); mobile phase, chloroform and methanol (9:1); flow rate, 0.3 ml/minute; column pressure, 1000 pounds per square inch; column temperature, 20°C; and detector, ultraviolet photometer at 254 nm. (B) Mass fragmentogram of TMS derivative of fraction 8. Instrument, LKB 9000s MID PM 9060S; column, 1.5% OV-101 on Chromosorb W HP 80 to 100 mesh (1 m × 4 mm i.d.); column temperature, 270°C; ionization current, 60 μA; ionization voltage, 20 eV; accelerating voltage, 3.5 kV; and ion source temperature, 290°C From Chino *et al.* (1974); *Science* **183**, 529. Copyright 1974 by the American Association for the Advancement of Science.

C. MASS FRAGMENTOGRAPHY

In order to obtain definitive data on the hormonal structure, we subjected fraction 8 to mass fragmentography. This rather new technique is based on a combination of gas chromatography and mass spectrometry with a computer system. Fraction 8 was first converted to a trimethylsilyl (TMS) derivative according to the method of Miyazaki *et al.* (1973), and then applied to a gas-chromatography/mass-spectrometry system. The ions m/e 425, 474, and 564 which are most characteristic of α-ecdysone TMS derivative (Miyazaki *et al.*, 1973) were monitored. As shown in Fig. 4B, the retention time of each of these three peaks coincided exactly with that (5 minutes) of the α-ecdysone TMS derivative, and the relative intensities of these three peaks were satisfactorily consistent with those of the standard spectrum of the α-ecdysone TMS derivative reported by Miyazaki *et al.* (1973).

V. Conclusions

The results of this study, together with our present knowledge on the insect molting hormone, lead to the conclusion that the prothoracic gland is indeed the site of α-ecdysone synthesis. Another research team in the United States working independently has recently reached the same conclusion that the prothoracic gland is mainly responsible for α-ecdysone synthesis (King *et al.*, 1974). They cultivated the prothoracic glands of the larvea of the tobacco hornworm *Manduca sexta* in cuture medium containing 10% whole egg ultrafiltrate, and used both a bioassay and a radioimmunoassay for determining hormonal activity. The average yield of hormone produced by a pair of glands was reported to be 30–50 ng of an α-ecdysone equivalent. In contrast, we expressed the amount of hormone as nanograms of β-ecdysone equivalent. Since β-ecdysone is known to be approximately twice as effective as α-ecdysone in the *Sarcophaga* test (Ohtaki *et al.*, 1967), the yield of hormone in our culture system should be 250–300 ng per pair of glands, if expressed as α-ecdysone equivalent. The chemical structure of the hormone produced by prothoracic glands of the tobacco hornworm was determined by thin-layer chromatography, gas chromatography, liquid chromatography, and mass spectrometry. This analysis has shown that the hormone in question is identical in structure to α-ecdysone.

Furthermore, a research team in Germany (Romer *et al.*, 1974) has recently presented evidence that the cultured prothoracic glands of *Tenebrio molitor* secrete only α-ecdysone. However, the chemical structure of the hormone was determined only by thin-layer chromatography.

Since both α-ecdysone and β-ecdysone occur naturally in insect tissues, one may assume that α-ecdysone is produced by the prothoracid glands and is subsequently converted to β-ecdysone in other tissues including the target organs. In fact, several investigators have provided evidence for this assumption. For example, King and Marks (1974) have observed that cultured leg regenerate of cockroach, a target organ of ecdysone, is capable of slowly converting α- to β-ecdysone. It has also been reported by Romer *et al.* (1974) that when prothoracic glands and oenocytes of the *Tenebrio* larvae are cultured together, the α-ecdysone produced by prothoracic glands is converted by the oenocytes to β-ecdysone.

Thus, in general it appears that the role of the prothoracic glands is to supply the insects with a prohormone, α-ecdysone, which is converted into a more active hormone, β-ecdysone, in tissues other than prothoracic gland.

Acknowledgment

Most of this study was done at the Biological Institute, University of Tokyo, where this author worked until 1973, in collaboration with S. Sakurai who was a predoctoral fellow, and Dr. T. Ohtaki of the National Institute of Health, Tokyo. Important chemical contributions, including mass fragmentography, were made by Dr. N. Ikekawa of the Tokyo Institute of Technology and Dr. H. Miyazaki of the Nippon Kayaku Company, Tokyo.

References

Agui, N., Kimura, Y., and Fukaya, M. (1972). *Appl. Entomol. Zool.* **7**, 71.
Bonner-Weir, S. (1970). *Nature (London)* **228**, 580.
Butenandt, A., and Karlson, P. (1954). *Z. Naturforsch. B* **9**, 389.
Chino, H., and Gilbert, L. I. (1971). *Insect Biochem.* **1**, 337.
Chino, H., Murakami, S., and Harashima, K. (1969). *Biochim. Biophys. Acta* **176**, 1.
Chino, M., Sakurai, S., Ohtaki, T., Ikekawa, N., Miyazaki, H., Ishibashi, M., and Abuki, H. (1974). *Science* **183**, 529.
Fukuda, S. (1940). *Proc. Imp. Acad. (Tokyo)* **16**, 414.
Fukuda, S. (1944). *J. Fac. Sci., Univ. Tokyo* **6**, 477.
Gilbert, L. I., and Chino, H. (1974). *J. Lipid Res.* **15**, 439.
Kambysellis, M. P., and Williams, C. M. (1971). *Biol. Bull.* **141**, 541.
Karlson, P., and Hoffmeister, H. (1963). *Hoppe Seyler's Z. Physiol. Chem.* **331**, 298.
King, D. S., and Marks, E. P. (1974). *Life Sci.* **15**, 147.
King, D. S., and Siddall, J. B. (1969). *Nature (London)* **221**, 955.
King, D. S., Bollenbacher, W. E., Borst, D. W., Vedeckis, W. V., O'Connor, J. D., Ittycheriah, P. I., and Gilbert, L. I. (1974). *Proc. Nat. Acad. Sci. U.S.* **71**, 793.
Locke, M. (1969). *Tissue & Cell* **1**, 103
Miyazaki, H., Ishibashi, M., Mori, C., and Ikekawa, N. (1973). *Anal. Chem.* **45**, 1164.
Nakanishi, K., Koreeda, M., Sasaki, S., Chang, M. L., and Hsu, H. Y. (1966). *Chem. Commun.* **24**, 915.
Nakanishi, K., Moriyama, H., Okauchi, T., Fujioka, S., and Koreeda, M. (1972). *Science* **176**, 51.
Ohtaki, T., Milkman, R. D., and Williams, C. M. (1967). *Proc. Nat. Acad. Sci. U.S.* **58**, 981.
Ohtaki, T., Milkman, R. D., and Williams, C. M. (1968). *Biol. Bull.* **135**, 322.
Romer, F., Emmerich, H., and Nowock, J. (1974). *J. Insect Physiol.* **20**, 1975.
Shaaya, E., and Karlson, P. (1965). *Develop. Biol.* **11**, 424.
Takemoto, T., Ogawa, S., and Nishimoto, N. (1967). *Yakugaku Zasshi* **87**, 1474.
Willig, A., Rees, H. H., and Goodwin, T. W. (1971). *J. Insect Physiol.* **17**, 2317.
Wyatt, S. S. (1956). *J. Gen. Physiol.* **39**, 841.

8

The Uses of Cell and Organ Cultures in Insect Endocrinology

EDWIN P. MARKS

I. Introduction

A few years ago while visiting the University of Sussex, I had the honor of having tea with Dame Honor B. Fell, former director of the Strangeways Laboratory and one of the founders of the science of organ culture. When she learned I was working with insects, she told me she once had a student, Joseph Frew, who attempted to culture the imaginal discs from blow flies. He worked for some time, but failing to determine why eversion of the discs occurred, became discouraged and eventually gave up research for a medical career. She was pleased to learn that Frew's paper (1928) is now regarded as a classic and 25 years ahead of its time. The work of Kopec (1922) on the role of the brain in insect metamorphosis was known at this time, but the pioneering work of Wigglesworth (1934) and Fukuda (1944) on insect hormones was yet to be done. Thus, in a real way, insect tissue culture and insect endocrinology have grown up together.

It was not until 1953 that the potential of organ culture as a tool in endocrinological research was finally realized by Schmidt and Williams (1953), who demonstrated the effect of molting hormone on differentiation in germinal cysts of insect testes. At about this time, Kuroda (1954) began working with imaginal discs from *Drosophila*, and shortly thereafter Kuroda and Yamaguchi (1955) demonstrated the effects of cultured endocrine tissues on these imaginal discs.

By this time, the advantages of tissue culture methodology were clear, and during the 15 years between 1955 and 1970, insect cell and organ culture became established tools for insect endocrinology. Today, they are used by numerous investigators. Three developments are largely responsible for this success: (1) the increasing availability of defined culture media; (2) the availability of highly purified hormone preparations; and (3) the development of microanalytical methods for the detection of minute amounts of hormonal compounds. The progress made in insect cell and organ culture in endocrine research during this period has been documented in at least two reviews (Marks, 1970; Berreur-Bonnenfant, 1972). Rather than attempting a comprehensive review of all work done with hormones and insect tissue culture since 1970, I will highlight the major developments that have occurred and identify some new areas where tissue culture methodology is making significant contributions.

Prior to 1970, almost all work done with insects *in vitro* involved either the effects of a gland extract or a purified hormone on a specific target tissue. However, with the recent development of microassay methods, two new areas are developing that hold exciting possibilities. The first of these involves the culture of isolated endocrine glands and the recovery of the hormone from the medium. The second involves the use of cultured target tissues to investigate the mode of action, metabolism, and degradation of hormonal compounds. In these kinds of studies, the ability to add to or remove materials from the cultures and to use varying combinations and amounts of tissue gives a much greater flexibility than is available with living insects.

II. Effect of Molting Hormone on Various Target Tissues

Much of the work done since 1970 involves the use of organ cultures to study the effects of hormones on a variety of insect tissues. This effort, in effect continuing and expanding the studies begun in the 1960's, when highly purified hormone preparations first became available, has involved a number of target tissues.

A. IMAGINAL DISCS

Hormonal control of development in insect imaginal discs was reviewed by Oberlander (1972a). He divided the work into five areas: (1) morphogenetic changes induced by ecdysone; (2) the competence of the disc to respond; (3) the effects produced by different ecdysone analogues; (4) the role of α-ecdysone; and (5) the role of various cofactors. Since 1972, many of the questions raised by Oberlander in his review have been partially resolved, and a number of new questions have been asked.

The role of β-ecdysone in the induction of morphological changes in wing discs of *Sarcophaga* was investigated by Ohmori and Ohtaki (1973) and Ohmori (1974). They found that the response to β-ecdysone could be divided into two stages: the synthesis and accumulation of the requisite materials; and the morphogenetic movements necessary for changes in shape. They reported that development of wing discs was dependent on the concentration of β-ecdysone and on the length of exposure, which supports the "covert-event/overt-response" hypothesis of Ohtaki *et al.* (1968). The entire synthetic phase (DNA, RNA, and protein synthesis) could be initiated by β-ecdysone, and morphogenic development was completely stopped when metabolic inhibitors were used to stop the synthetic activities. The studies of Raikow and Fristrom (1971) and Chihara *et al.* (1972) on the development of *Drosophila* imaginal discs *in vitro* also showed that β-ecdysone initiated the synthetic phase and that it was more than 200 times as effective as α-ecdysone in doing so. Chihara *et al.* also found that massive doses of *Cecropia* juvenile hormone inhibited the initiation of evagination by β-ecdysone. They concluded from their dosage studies that β-ecdysone was the morphogenic hormone in *Drosophila* and that α-ecdysone was probably a prehormone.

Fristrom *et al.* (1973) worked with the same system and found that the effects of β-ecdysone were concentration-dependent and that low concentrations of hormone required longer exposure periods than high concentrations. They also found that inhibitors of RNA and protein synthesis prevented disc evagination, but inhibitors of DNA synthesis alone did not. Ohmori (1974), like Fristom *et al.* (1973), found that with very high concentrations of β-ecdysone, abnormalities and inhibition of evagination occurred. The development of *Drosophila* wing discs was also studied by Mandaron (1973). Mandaron reported that explants taken from late (day 6.5) third-instar larvae differentiated almost completely under the influence of α-ecdysone and produced a partially sclerotized cuticle. In addition, he found that only α-ecdysone and inokosterone would produce this effect and that the presence of even minute quantities of β-ecdysone inhib-

ited and introduced abnormalties into the developmental process. Since
β-ecdysone alone did not initiate development, Mandaron ruled out the
possibility that the activity of α-ecdysone was related to its conversion
to β-ecdysone by the tissue. More recently, Mandaron (1974) studied
the mechanism of disc evagination under the influence of α-ecdysone and
a variety of substances that affect morphogenesis. Cytocholasin B and
concanavalin A did not affect the binding of ecdysone though they effectively prevented evagination of the disc and the formation of cuticle.
Mandaron concluded that morphogenesis resulted from changes in the
physicochemical properties of the plasma membrane caused by the molting hormone. It is difficult to reconcile the conflict between the findings
of Mandaron and those of Chihara et al., Fristrom, and Ohmori concerning the role of α-ecdysone, but it is becoming increasingly evident that
ecdysone is secreted by the prothoracic gland in the α form and that
it is readily converted to the β form by the oenocytes, fat body, and
epidermal tissues (see Section III). Since the imaginal discs used by
Mandaron had probably received some ecdysone exposure in vivo before
explantation and since Chihara et al. (1972) and Ohmori (1974) found
that overdosing with β-ecdysone caused partial failure of eversion, the
inhibition of α-ecdysone activity by β-ecdysone may have represented
hyperecdysonism. Other aspects of this apparent conflict remain
unexplained.

The effect of ecdysones on lepidopteran imaginal discs has received considerable attention since 1972. Working with wing discs from Galleria,
Oberlander (1972b) reported that DNA synthesis initiated by α-ecdysone
was inhibited by the presence of both β-ecdysone and 22-isoecdysone.
From this and earlier work, Oberlander hypothesized that α-ecdysone initiated morphogenic processes by stimulating DNA synthesis and that the
momemtum of the process was maintained by β-ecdysone as the conversion from the α to the β form took place. Then, the increasing titer of
β-ecdysone shut down DNA synthesis and stimulated RNA and protein
synthesis. More recently, Bergtrom and Oberlander (1975) showed that
the initiation of RNA synthesis by α- and β-ecdysones may have multiple
and simultaneous, but independent, effects on a target tissue.

Oberlander and Tomblin (1972), working with Plodia wing discs, found
that very low concentrations of β-ecdysone (0.05 μg/ml) induced morphogenesis and tracheal migration, and higher concentrations (2–5 μg/ml)
induced the deposition of tanned, pupa-type cuticle. Suppression of the
ecdysone-induced cuticle deposition was obtained with massive doses of
juvenile hormone. Subsequently, Oberlander et al. (1973), while studying
the effects of several ecdysone analogues on lepidopteran wing discs,
found that though α-ecdysone stimulated cuticle deposition poorly if at

all, a combination of α- and β-ecdysones was more effective than β-ecdysone alone.

Dutkowski and Oberlander (1973) studied the effect of fat body on β-ecdysone-induced cuticle deposition in cultured *Plodia* wing discs. They concluded that a "fat body factor" was produced that stimulated and regulated the effects produced by the application of exogenous ecdysone. Also, Benson and Oberlander (1974) studied the effect of α- and β-ecdysones on the synthesis of protein by cultured *Galleria* wing discs. They found that protein synthesis could be increased by the application of either α- or β-ecdysone, but the former had a somewhat greater effect. The presence of fat body synergized the stimulation of protein synthesis induced by the presence of α-ecdysone.

The ecdysone-stimulated uptake and incorporation of labeled D-glucosamine by cultured imaginal discs from *Plodia* was studied by Oberlander and Leach (1975). Stimulation of uptake of the labeled material was produced by either α- or β-ecdysone, but localization of the incoporated material at the epidermal surface occurred only when β-ecdysone was present. Most of the localized label was removed by treatment with chitinase. Concentrations of β-ecdysone as low as 0.005 μg/ml stimulated uptake but were subthreshold for localization of the material at the epidermal surface. Agui and Fukaya (1973), working with imaginal discs from *Mamestra*, found that concentrations of β-ecdysone as low as 0.03 μg/ml produced tracheal migration and evagination of the wing discs. They also found that longer incubation periods were required when concentrations of the hormone below 0.75 μg/ml were used. Similarly, when 1 μg/ml doses of hormone were used, the period of incubation required with α-ecdysone was four times as long as when β-ecdysone was used. High concentrations of β-ecdysone caused abnormalities and degeneration of the organs, but a developmental pattern much like that observed *in vivo* was produced by α-ecdysone acting over a prolonged period of 4 days. When wing discs were cocultured with five pairs of prothoracic glands, similar evagination occurred. When larger numbers of prothoracic glands were used, deleterious effects were obtained, apparently the result of overdose with the prothoracic gland hormone. Interestingly enough, Agui and Fukaya (1973) did not report any evidence of cuticle deposition by the wing discs, even in the presence of large amounts of β-ecdysone. Nevertheless, the overall impression gained from the imaginal disc studies is that morphogenesis requires only minute doses of ecdysone and that larger doses may induce abnormal development and premature cuticle deposition. Furthermore, many, but not all, effects attributed to α-ecdysone can be produced by β-ecdysone when given in minute but prolonged dosages.

122   EDWIN P. MARKS

B. OTHER EPIDERMAL STRUCTURES

In an epidermal system, Wyatt and Wyatt (1971) used minced pupal wing tissue from *Cecropia* and showed that β-ecdysone was more effective in stimulating RNA synthesis than was α-ecdysone. In addition, Agui *et al.* (1972) and Agui (1973) studied the effect of cultured prothoracic glands and β-ecdysone on the process of molting in cultured fragments of the larval body wall and found that the prothoracic glands released an active agent that induced ecydsis. Subsequently, Agui (1974) developed a quantitative bioassay for molting hormone action *in vitro* by using body wall fragments. The results showed a correlation between the concentration of the hormone used and the length of exposure to it.

Using an *in vitro* system employing cultured regenerating legs from the cockroach *Leucophaea maderae*, Marks and Leopold (1971) studied the induction of cuticle deposition by β-ecdysone. They found that the response to the hormone depended on both the age of the donor and the amount of the exogenous hormone used. In a later study, Marks (1972) showed that the length of exposure of the tissue to the hormone was as critical as the amount of hormone used and that the repetition of small, subthreshold doses eventually produced a response. These findings, like those of Fristrom *et al.* (1973) and Ohmori (1974), gave strong support to the covert-event/overt-response hypothesis of ecdysone action proposed by Ohtaki *et al.* (1968). Dutkowski and Oberlander (1973) obtained similar results. Marks (1973a) subsequently used the leg regenerate system to compare the dose–response patterns obtained with cuticle deposition with those obtained with the differentiation of epidermal cells into setae; both were induced by β-ecdysone and had about the same concentration threshold, but they responded differently to increase in concentration of and exposure to the hormone. Moreover, the difference in response occurred in such a way that differentiation of the setae was complete before the cuticle was deposited. When the hormone concentration was increased too rapidly in the early stages of development, cuticle deposition occurred prematurely, and no setae were formed. Similar effects of hyperecdysonism were reported by Bullière (1973) in cultured legs from *Blaberus* embryos.

The cockroach leg regenerate system was also used to investigate the roles of α- and β-ecdysone in the induction of cuticle deposition. In his initial studies, Marks (1973b) found that even when large doses of α-ecdysone were used to initiate cuticle deposition, the response was often only partial, and the frequency of formation of setae was low. When the formation was analyzed by time-lapse photomicrography, it was apparent that when the explants were treated with α-ecdysone, cell activation occurred earlier than when treated with β-eceysone, but that the terminal stages of cuticle formation often failed completely. In a subsequent study, Marks (1973c) showed that with β-ecdysone, the frequency of response could be increased either by increasing the concentration of the hormone or the length of exposure to it. With α-ecdysone, an increase in the concentration of the hormone had little effect, and the frequency of response was almost entirely time-dependent. From this, he hypothesized that the response produced by α-ecdysone was caused by the slow conversion to β-ecdysone by the target tissue and that α-ecdysone per se was probably inactive. Recently, the conversion of α- to β-ecdysone by cultured leg regenerates was confirmed at the chemical level by King and Marks (1974). In addition to the demonstration by Marks and Leopold (1971) that β-ecdysone induced the deposition of a chitin-bearing cuticle, Marks and Sowa (1974) demonstrated that the chitinous portion of the cuticle was synthesized from soluble precursors in cultured leg regenerates and that the synthetic process occurred only when ecdysone was present.

C. TESTES AND OVARIES

A number of researchers have studied the effects of ecdysone on other cultured organs. Kambysellis and Williams (1971, 1972) demonstrated that ecdysone exerted an effect on the initiation of spermatogenesis in cultured testes of *Samia* by permitting the passage of a "macromolecular factor" (MF) through the testis wall. Without the presence of MF, ecdysone apparently had no effect of its own. They also found that coculture with either "activated" or "inactive" prothoracic glands together with an "active" brain could take the place of ecdysone in this system.

The effects of ecdysones on the development of cultured ovaries of several insects has also been studied. Both Laverdure (1971, 1972), working with *Tenebrio*, and Shibuya and Yagi (1972), working with *Galleria*, found that β-ecdysone was necessary for continued development of ovaries from late-instar larvae. Indeed, Laverdure found that the addition of juvenile hormone inhibited ecdysone-induced development. Subsequently, Laverdure (1975) found that cultured ovaries from adult *Tenebrio* took up vitellogenic proteins from the culture medium but that juvenile hormone did not appear to affect this process or that oocyte development. In contrast, Adams and Eide (1972) and Ittycheriah *et al.* (1974) found that oocyte development in adult *Diptera* was enhanced by the presence of juvenile hormone, and Ittycheriah *et al.* found that ecdysone had a deleterious effect. Leloup (1970) studied the effect of coculturing the neuroendocrine complex and the ring glands with prepupal ovaries from *Calliphora*. Like Laverdure (1972), she found that the presence of both the brain and prothoracic glands was necessary for differentiation in the ovarian tissues. Similarly, Ittycheriah *et al.* (1974) found that the "most normal" development of cultured mosquito ovaries occurred when they were cocultured with the neuroendocrine complex.

As a generalization, ecdysone is necessary for the development of larval ovaries, and juvenile hormone regulates egg maturation in adult ovaries. However, the situation is apparently more complex than this. The role of juvenile hormone in stimulating oocyte development and vitellogenic uptake has yet to be satisfactorily demonstrated *in vitro*. In addition, the report of Fallon *et al.* (1974) that β-ecdysone induced vitellogenic synthesis in *Aedes* and the discovery by King and Marks (1974) that β-ecdysone is present in the blood of mated female *Leucophaea* suggest that ecdysone may also be involved in regulating reproduction. Furthermore, the neuroendocrine system is probably involved in the control mechanism as well. Many attempts to study ovarian development and egg maturation *in vitro* have either failed or have produced inconclusive results. In all likelihood, there are many reasons for this. The control of these processes is more complex than previously suspected. Also, the control mechanisms appear to differ in detail from one group of insects to another so that comparisons are difficult.

Many technical problems remain to be solved in working with gonads *in vitro*, and the working out of the various control mechanisms through the use of organ culture may prove to be laborious. However, the rewards in terms of understanding the processes involved should more than compensate for the efforts expended.

D. OTHER ORGANS

Morphological changes in response to ecdysone stimulation have also been demonstrated in pupal hindgut tissue of *Manduca* (Judy and Marks, 1971), in the ventral nerve chain of *Galleria* (Robertson and Pipa (1973), and in the cultured silk glands of *Bombyx* (Y. Chinzei, personal communication). In all three cases, the responses (Tissue reorganization in *Manduca*, shortening of the interganglionic connective in *Galleria*, and the degeneration of the larval silk gland in *Bombyx*) were precisely those that occurred in the intact insect. Chinzei's work with the silk gland of *Bombyx* is of particular interest since we know little about the role of hormones in the induction of cytolysis and tissue breakdown. This *in vitro* system, therefore, provides a long-needed opportunity to study these processes in detail. Most other studies of the effects of hormones on target organs have been continuations and refinements of work begun before

1970. Since that time, and especially in the last two years, the area of biosynthesis and the study of release of hormones by cultured glands has developed rapidly and has produced spectacular results.

III. Hormone Production by Cultured Glands

A. CORPORA ALLATA

Since 1970, the original work of Röller and Dahm (1970) on the secretion of juvenile hormone by incubated corpora allata has been greatly expanded and refined by Judy et al. (1973a,b). When they incubated corpora allata from *Manduca* and *Schistocerca*, they were able to extract the secreted hormone from the culture medium. Then when they used microanalytical techniques to establish the chemical configuration, they were able to demonstrate that there are three distinct, natural juvenile hormones that occur in different numbers and combinations in different insects. The same research group used radiotracer techniques to study the biosynthetic pathways involved in the formation of these compounds (Schooley et al., 1973). From this work, it was apparent that the organ culture technique has three advantages over the extraction of the hormone from whole insects or from excised glands: (1) The amount of the hormone recovered from each insect is up to 100 times as great when the glands are cultured; (2) the isolation and cleanup of the hormone from the culture medium is much easier than from the whole insect; and (3) radiolabels are more easily incorporated into hormones synthesized *in vitro*. As a result, the methods developed in these studies have been adopted for use with other insects (Müller et al., 1974) and other endocrine products.

B. PROTHORACIC GLANDS

Earlier organ culture studies have thrown new light on the old question of the role of the prothoracic gland in the biosynthesis of molting hormone (see Marks, 1970). Recently two approaches to this problem have been made. The first involved the use of a bioassay to detect the secretory product. This approach was followed by Kambysellis and Williams (1972), Agui et al. (1972), Agui and Yagi (1973), and Agui (1974, 1975). These investigators were able to identify the prothoracic gland as the source of the molting hormone. Indeed, Agui (1974) compared the effects of α- and β-ecdysones on wing discs cocultured with oenocytes and found that α-ecdysone cultured in the presence of oenocytes produced the same hormonal effect as β-ecdysone. Similarly, active prothoracic glands (when cocultured with oencytes) produced effects on the wing discs of *Mamestra* that were identical to those produced by β-ecdysone. Yet, without chemical analysis, neither Kambysellis and Williams, nor Agui, were able to identify with certainty the product of the prothoracic gland as α-ecdysone. For further discussion of Agui's work, see Chapter 9.

The second or microanalytical approach was employed by King et al. (1974), Chino et al. (1974), King and Marks (1974), and Romer et al. (1974). These researchers cultured the prothoracic glands of different insects, isolated the secretory product from the culture medium, and identified it by chemical means. In each instance, the product of the prothoracic gland was identified as α-ecdysone. The work done by Chino and his co-workers with the prothoracic glands of *Bombyx* is discussed in detail in Chapter 7. King et al. (1974), working with *Manduca*, and King and Marks (1974), working with *Leucophaea*, showed that cultured prothoracic glands were unable to convert α-ecdysone to β-ecdysone though several other tissues could. In addition, they confirmed the identification of α-ecdysone by mass spectroscopy. Romer et al. (1974) demonstrated that cultured prothoracic glands from late-instar *Tenebrio* larvae can synthesize α-ecdysone but not β-ecdysone from labeled cholesterol. They also showed that when oenocytes and prothoracic glands were cocultured, α-ecdysone produced by the prothoracic glands was converted to β-ecdysone.

The extent to which tissues other than the prothoracic gland can synthesize ecdysone is not entirely certain (see Romer et al., 1974). Certainly, a number of tissues are able to convert α- to β-ecdysone and some may be able to store it for eventual use in the adult insect (see Fallon et al., 1974; King and Marks, 1974). The answers to these questions will probably be worked out in the near future by using organ culture techniques.

C. NEUROENDOCRINE GLANDS

It has been apparent for some time that neurosecretory cells survive for long periods *in vitro*. However, the question whether synthesis and release of neurohormones occur in such cells has been difficult to answer. Basically, two approaches have been tried.

Morphological and histochemical criteria have been the most widely used approach for following the buildup and disappearance of stainable material, but interpretation of the results has often proved difficult (see Marks, 1970). Leloup and Marks (1973a,b) and Borg and Marks (1973), in their studies of cultured *Manduca* brains, found that the combination of histochemical and electron microscope evidence indicated the neurosecretory cells in the cultured brains were both synthesizing and releasing neurosecretory material. Seshan *et al.* (1974) studied cultured neurosecretory cells from *Periplaneta* and found that these isolated cells maintained electrical activity for as long as 6 weeks *in vitro*. Electron micrographic examination of the cells showed that the axons were filled with electron-dense, membrane-bound neurosecretory granules. Marks *et al.* (1972) studied the effect of β-ecdysone on the median neurosecretory cells of cultured cockroach brains. They found that Victoria blue-stainable material accumulated in untreated brains for as long as 3 days after explanation and that the addition of β-ecdysone to the culture medium caused the discharge of the accumulated material. In spite of the valuable information obtained from the morphological approach, we still know little about the synthesis, storage, and release of neurohormones per se, because these processes can be detected only by bioassay or perhaps by chemical analysis, when more is known about their structure.

The second approach involves the use of some kind of bioassay to identify and quantitate the particular neurohormone being studied. Such studies may involve either coculture of the neuroendocrine tissue and target organ or the treatment of cultured target organs with extracted materials. Attempts to demonstrate the activation of cultured prothoracic glands by the prothoracotropic neurohormone were made by Kambysellis and Williams (1972), Gersch and Braüer (1974), and Agui (1975), each of whom used different *in vitro* techniques. Gersch and Braüer treated the cultured glands of *Periplaneta* with extracts of corpora cardiaca that contained "activation factor I," and assayed the prothoracic gland by measuring the incorporation of [5-^{3}H]uridine into the RNA fraction of the cells. In contrast, Agui used a three-level bioassay in which he cocultured "active brains" and "inactive prothoracic glands" with fragments of the body wall. Molting of the body wall occurred only when all three components were present. All these experiments indicated that the prothoracic glands were probably being activated to produce molting hormone, but the results would be more convincing if the product of the activated prothoracic gland had been extracted from the culture medium and identified by chemical means. The ultimate goal would be to extract and determine the structure of the neurohormone itself, but, unfortunately, adequate methodology for this has not yet been developed.

An attempt to bridge the gap between information about the neurosecretory system obtained from morphological studies and that obtained by bioassay was made by Marks *et al.* (1973). In this study, the production of neurosecretory material by cultured cockroach brains was studied by selective staining and by electron photomicrography. The results were then compared with information obtained by bioassay for the hindgut-stimulating neurohormone. The appearance of morphological evidence of neurosecretion correlated closely with a rise in bioassayable activity. In an additional study, Sowa and Borg (1975) showed that when dispersed cockroach brains were fractioned by density gradient ultracentrifugation, the fraction that contained an almost pure accumulation of identifiable neurosecretory granules also possessed virtually all the neurohormonal activity in the bioassay.

Holman and Marks (1974) showed that the neurohormone synthesized *in vitro* by intact cockroach neuroendocrine complexes passed into the corpus cardiacum from which it was released into the culture medium. In a subsequent study, Marks and Holman (1974) found that the presence of the corpora cardiaca was necessary for release of the accumulated hormone from cultured brains but that intact nervi corporis cardiaci (NCC) were not. Furthermore, when the NCC were not intact, the hormone released by the brain was sequestered from the medium by the corpora cardiaca and retained. The nature of the mechanism that controlled the release from the cardiaca into the medium is currently being investigated.

With the possible exception of Seshan's work with isolated neurosecretory cells, all the preceding studies fall into the category of organ culture. However, an increasing number of investigators are now turning to the use of cell cultures and the study of endocrine action at the cellular and subcellular levels.

IV. Hormonal Effects in Cell Cultures

Most studies that have been discussed thus far have involved multicellular organotypic responses. However, all organotypic responses are the result of hormonal effects on individual cells. In fact, responses such as the synthesis of macromolecular substances, the metabolic conversion of prohormones, and the activation of receptor sites should occur as readily in cell cultures as in organ cultures. Cell cultures lend themselves readily to quantitation, and large numbers of cells of a limited number of types can be produced with a minimum of effort. Since 1970, both primary cultures and cell strains have come into increasing use for the study of hormonal effects on nucleotide synthesis and mitosis.

Earlier studies showed that mitosis persisted in hormone-free primary cultures but apparently was inhibited by treatment with β-ecdysone (see Marks, 1970). Mitsuhashi and Grace (1970), working with established cell strains from *Antherea*, also found that high levels of β-ecdysone (5.0 μg per ml) produced an inhibitory effect on mitosis; however, lower levels (0.1 μg per ml) did not and may have had an enhancing effect on mitosis.

On the other hand, in a similar experiment using *Drosophila* cells, Cour-geon (1972a) found that the application of β-ecdysone, even at very low levels (0.006 μg per ml), produced a decrease in mitosis, but α-ecdysone at much higher levels (0.3–1.0 μg per ml) stimulated mitosis. However, at a dose level of 1.0 μg per ml, this stimulation ceased after the third day and was replaced by an inhibitory effect. This long-term response suggested a possible conversion of α- to β-ecdysone by the cells. At the same conditions, Courgeon (1972b) found changes in cellular activity and conformation that she attributed to cell differentiation, initiated by the molting hormone. She also found that β-ecdysone was 500 times more effective than α-ecdysone in producing this response. Similar results were reported by Eide and Adams (1973) who used a cell line derived from *Musca*. Reinecke and Robbins (1971) studied the effect of ecdysone on nucleotide synthesis in cell lines derived from *Antherea*. They found that β-ecdysone (at a level of 2 μg per ml) caused a decrease in cell number that was accompanied by a decrease in DNA synthesis and an increase in RNA synthesis.

Cohen and Gilbert (1972) investigated the effects of insect hormones on RNA and protein synthesis in cell lines derived from *Antheraea* and *Trichoplusia*. They found that neither α- nor β-ecdysone had any appre-ciable effect on these processes and that no α-ecdysone and very little β-ecdysone was taken up by the cells. However, juvenile hormone proved to be a potent inhibitor of both RNA and protein synthesis. The authors attributed the failure of β-ecdysone to produce an effect to the inability of the cells to take up the hormone, and the effect of the juvenile hormone to toxic effects that produced swelling and lesions in the plasma mem-brane. In a subsequent study, Cohen and Gilbert (1973) found that juvenile hormone treatment of cell cultures from *T. ni* and cultured ger-minal cysts from *Hyalophora* caused disaggregatin of the polysomes. This response occurred only in intact cells, indicating that the hormone prob-ably acted at the level of the cell membrane.

There are discrepancies in the results obtained by the various investi-gators, but there is reasonable correlation between the results obtained with cultured organs and cultured cells (see Wyatt and Wyatt, 1971). The use of gross cell morphology as a criterion for cellular response makes interpretation difficult, but such cultures are proving to be nearly ideal for electron photomicrography, radiotracer studies, and microchemical analyses. In this context, it is interesting that no one has reported a hor-mone-producing insect cell line though vertebrate cell lines that produce various pituitary hormones have been known for some years (Dannies and Tashjian, 1973). The work of Seshan *et al.* (1974) suggested that efforts made along this line might be productive. Hormone-producing cell

130 EDWIN P. MARKS

lines would be extremely useful for the study of biosynthetic pathways and even for mass production of complex peptide hormones.

V. Conclusions

Since 1970, cell culture and organ culture have become more and more important in insect endocrinology, and they have become the materials of choice for some types of study. However, the usefulness of this methodology is still limited by technical problems. For example, little work has been done with juvenile hormone(s) *in vitro*, largely because methods of applying known amounts to the explants within physiological ranges are only now being worked out (see Cohen and Gilbert, 1972, 1973). Products of incubated tissues may differ from those produced *in vivo* because of altered availability of precursors or substrates or because of ruptured membranes and altered permeabilities (see Müller *et al.*, 1974). Suitable bioassays and microchemical assays are not yet available for most of the neurohormones. Suitable culture media are still not available for tissues from many insect species. Methods of deriving cell lines from physiologically significant tissues still have to be worked out in most cases. Finally, the tissue of origin is rarely known for insect cell lines. However, the lines derived from cockroach hemolymph by Landureau and Grellet (1975), in which the differentiation of typical hemocytes can be induced, represent exceptions in this regard and demonstrate the possibilities inherent in this particular approach. In the future, established lines of specialized cells will likely be important in the use of cell culture in insect endocrinology. The usefulness of organ culture in this regard is already apparent.

References

Adams, T., and Eide, P. (1972). *Gen. Comp. Endocrinol.* 18, 12.
Agui, N. (1973). *Appl. Entomol. Zool.* 8, 236.
Agui, N. (1974). *Appl. Entomol. Zool.* 9, 256.
Agui, N. (1975). *J. Insect Physiol.* 21, 903.
Agui, N., and Fukaya, M. (1973). *Appl. Entomol. Zool.* 8, 73.
Agui, N., and Yagi, S. (1973). *Appl. Entomol. Zool.* 8, 239.
Agui, N., Kimura, Y., and Fukaya, M. (1972). *Appl. Entomol. Zool.* 7, 71.
Benson, J., and Oberlander, H. (1974. *Insect Biochem.* 4, 423.
Bergtrom, G., and Oberlander, H. (1975). *J. Insect Physiol.* 21, 39.
Berreur-Bonnenfant, J. (1972). In "Invertebrate Tissue Culture" (C. Vago, ed.), Vol. 2, pp. 181–242. Academic Press, New York.
Borg, T., and Marks, E. (1973). *J. Insect Physiol.* 19, 1913.

Bullière, F. (1973). *J. Insect Physiol.* **19**, 1465.
Chihara, C., Petri, W., Fristrom, J., and King, D. (1972). *J. Insect Physiol.* **18**, 1115.
Chino, H., Sakuri, S., Ohtaki, T., Ikekawa, N., Miyazaki, H., Ishibashi, M., and Abuki, H. (1974). *Science* **183**, 529.
Cohen, E., and Gilbert, L. (1972) *J. Insect Physiol.* **18**, 1061.
Cohen, E., and Gilbert, L. (1973). *J. Insect Physiol.* **19**, 1857.
Courgeon, A. (1972a). *Nature (London), New Biol.* **238**, 250.
Courgeon, A. (1972b). *Exp. Cell Res.* **74**, 327.
Dannies, P., and Tashjian, A. (1973). *In* "Tissue Culture: Methods and Application" (P. F. Kruse, Jr. and M. K. Patterson, Jr., eds.), pp. 561–569. Academic Press, New York.
Dutkowski, A., and Oberlander, H. (1973). *J. Insect Physiol.* **19**, 2155.
Eide, P., and Adams, T. (1973). *Proc. N. Dak. Acad. Sci.* **27**, 10.
Fallon, A., Hagedorn, H., Wyatt, G., and Laufer, H. (1974). *J. Insect Physiol.* **20**, 1815.
Frew, J. G. H. (1928). *J. Exp. Biol.* **6**, 1.
Fristrom, J., Logan, W., and Murphy, C. (1973). *Develop. Biol.* **33**, 441.
Fukuda, S. (1944). *J. Fac. Sci., Imp. Univ. Tokyo Sect. IV* **6**, 477.
Gersch, M., and Braüer, R. (1974). *J. Insect Physiol.* **20**, 735.
Holman, G., and Marks, E. (1974). *J. Insect Physiol.* **20**, 479.
Ittycheriah, P., Marks, E., and Quraishi, S. (1974). *Ann. Entomol. Soc. Amer.* **67**, 595.
Judy, K., and Marks, E. (1971). *Gen. Comp. Endocrinol.* **17**, 351.
Judy, K., Schooley, D., Dunham, L., Hall, M., Bergot, J., and Siddall, J. (1973a). *Proc. Nat. Acad. Sci. U.S.* **70**, 1509.
Judy, K., Schooley, D., Hall, M., Bergot, J., and Siddall, J. (1973b). *Life Sci.* **13**, 1511.
Kambysellis, M., and Williams, C. M. (1971). *Biol. Bull.* **141**, 527.
Kambysellis, M., and Williams, C. M. (1972). *Science* **175**, 769.
King, D., and Marks, E. (1974). *Life Sci.* **15**, 147.
King, D., Bollenbacher, W., Borst, D., Vedeckis, W., O'Connor, J., Ittycheriah, P., and Gilbert, L. (1974). *Proc. Nat. Acad. Sci. U.S.* **71**, 793.
Kopec, S. (1922). *Biol. Bull.* **42**, 323.
Kuroda, Y. (1954). *Jap. J. Genet.* **29**, 163.
Kuroda, Y., and Yamaguchi, Y. (1955). *Jap. J. Genet.* **31**, 97.
Landureau, J., and Grellet, P. (1975). *J. Insect Physiol.* **21**, 137.
Laverdure, A. (1971). *Gen. Comp. Endocrinol.* **17**, 467.
Laverdure, A. (1972). *J. Insect Physiol.* **18**, 1477.
Laverdure, A. (1975). *J. Insect Physiol.* **21**, 33.
Leloup, A. (1970). *Annee Biol.* **9**, 207.
Leloup, A., and Marks, E. (1973a). *Proc. Int. Colloq. Invertebr. Tissue Cult., 3rd, 1971* pp. 63–65.
Leloup, A., and Marks, E. (1973b). *Ann. Endocrinol.* **34**, 766.
Mandaron, P. (1973). *Develop. Biol.* **31**, 101.
Mandaron, P. (1974). *Wilhelm Roux' Arch. Entwicklungsmech. Organismen* **175**, 49.
Marks, E. (1970). *Gen. Comp. Endocrinol.* **15**, 289.
Marks, E. (1972). *Biol. Bull.* **142**, 293.
Marks, E. (1973a). *Biol. Bull.* **145**, 171.
Marks, E. (1973b). *Proc. Int. Colloq. Invertebr. Tissue Cult., 3rd, 1971* pp. 221–232.
Marks, E. (1973c). *Gen. Comp. Endocrinol.* **21**, 472.

Marks, E., and Holman, G. (1974). *J. Insect Physiol.* **20**, 2087.

Marks, E., and Leopold, R. (1971). *Biol. Bull.* **140**, 73.

Marks, E, and Sowa, B. (1974). *In* "Mechanism of Pesticide Action" (F. Kohn, ed.), Amer. Chem. Soc. Symp. Ser., pp. 145–155. Academic Press, New York.

Marks, E., Ittycheriah, P., and Leloup, A. (1972). *J. Insect Physiol.* **18**, 847.

Marks, E., Holman, G., and Borg, T. (1973). *J. Insect Physiol.* **19**, 471.

Mitsuhashi, J., and Grace, T. (1970). *Appl. Entomol. Zool.* **5**, 182.

Müller, P., Masner, K., Trautman, K., and Maag, R. (1974). *Life Sci.* **15**, 915.

Oberlander, H. (1972a). *In* "Results and Problems in Cell Differentiation" (H. Urspring and R. Nöthinger, eds.), Vol. 5, pp. 155–172. Springer-Verlag, Berlin and New York.

Oberlander, H. (1972b). *J. Insect Physiol.* **18**, 223.

Oberlander, H., and Leach, C. (1975). *Insect Biochem.* **5**, 235.

Oberlander, H., and Tomblin, C. (1972). *Science* **177**, 441.

Oberlander, H., Leach, C., and Tomblin, C. (1973). *J. Insect Physiol.* **19**, 993.

Ohmori, K. (1974). *J. Insect Physiol.* **20**, 1697.

Ohmori, K., and Ohtaki, T. (1973). *J. Insect Physiol.* **19**, 1199.

Ohtaki, T., Milkman, R., and Williams, C. M. (1968). *Biol. Bull.* **135**, 322.

Raikow, R., and Fristrom, J. (1971). *J. Insect Physiol.* **17**, 1599.

Reinecke, J., and Robbins, J. (1971). *Exp. Cell Res.* **64**, 335.

Robertson, J., and Pipa, R. (1973). *J. Insect Physiol.* **19**, 673.

Röller, H., and Dahm, K. (1970). *Naturwissenschaften* **57**, 454.

Romer, F., Emmerich, H., and Nowock, J. (1974). *J. Insect Physiol.* **20**, 1975.

Schmidt, E., and Williams, C. M. (1953). *Biol. Bull.* **105**, 174.

Schooley, D., Judy, K., Bergott, J., Hall, M., and Siddall, J. (1973). *Proc. Nat. Acad. Sci. U.S.* **70**, 2921.

Seshan, K., Provine, R., and Levi-Montalcini, R. (1974). *Brain Res.* **78**, 359.

Shibuya, I., and Yagi, S. (1972). *Appl. Entomol. Zool.* **7**, 97.

Sowa, B., and Borg, T. (1975). *J. Insect Physiol.* **21**, 511.

Wigglesworth, V. B. (1934). *Quart. J. Microsc. Sci.* [N.S.] **77**, 191.

Wyatt, S., and Wyatt, G. (1971). *Gen. Comp. Endocrinol.* **16**, 369.

9

Attempts to Establish Insect Endocrine System *in Vitro*

NORIAKI AGUI

I. Introduction

In 1915, Goldschmidt cultured the spermatocysts of the cecropia silk-moth in blood from the same species and observed the process of spermatogenesis *in vitro*. This was the first attempt at insect organ culture.

Organ culture methodology has many advantages for the study of the complicated physiological phenomena concerning growth, development, and differentiation of insects. Investigators can analyze these phenomena by this technique under simpler conditions than *in vivo*. This method provides a particularly good tool for the study of the various insect hormones which control the molting and metamorphosis of insects. The progress of insect organ culture in endocrine research has been reviewed by Marks (1970), Brooks and Kurtti (1971), Berreur-Bonnenfant (1972), Demal and Leloup (1972), and Oberlander (1972a).

Two important problems are faced when the techniques of organ culture are applied to the study of insect endocrinology. The first problem is to keep the target organs functional for long periods similar to those *in vivo*, and the second is to keep endocrine organs active for long periods. If both problems could be solved, investigators would be able to coculture the various target organs with endocrine glands *in vitro* for the study of interactions between these organs. Also, they would be able to apply purified insect hormones directly to target organs to investigate the effect of the hormones on the target organs *in vitro*. Furthermore, the accumulation of data concerning the interaction between target organs and endocrine glands *in vitro* would facilitate the analysis of complicated phenomena that take place in insect endocrine systems *in vivo*.

In this paper, the author has described current insect endocrine research and applications of the culture of target organs and endocrine glands *in vitro*, as well as the possibility of establishing insect endocrine systems *in vitro*.

II. Effects of Hormones on the Cultured Target Organs *in Vitro*

A. IMAGINAL DISC

The imaginal discs have been utilized by many investigators as suitable materials for studying effects of hormones on the growth and differentiation of insects because these organs develop and differentiate rapidly under hormonal control. Kuroda and Yamaguchi (1956) reported that the morphogenetic development of the eye-antennal discs of last instar larvae *Drosophila* was strongly enhanced *in vitro* if these discs were cultured with cephalic complex which included the cephalic ganglion, ventral ganglion, and ring gland. Following this investigation, many workers have attempted to clarify the effects of the brain ring gland complexes on the morphogenesis of the cultured imaginal discs (i.e., eye-antennal, wing, and leg discs) in dipterous insects (Gottschewski, 1958, 1960; Horikawa, 1958, 1960; Horikawa, and Sugihara, 1960; Schneider, 1964; Fujio 1962; Burdette *et al.*, 1968; Courgeon, 1969; Sengel and Mandaron, 1969). However, morphogenetic stimulation of the entire brain ring gland complex of the cultured discs does not permit determination of the hormones involved because the cultured brain ring gland complex may secrete brain, juvenile, and molting hormones during cultivation. Recently, there have been a number of reports on the effects of the purified insect hormone ecdysone, on the morphogenesis of the cultured dipterous imaginal discs (Sengel and Mandaron, 1969; Kuroda, 1969; Ohmori and Ohtaki, 1973). According to Mandaron (1973), α-ecdysone and inokosterone were able to initiate and sustain metamorphosis of the cultured imaginal discs of *Drosophila*, whereas β-ecdysone suppressed the continued development of the discs. In view of this finding Mandaron (1973) concluded that only α-ecdysone intrinsically may act on the target organ *in vivo*. Labeled α-ecdysone may be rapidly converted into β-ecdysone if injected into larvae or pupae of the blowfly *Calliphora vicina* (King and Siddall, 1969) or larvae of *Bombyx mori* (Moriyama *et al.*, 1970). The conversion of α-ecdysone to β-ecdysone takes place rapidly in many organs isolated from the tobacco hornworm *Manduca sexta* (King, 1972) and the cockroach *Leucophaea maderae* (King and Marks, 1974). Furthermore, Borst *et al.* (1974) found that only β-ecdysone was detected in the methanol extracts of various stages of *Drosophila*, and α-ecdysone may be present in small amounts in mid-third instar larvae. The amount of β-ecdysone is comparable to the amount required to induce morphogenesis of imaginal discs cultured *in vitro*. From this finding the authors concluded that β-ecdysone is the molting hormone responsible for metamorphosis in *Drosophila*. The answer to the problem of whether α-ecdysone or β-ecdysone is responsible for the morphogenesis of imaginal discs in *Drosophila* requires further experimentation.

As to the effects of the juvenile hormone on the cultured discs, Chihara *et al.* (1972) and Chihara and Fristrom (1973) reported that the evagination of mass-isolated *Drosophila* imaginal discs was induced *in vitro* by β-ecdysone but that the β-ecdysone-induced development was reversibly inhibited by cecropia C 18 juvenile hormone (90 µg/ml). In the cultivation of imaginal discs of lepidopterous insects, Oberlander and Fulco (1967) were the first to report that the wing discs from matured larvae of the wax moth *Galleria mellonella*, partially metamorphosed *in vitro* after the addition of crystalline α-ecdysone (3 µg/ml) to chemically defined Grace's medium. Subsequently, it was demonstrated that even younger wing discs from the final instar larvae responded to α-ecdysone and its analogues, including β-ecdysone *in vitro* (Oberlander, 1969). Furthermore, different action of α-ecdysone and β-ecdysone has been reported on the wing discs of *G. mellonella* that synthesized DNA during the first 24 hours of culture in the presence of α-ecdysone, but ecdysone analogues (β-ecdysone or inokosterone) inhibited α-ecdysone-induced DNA synthesis *in vitro* (Oberlander, 1972b). In these findings he proposed that α-ecdysone initiates metamorphosis, including DNA synthesis, and that subsequent conversion of α-ecdysone to β-ecdysone inhibits DNA synthesis without interfering with continued morphogenesis. Agui *et al.* (1969b) tried to ascertain that the molting hormone β-ecdysone affects the morphogenesis of wing discs from the different stages of the last instar larvae of the rice stem borer *Chilo suppressalis in vitro*. Wing discs taken from diapausing larvae or nondiapausing larvae within 24

hours after the final molting did not change when cultured with 3 μg/ml
β-ecdysone in Mitsuhashi's CSM-2F medium (1968). On the other hand,
more developed wing discs from nondiapausing larvae at 2–3 days of
the final instar metamorphosed about 48 hours after the onset of culture.
The most remarkable changes of the wing discs were caused by β-ecdysone when the cultured material was derived from larvae 3–4 days
of the final instar—namely, the wing tissue expanded rapidly and
ruptured the hypertrophied wing pouch. Nevertheless, the discs could not
develop *in vitro* without β-ecdysone. Wing discs of the later stage, about
1–2 days before pupation, grew *in vitro* in the absence of β-ecdysone.
Probably, the metamorphosis of the wing discs which had passed this
critical stage did not require the control of the molting hormone. Similar
results were obtained from the cultured pharate adult testes of *Drosophila* (Kuroda, 1974) and the cultured lepidopterous wing discs of the wax
moth *G. mellonella* (Oberlander, 1969) and cabbage armyworm larvae
Mamestra brassicae (Agui and Fukaya, 1973).

The successive development of *Mamestra* wing discs *in vivo* is illustrated in Fig. 1; in the wing discs of 1-day-old larvae, a tracheal mass
covered a larger area of the wing than did the transparent tissue. The
volume of wing tissue increased, and tracheoles passed into the channels

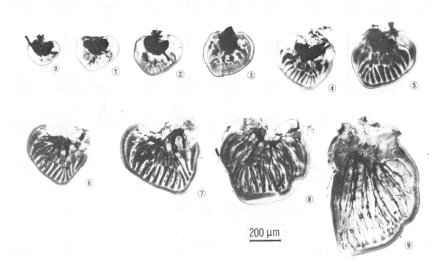

Fig. 1. Successive daily development of wing discs *in vivo* at 25°C. 0–9 age in days
of last instar larvae of *Mamestra brassicae*. From Agui and Fukaya (1973).

when the larvae were 3 or 4 days old. At the prepupal stage, the wing discs began to develop quickly; when pupal ecdysis occurred, the wing discs were exposed as external organs by evagination. When the morphogenetic effects of 0.3 μg/ml β-ecdysone on *Mamestra* wing discs from the 4-day last instar larvae were examined, it was apparent that the development of wing discs began 2 days after culture (see Fig. 2). After 4 days of culture, striking development occurred, but degeneration of the developed tissue was started on the fifth day of cultivation. In another experiment, the relationship between the morphogenetic development of wing discs and hormone concentration was examined under *in vitro* condition. The effects of different concentrations of β-ecdysone on *Mamestra* wing discs from the 3- or 4-day-old larvae are summarized in Fig. 3. High concentrations of the hormone caused the early development of wing discs but also accelerated the degeneration of the organs. At a concentration of 0.15 and 0.3 μg/ml, most of the wing discs evaginated by bursting the pouch membrane and then grew to form a winglike shape. When the hormone concentration dropped as low as 0.03 μg/ml, the wing discs took about 9 days to develop. Moreover, the effect of a molting hormone such as α-ecdysone, β-ecdysone, ponasterone A, cyasterone, or rubrosterone, which was dissolved in Grace's medium at a concentration of 1 μg/ml on wing discs of 3-day-old larvae was examined. The activity of five

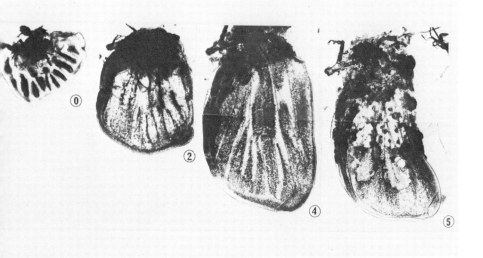

Fig. 2. Morphogenetic patterns of *Mamestra* wing discs for a 4-day-old last instar larva cultivated in the presence of β-ecdysone (0.3 μg/ml) for 0, 2, 4, and 5 days. Note the striking development at 4-day cultivation and the initiation of degeneration at 5-day cultivation. ×45. From Agui and Fukaya (1973).

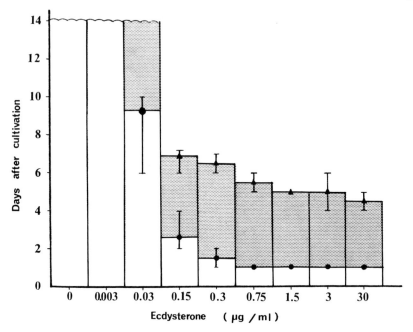

Fig. 3. Relationship between morphogenetic development of *Mamestra* wing discs and hormone concentration *in vitro*. ●, mean days required for initiation of development of wing discs; ▲, mean days before occurrence of degeneration of developed wing discs. The vertical lines represent the ranges of the results. From Agui and Fukaya (1973).

kinds of molting hormones assayed from *Mamestra* wing discs ranged as follows: ponasterone A = cyasterone = β-ecdysone > α-ecdysone > rubrosterone. Of these hormones only α-ecdysone at this concentration could induce development of wing discs similar to that observed *in vivo*, although it required about 4 days to do so. Elongation of wing tissues, increase in pouch volume, and extention of tracheoles into the wing cavity all took place normally in the medium containing 1 μg/ml α-ecdysone. However, degeneration of the developed wing discs occurred 6 days after the molting hormone treatment except in the case of rubrosterone. Indeed, application of rubrosterone barely initiated development of wing discs. Similar phenomena were observed when wing discs were cultured in the presence of prothoracic glands of *Mamestra*. The initiation period of wing disc development was shortened with an increase in the number of prothoracic glands, but they degenerated quickly. On the other hand, in cultures with a moderate number of prothoracic glands, the wing discs developed normally, just as *in vivo*.

On the basis of these results, it was concluded that the action of molting hormone was necessary for the growth and development of imaginal discs, although the effect of hormone on *in vitro* growth and development differed according to the insect species, imaginal discs, and stage of the insect. The hormones intrinsically responsible for metamorphosis of imaginal discs will be discussed further.

B. INTEGUMENT

From experimental morphological investigation, it is well known that the molting of insect integument is induced by the action of the molting hormone ecdysone and that the pattern of ecdysis is controlled by a quantitative relationship between molting hormone and juvenile hormone. Therefore the organ culture of isolated integuments has drawn the attention of many insect endocrinologists who hope to clarify the many unknown functions of the hormone by means of these techniques.

Miciarelli *et al.* (1967) first reported that the integument of the desert locust *Schistocerca gregaria* larva was able to molt *in vitro*. The synthesis of a new cuticle was induced *in vitro* by the cultivated integument of the fourth instar nymph if taken after the stage of multiplication of intact epidermal cells. However, in this case the role played by the molting hormone was not clarified. Agui *et al.* (1969a) succeeded in inducing *in vitro* ecdysis of the integument taken from the diapausing rice stem borer larva *C. suppressalis* in the presence of β-ecdysone. Recently this author induced ecdysis of the larval integument in the cabbage armyworm *M. brassicae* and the European corn borer *Ostrinia furnacalis* by the addition of the molting hormone *in vitro* (N. Agui, unpublished). As to the culture method used in this experiment, fragments of *Chilo* integuments, each of which was cut as large as 3×4 mm, were added immediately after being rinsed with physiological saline solution to the medium containing the molting hormone. When the explants were added to the culture they were soaked in the medium and their outer surfaces subjected to air so that integument fragments were floating on the medium. At 24 hours after the start of cultivation, the integuments in the medium containing hormone began ecdysis (see Fig. 4). This ecdysis occurred in either of three different media (Ringer–Tyrodes's solution, Grace's insect TC medium, and CSM-2F medium) if the hormone was dissolved in these media above the threshold concentration (β-ecdysone: ED_{50}, 0.04 μg/ml). Microscopic examination revealed that the epidermis of the molted integument was much thickened, that the nuclei of the cells deeply stained with hematoxylin, and that new cuticle was deposited, whereas, the integument cultivated in the β-ecdysone-free medium showed no change in epidermal cells. This author also investigated the effects of various concentrations

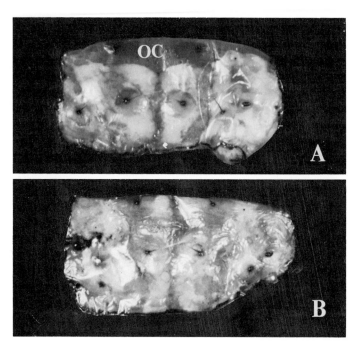

Fig. 4. *Chilo* integument cultured in a medium with (A) or without β-ecdysone (B) for 3 days. Old cuticle (OC). ×20. From Agui (1973).

of β-ecdysone, in different media, on the molting rate of *Chilo* integument *in vitro*. As shown in Table I, the molting rate of the integuments cultured on three different media (Ringer–Tyrode's solution, Grace's medium, and CSM-2F medium) correlated with the hormone concentration. On the other hand, the time required for molting tended to become shorter with increased concentration of the hormone. Moreover, when the effects of the three different media on both molting time and rate were examined, there was little difference between Ringer–Tyrode's solution and Grace's medium, whereas in CSM-2F the molting time and rate were prolonged and lower than those of the other media. In order to examine the difference between the data obtained from 2- and 3-day-old cultures, a regression line was drawn from the relation between the hormone concentration and molting rate in each medium by means of the standard probit method. It was ascertained that there was no significant difference between the regression line of Ringer-Tyrode's solution and Grace's medium at 2 days and CSM-2F medium at 3 days. Therefore, when the hormonal activity of the unknown agent is to be bioassayed, results should be recorded 2 days after onset of culture in Ringer–Tyrode's

TABLE I

Effects of Various Concentrations of Ecdysterone and Different Media on the Molting Rate of the Cultivated *Chilo* **Integuments** *in Vitro*[a]

Medium[b]	Concentration of ecdysterone (μg/ml)	No. of cultured integuments	Molting rate (%) after cultivation for			
			1 day	2 days	3 days	4 days
R–T	0.15	28	14.3	96.4	96.4	100.0
	0.075	28	3.6	46.4	78.0	78.0
	0.0375	27	0.0	10.8	29.6	51.9
	0.0	29	0.0	0.0	0.0	0.0
G–M	0.15	30	16.7	96.6	96.6	100.0
	0.075	29	3.4	62.0	73.3	89.7
	0.0375	30	0.0	0.0	36.6	46.7
	0.0	30	0.0	0.0	0.0	0.0
CSM–2F	0.15	30	0.0	70.0	96.6	100.0
	0.075	30	0.0	46.0	63.3	90.0
	0.0375	30	0.0	10.0	20.0	33.3
	0.0	30	0.0	0.0	0.0	0.0

[a] From Agui (1973).
[b] R–T, Ringer–Tyrode; G–M, Grace's medium.

solution or Grace's medium, and 3 days after in the case of CSM-2F medium.

In practice, when agents with molting activity are being assayed, the following procedures should be followed. First, β-ecdysone is dissolved in each medium at three different concentrations, for instance 0.15, 0.075, and 0.0375 μg/ml. Next, when Ringer–Tyrode's solution or Grace's medium is used as the culture medium the molting rates are recorded 2 days after cultivation; when CSM-2F medium is used they should be recorded 3 days after. Third, the regression equation or regression line is calculated from the relationship between concentration and molting rate which is established by observation. Finally, the activity of the unknown agent can be estimated as a value equivalent to β-ecdysone from either the regression equation or regression line.

The interesting fact was that the cultured integument could ecdyse several times if molting hormone was applied to the integument cyclinically *in vitro*. Subsequently, the author examined the relationship between induction of ecdysis and duration of hormone treatment. It was ascertained that if the integuments were subjected to 0.3 μg/ml β-ecdysone for 3–7 hours and then returned to the hormone-free medium, the integuments that were treated with hormone for more than 5 hours ecdysed 1 day after that treatment. In another experiment, the integuments were cultured for 1–4 hours in a medium containing 0.3 μg/ml β-ecdysone and then cultured in the hormone-free medium for 5 days. The same integuments were again treated with β-ecdysone for 2–5 hours. The total duration of the hormone treatment was 6 hours, and after the second treatment they were again returned to hormone-free medium. Almost all integuments ecdysed 1 day after treatment (N. Agui, unpublished). This result suggests that the action of β-ecdysone on the cultured integument is not a trigger action, but is cumulative. Evidence that ecdysone does not provide a trigger mechanism, but that it is required as a sustained stimulus for a certain critical period to ensure metamorphosis of cultured imaginal discs or to regenerate tissue *in vitro* was provided by observations of other investigators (Oberlander, 1969; Marks, 1972, 1973; Ohmori, 1974).

RNA, DNA, and protein synthesis of the cultured integuments were examined by incorporation of radioactive precursors (N. Agui, unpublished). RNA synthesis of the integuments treated with and without hormone was at a maximum 48 hours after cultivation. RNA synthesis in hormone-treated integuments was higher than in the controls. RNA synthesis decreased quickly 48 hours after onset of the culture. There was no significant difference between DNA synthesis of the hormone-treated integuments and that of the control. The protein synthesis of the hormone-treated integuments decreased significantly 48 hours after onset of the culture, but the synthesis increased after ecdysis and later decreased again.

C. GONAD

Insect gonads are also suitable for studying the insect endocrine system because their growth and development are greatly influenced by hormones. As already mentioned, Goldschmidt (1915) was able to culture the spermatocysts of the cecropia silkmoth in the hemolymph of this species. In cultured testes, Schmidt and Williams (1953) showed that spermatocysts from *Hyalophola cecropia* and *Samia cynthia* in a hanging drop of blood obtained from pupating larvae or developing adults of the same species developed promptly into spermatids. After this study, some investigators tried to culture gonads in media which did not contain insect hemolymph (Lender and Duveau-Hagége, 1962, 1963a,b; Mitsuhashi, 1965). Later, Yagi et al. (1969) reported that remarkably rapid development of both testes and spermatocysts of the diapausing rice stem borer larvae *C. suppressalis* occurred through direct application of β-ecdysone. Similar results were obtained in the diapausing slug moth pharate pupae *Monema flavescens*. The spermatocysts in intact testes developed into well-differentiated spermatids with the addition of β-ecdysone, and naked spermatocysts were more sensitive than the intact spermatocysts (Takeda 1972a,b). On the other hand, Kambysellis and Williams (1971a,b) reported that the naked spermatocysts taken from the diapausing pupae of the silkworns *H. cecropia* and *S. cynthia* were insensitive to α- and β-ecdysone *in vitro* but that spermatogenesis was promptly acceletated when a macromolecular factor contained in hemolymph was added to ecdysone-free medium. The same authors found that spermatocysts within intact testes responded only when both the macromolecular factor and ecdysones were present in the medium. As to the role of α- or β-ecdysone, they considered that its sole function was to alter the penetrability of the testis wall thereby facilitating the entry of the marcromolecular factor. Recently, Fukushima and Yagi, (1975) obtained interesting results on the role of ecdysones on gonads. The testes from the cabbage armyworm *M. brassicae* were cultured in Grace's medium devoted of insect hemolymph but containing dissolved α-ecdysone (0.5–5 μg/ml) or cocultured with the active prothoracic glands. Spermatogenesis was promptly promoted in intact testes 7 days after cultivation. But if β-ecdysone (0.5–30 μg/ml) was applied to them, there was no change in the cultured testes or in the controls cultured in the medium without hormone. The results suggest that only α-ecdysone may act directly on the peritoneal sheath of the testes or the spermatocysts in the testes, and, consequently, may promote spermatogenesis without the macromolecular factor.

As for the effect of juvenile hormone on spermatogenesis *in vitro*, the cultured spermatocysts from *Bombyx* larvae at the third to fourth day of the fourth instar showed definite elongation in Grace's medium containing insect hemolymph, but the development was completely inhibited when they were cultured in medium containing 50 μg/ml cecropia C 18 juvenile hormone or in medium conditioned with precultivation of only one corpus allatum of *Bombyx* larvae at day 0 of the last instar (Yagi and Fukushima, 1975) (see Table II). Attempts are now being made to establish the *in vitro* system for assay of the activity of juvenile hormone or corpora allata by the described methods.

Studies on the effects of hormones on female gonads (ovary) were also carried out at the same time as studies on the effects of hormone on male gonads. Lender and Laverdure (1967) studied the process of vitellogenesis on the cultured ovaries of the mealworm *Tenebrio molitor* L. When the explants of a cephalic complex were added to the cultures, development of the oocytes and deposition of glycogen and vitelline droplets in the cytoplasm occurred. Furthermore, Leloup and Demal (1968) reported

TABLE II
Effects of Precultured Corpora Allatum or Prothoracic Ganglion on Spermiogenesis of the Silkworm *B. mori in Vitro*

No. of pre-cultured organs[a]		No. of experi-ments	Spermiogenesis in the medium[b]				
			−	±	+	+ +	+ + +
PG(control)	2	12	0	0	2	5	5
CA	1	12	12	0	0	0	0
CA	2	12	12	0	0	0	0
CA	5	8	8	0	0	0	0
CA	10	8	8	0	0	0	0

[a] PG, prothoracic ganglion; CA, corpus allatum.
[b] (−) Spherical spermatocysts only; (±) mixture of spherical spermatocysts and pyriform spermatocysts; (+) elongated spermatocysts are less than 10%; (++) elongated spermatocysts are more than 10% but less than 50%; (+++) elongated spermatocysts are more than 50%. From Yagi and Fukushima (1975).

that the development of cultured ovaries from *Calliphora* pupae was promoted when they were cocultured with the complex of ring gland and abdominal imaginal discs. In the cultured ovary of *Iphita limbata* from the fifth instar larva, the development of and yolk deposition of oocytes were stimulated if the ovaries were cultured with both brain and corpora allata from the copulated female in medium containing fat-body extract from females (Ittycheriah and Stephanos, 1969). Development and differentiation of cultured *Tenebrio* ovaries occurred when ecdysone (2–3 μg/ml) was added to the culture medium (Laverdure, 1969). Similar results were obtained with ovaries from *G. mellonella* larvae when β-ecdysone (3 μg/ml) was added to the culture medium. However, ovaries taken from earlier instar stages did not respond to such treatment (Shibuya and Yagi, 1972).

It has generally been considered that juvenile hormone induces the synthesis of vitellogenic protein and yolk deposition on oocytes *in vivo*. Adams and Eide (1972) reported that when ovaries from allatectomized houseflies *Musca domestica* were cultured in a medium containing 200–1800 μg of a juvenile hormone analogue (mixed isomers of methyl-10,11-epoxy-7-ethyl-3,11-dimethyl-2,6-dodecadionoate), egg chamber growth was induced. From an endocrinological viewpoint, this concentration is too high in comparison with the juvenile hormone titer in the blood of intact insects. How does the juvenile hormone dissolve in the culture medium and how does the dissolved juvenile hormone efficiently incorporate into the target organ?

D. REGENERATE TISSUE

Marks and Reinecke (1965) attempted to test the effect of various endocrine glands on the leg regenerate tissue from *Leucophaea, in vitro*. They removed the two middle legs at the trachanterofemoral joint of the fifth instar larvae and 8 days later dissected regenerate tissues and cultured them in a Rose multipurpose chamber under dialysis strips.

Only the regenerate tissue taken at 8 days after amputation was affected by endocrine glands such as prothoracic gland *in vitro*, but tissue at earlier stages of regeneration did not show any change. Tissue isolated after 8 days grew with or without endocrine glands. Subsequently, Marks (1968) reported that corpora allata, corpora cardiaca–corpora allata complex, and prothoracic ganglion promoted the stimulating effect of the prothoracic gland on the growth of regenerated tissue. On the other hand, the brain inhibited the effect of the prothoracic gland on the growth of regenerate tissue (Marks, 1968). In addition, the secreted materials from cultured prothoracic glanglion promoted the growth of nerve tissue on the cultivated leg regenerate tissue, but the production of materials at the cultured prothoracic ganglion was inhibited by the cocultured prothoracic gland (Marks, 1969). When β-ecdysone (2.5 μg/ml) was added to the culture medium instead of endocrine glands, deposition of cuticle could be induced *in vitro* in the regenerate tissue 22 days after amputation (Marks and Leopold, 1970). A high dose (2–25 μg/ml) given over a long period (7–10 days) resulted in the deposition of heavy cuticle complete with well-defined setae; smaller doses produced a thinner cuticle on which droplets of chitin-bearing material accumulated. The lowest concentration that induced any deposition was 0.1 μg/ml given over a period of 7 days. Large doses given 1 week apart resulted in the production of multiple cuticles; small doses given at shorter intervals gave a single cumulative response. Time—dosage studies showed that the concentration of hormone and the length of exposure make roughly equal contributions to the effect of the dose (Marks, 1972).

III. Secretory Activity of the Cultured Endocrine Glands *in Vitro*

A. RING GLAND AND PROTHORACIC GLAND

It has been generally accepted that the prothoracic gland and its analogous organs are responsible for ecdysone synthesis. Although there is indirect evidence that the prothoracic gland secretes ecdysone, clear support of this theory is lacking. However, this important problem in insect endocrinology has been solved by the use of organ culture. Several attempts have been made to learn the effect of the brain ring gland complex on the development of imaginal discs *in vitro* as mentioned in Section I,A.

Marks (1971) reported that when a single pair of prothoracic glands was placed in a Rose chamber, the tissue died within a few weeks without any visible sign of activity. However, when the number of glands was increased to 8 or 10 pairs per chamber, cell migration and monolayer formation with mitotic activity persisted for several weeks. The most successful preparation consisted of one leg regenerate and two pairs of prothoracic glands. Thus, he considered that the survival of the glands seemed to depend upon "tissue mass effect" and that the monolayer around the cultured prothoracic gland grew rapidly by mitotic cell division and by migration from the explant for up to 60 days. Agui *et al.* (1972) tried to culture the prothoracic gland taken from the cabbage armyworm *M. brassicae* L. The prothoracic glands were maintained well in both CSM-2F medium and Grace's medium for long periods. Some of the prothoracic gland cells appeared extraordinarily swollen a few days after the initiation of the culture. Numerous oily droplets were successively released from the cultivated prothoracic gland for the first 2–3 days. The size of the droplets ranged from 10 to 80 μm in diameter. These droplets gradually divided into minute particles and finally disappeared. But there was no cell migration and no formation of monolayer cell sheet from cultured prothoracic gland in *M. brassicae*. When fragments of *Chilo* integuments were cultured with active prothoracic glands, a high rate of molting was obtained from both CSM-2F medium and Ringer–Tyrode's solution (R–T sol). *Chilo* integuments were cultured in 0.25 ml R–T sol in which 60 active prothoracic glands had been precultured for 48 hours. Molting was induced at a high rate, comparable to the result obtained from the application of 0.75 μg β-ecdysone/0.25 ml medium. The result seems to suggest that a hormonal agent is released from the active prothoracic glands *in vitro* which induces the ecdysis of integuments. On the other hand, no *Chilo* integument molted in R–T sol conditioned with the prothoracic glands of 30-day-old diapausing pupae, although some oily droplets were released from explants. Therefore, it may be assumed that the prothoracic gland taken from diapausing pupae could not be activated in R–T sol during cultivation. Furthermore, 10 fragments of *Chilo* integuments were cultured in 0.25 ml of R–T sol for 72 hours in the presence of 50 active prothoracic glands. At the end of a 72-hour cultivation, the rate of molting reached 90%; the prothoracic glands were then transferred to fresh R–T sol immediately after rinsing with physiological saline solution. Molting was not induced within 72 hours of cultivation in the new medium, while fresh *Chilo* integuments cultured in the old medium molted at the rate of 70% for 72 hours of cultivation. An amount of hormonal agent adequate to induce ecdysis of the *Chilo* integuments seems to be released into the medium for the first 72 hours, and thereafter the glands do not release the agent in sufficient quantity to induce ecdysis of the *Chilo* integuments.

The successive changes of ecdysone titer and activity of cultured prothoracic gland during metamorphosis in the cabbage armyworm were examined by using both the organ culture of the prothoracic gland and the *Chilo* integument assay (Agui and Yagi, 1973; Agui, 1973). As shown in Fig. 5, *in vitro* activity of prothoracic glands from the larvae increased considerably for day 2; the greatest activity was found at day 9 (prepupal stage) when the activity was calculated to be 0.003 μg dose/1 PG/0.1 ml equivalent to β-ecdysone. As for the ecdysone titer, the maximum titer was also obtained at day 9; it was estimated to be about 0.1 μg/gm weight. The ecdysone titer decreased rapidly in the 12-day-old larvae, although *in vitro* activity of prothoracic glands was still maintained at this stage. The differences between *in vitro* prothoracic gland activity and ecdysone titer suggest that the inactivation of ecdysone *in*

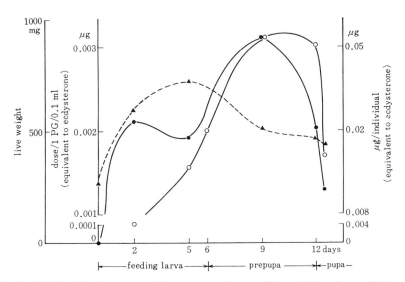

Fig. 5. Relationship between endogeneous ecdysone titer and *in vitro* activity of prothoracic glands (PG) of *Mamestra*. ●——●, ecdysone titer; ○——○, *in vitro* PG activity; ▲----▲, live weight. From Agui and Yagi (1973).

vivo occurs rather rapidly at this stage. This was supported by the fact that the exogeneous ecdysone was immediately inactivated, namely, when 2 μg of β-ecdysone was injected into each of the prepupae of those stages, the activity of the hormone decreased less than one-tenth or so 24 hours after the treatment (S. Yagi, personal communication). The results essentially correspond to the experimental results of Karlson and Bode (1969): the enzyme that inactivates ecdysone was nearly a mirror image of the ecdysone titer in *Calliphora* larvae during metamorphosis. It was revealed not only that ecdysone titer but also *in vitro* prothoracic gland activity declined and showed low levels for a while after pupation. These results indicate that the amount of the molting hormone liberated from the prothoracic gland decreased at this stage because the prothoracic gland may not yet have been reactivated by the brain. Indeed, the onset of brain reactivation for adult dvelopment starts about 3 days after pupation in *Mamestra* (T. Honda, personal communication).

Hormone production in the prothoracic gland has been described in detail in Chapter 7. Briefly, in 1971. Willig *et al.* reported that incubation of ^{14}C-labeled cholesterol with brain–ring gland complexes isolated from the 7-day-old larvae of blowfly *Calliphora erythrocephala* produced substances that cochromatographed with authentic α- and β-ecdysone. The cholesterol metabolites from the brain-ring gland complex did not cocrystallize with one another and their hormonal activity was initiated after enzymatic hydrolyzation by esterase or glucosidase. It is impossible to determine the chemical structure of these prothoracic gland secretions because (1) the cultured brain-ring gland complex consists of organs such as brain, corpus allatum and cardiacum, and prothoracic gland, and (2) there has not been adequate chemical analysis. Recently, identification of hormones secreted by the prothoracic gland has been reported simultaneously by Chino *et al.* (1974) and King *et al.* (1974). Both groups cultured lepidopterous prothoracic glands (Chino *et al.* used prothoracic glands from *B. mori* at spinning larvae, and King *et al.* used prothoracic glands from *M. sexta* at early prepupae), and identified the recovered secreted material as α-ecdysone. Also, it was found that the cultured prothoracic gland taken from the cockroach *L. maderae* (King and Marks, 1974) and the mealworm *T. molitor* (Romer *et al.*, 1974) secreted α-ecdysone.

B. CORPUS ALLATUM

Röller and Dahm (1970) cultured the complexes brain–corpora cardiaca–corpora allata from the last instar larvae of the greater wax moth

G. mellonella, in the Rose multipurpose tissue chambers filled with 1 ml Grace's medium supplemented with 6% heat-treated hemolymph from chilled diapausing pupae of *H. cecropia*. The three pairs of corpora allata that separated from the cultured complexes were implanted into full-grown, last instar larvae of *G. mellonella*, causing the animal to molt into a larval–pupal intermediate, while implantation of the remaining brain–cardiacum units had no visible effect. This result demonstrated that corpora allata kept for 6 or 7 days *in vitro* still have the potential to release juvenile hormone when implanted into larvae. Further, 50 brain–cardiaca–allata complexes were maintained in culture for 7 days. After the third and seventh days the medium was replaced and extracted three times with diethyl ether to purify the active principle which was identified to be the same as the cecropia C 18 juvenile hormone. These results provide the first direct evidence that juvenile hormone is produced and secreted by corpora allata. Two new natural insect juvenile hormones were isolated from the organ culture of corpora allata of the tobacco hornworm moth *M. sexta* and purified by high-resolution liquid chromatography (Judy *et al.*, 1973). Successively, using organ culture, chromatography, and other microchemical techniques, Schooley *et al.* (1973) demonstrated the efficient incorporation *in vitro* of several radiolabeled precursors—acetate, mevalonate, and propionate—into the juvenile hormone of *M. sexta*. Based on the position of the labeled atoms in the precursors and on the position of the incorporation obtained from label-distribution data, a scheme for juvenile hormone biosynthesis has been advanced.

C. BRAIN

The insect brain is not only the center of the nervous system but also the organ that secretes various physiologically active substances from a number of groups of neurosecretory cells. Thus, organ culture of the brain has drawn the attention of many insect endocrinologists because many unknown functions of the intact brain may be clarified by means of this technique.

Leloup and Gianfelici (1966) first observed that stainable material accumulated in the neurosecretory cells of the cultured brain from *Calliphora*. After these experiments, accumulations of neurosecretory material in the cultured brain were reported by many investigators (Schaller and Meunire, 1967; Gianfelici, 1968; Courgeon, 1969). Marks (1971) reported that, in the cultivation of the brain–corpora–cardiaca complex of the cockroach *L. maderae*, stainable material did not accumulate in the neu-

150   NORIAKI AGUI

rosecretory cells but passed to the corpora cardiaca, but if corpora cardiaca were removed from the complex, the material accumulated in the neurosecretory cells and axon tracts. In addition, β-ecdysone facilitated the release of accumulated neurosecretory materials from the neurosecretory cells of the isolated brain (Marks *et al.*, 1972).

In vitro culture of whole brain helped clarify the intrinsic relationships between the brain and prothoracic gland. The effect of cultured brain on the inactive prothoracic gland in Lepidoptera *in vitro* was first reported by Kambysellis and Williams (1971b): inactive prothoracic glands from diapausing *Cynthia* silkworm pupae were activated by culturing with *Cynthia* active brain; the hormone released from the activated prothoracic glands promoted spermatogenesis of cocultured testes *in vitro*. Agui (1975) has attempted to examine morphologically and functionally how inactive prothoracic gland can be activated by brains taken from either diapausing or nondiapausing *M. brassicae in vitro*. Figure 6 shows the morphological changes in the cultured brain from nondiapausing *Mamestra* pupae just after pupation. The cultured brain remained in good condition for a week, but considerable degeneration occurred after 2 weeks of cultivation despite the increase in volume of the

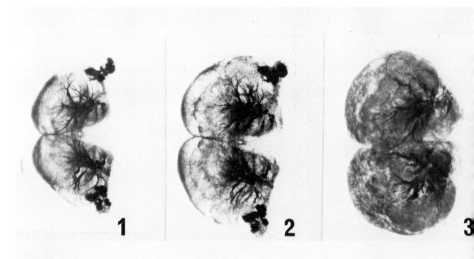

Fig. 6. Morphological changes of the cultured brain from nondiapausing *Mamestra brassicae* pupae immediately after pupation: (1) immediately after the onset of culture; (2) 1 week after the onset of culture; (3) 2 weeks after onset of culture ×78. From Agui (1975).

whole brain. The effects of nondiapausing pupae brains on prothoracic glands were investigated by means of *Chilo* integument assay *in vitro*, mentioned previously. When the diapausing prothoracic glands taken 0, 7, 40, and 70 days after pupation were cultured with nondiapausing brains from day 0 after pupation, greatest activity of the prothoracic glands was observed from the combination of diapausing prothoracic gland from 40-day-old pupae and nondiapausing brain (Fig. 7). The diapausing prothoracic glands taken 70 days after pupation could not be activated by the active brain. On the other hand, in the prothoracic gland from diapausing pupae just after pupation, there was no difference in the molting hormone activity irrespective of whether it was cultured alone or with active brain. Interestingly, prothoracic glands from diapausing pupae just after pupation showed higher hormonal activity than the prothoracic glands from nondiapausing pupae at the same stage when cultured with or without active brains.

Next, the effects of diapausing pupae brains on the prothoracic glands *in vitro* were examined. The prothoracic glands from nondiapausing

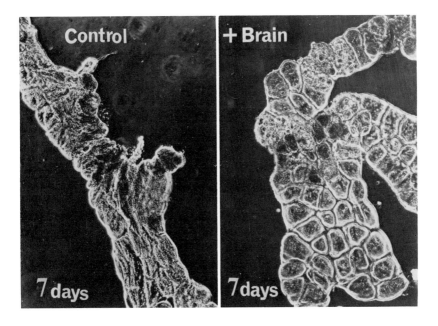

Fig. 7. The prothoracic gland activated by the active brain *in vitro*. Left side, the cultured prothoracic gland without brains on the seventh day of cultivation. Right side, the prothoracic gland activated by cocultured active brains on the seventh day of cultivation. The cultured prothoracic glands were taken from diapausing *Mamestra brassicae* pupae 40 days after pupation. ×260. From Agui (1975).

pupae just after pupation showed higher hormonal activity when cocultured with brains from diapausing pupae just after pupation as compared with the cultivation of nondiapausing prothoracic gland alone. The prothoracic glands of 30-day-old diapausing pupae were strongly activated by the brains of 25-day-old diapausing pupae. The observation that the brain from diapausing pupae had the ability to activate the inactive prothoracic gland *in vitro* is very interesting. In the control experiment, neither diapausing prothoracic glands of 7- or 70-day-old pupae were activated when cocultured with fat-body or abdominal ganglia from 0-day-old diapausing pupae.

I have attempted to investigate the relationship of the morphological changes of the neurosecretory cells by the application of molting hormone to the ability of the prothoracic glands to secrete activating hormone (prothoracotropic hormone). The stainability of the neurosecretory cells with paraldehyde–fuchsin increased when the brain was cultured in CSM-2F medium alone. If α- or β-ecdysone were added to the culture medium or the five active prothoracic glands were cocultured with diapausing brain, the neurosecretory cells showed stronger stainability than those in CSM-2F medium alone. Moreover, the neurosecretory cells treated with these hormones became tear-shaped. Also, the axon tract from neurosecretory cells manifested good stainability and chiasma-like shape.

Next, the prothoracotropic hormone released into the medium from the precultured brain by treatment with ecdysone was assayed in the following manner: Five brains from diapausing *Mamestra* pupae were cultured for 2 days in 0.02 ml CSM-2F medium containing 0.3 μg/ml β-ecdysone and were then transferred to a hormone-free medium for another 2 days in order to allow secretion of the prothoracotropic hormone. Then the brains were removed from the culture and only the conditioned medium was harvested. For the assay of prothoracotropic hormone, the inactive prothoracic glands from diapausing pupae were cultured for 4 days in this medium and then the five *Chilo* integuments were cocultured for 4 days with the prothoracic glands in order to assay the molting hormone activity. The molting rate was 8/15 when the above treatment was administered. However, if ecdysone was not applied and the medium was conditioned only with diapausing brain for 2 or 4 days, the molting rate was 2/15 or 0/15. From these results it was concluded that the brains from diapausing pupae could be activated by molting hormone *in vitro* and that when the brains conditioned with ecdysone were transferred to hormone-free medium they released the prothoracotropic hormone. Thus, these results suggest that there may be some feedback regulation between the brain and the prothoracic gland.

D. OENOCYTE

It has been known that oenocytes are in some way connected with humoral activity and that the morphological changes of these organs are linked with the molting cycle of the insect. Accordingly, attempts have been made to clarify the role of these organs, but their function is still far from clear. In 1969, Locke proposed that the great amount of endoplasmic reticula in oenocytes suggests the occurrence of steroid metabolism and that the oenocyte might be the site of ecdysone synthesis. Recently, Romer (1971) reported that there was high molting activity in the oenocyte of the mealworm prepupae *T. molitor*, but the actual role of the oenocyte *in vivo* is unknown to date. On the other hand, based on results of *in vitro* incorporation of ^{14}C-acetate, Diehl (1973) reported that oenocytes which were associated with fat-body tissue in the desert locust larvae *S. gregaria*, synthesized paraffin during short-term incubation (1 hour) in buffered Ringer's solution.

Agui (1974) carried out a series of experiments which were designed to examine the effects of the prothoracic glands and oenocytes from the last instar larvae of the cabbage armyworn *M. brassicae* on the morphogenesis of the wing discs of the same host *in vitro*. It was found that the oenocytes alone barely affected the wing discs, but the latter showed remarkable development when cocultured with prothoracic glands and oenocytes. These results seem to indicate that oenocytes are activated by prothoracic gland hormone. Subsequently, the author examined the effect of oenocytes and α- and β-ecdysone on the development of wing discs. The wing discs cocultured with oenocytes in the medium containing α-ecdysone also developed more than those treated with α-ecdysone alone. Another interesting fact was that there was no difference between wing discs treated with oenocyte plus β-ecdysone, and β-ecdysone alone. Although many hypotheses may be postulated, the above results seem to suggest that α- and β-ecdysone may have different and distinct roles in oenocyte activation and that the agent secreted from the oenocyte does not directly affect the development of cocultured wing discs but acts synergistically with α-ecdysone. Recently, Romer *et al.* (1974) reported that the isolated abdominal oenocytes from *Tenebrio* larvae mainly synthesized β-ecdysone and only a little α-ecdysone. Moreover, when prothoracic glands and oenocytes were cultured together, the α-ecdysone derived from the prothoracic gland was oxidized to β-ecdysone by the oenocytes. In addition, they mentioned that the abdominal oenocytes synthesized ecdysone independently but did not give any explanation for the discrepancy of their results with those of most experimental morphological studies that have shown a clear-cut predominance of the prothoracic gland over the abdomen in the control of the molting process. They felt that the prothoracic gland may produce not only α-ecdysone but also the unknown activator for the oenocytes.

IV. Feedback Regulation on Neuroendocrine and Endocrine Gland *in Vitro*

In general, the protocephalic corpus cardiacum–allatum system in insects may be regarded as analogous to the hypothalomo–hypophyseal system in vertebrates. The possibility that the insect neurosecretory system is subject to feedback regulation during metamorphosis has been suggested in earlier reviews (Williams, 1950; Bodenstein, 1954), but this suggestion has not been studied extensively. Recently, some experimental results regarding feedback mechanisms between the neuroendocrine system and the secretory organs *in vivo* and *in vitro* have been reported by insect endocrinologists. Steel (1973, 1975) observed that when the fifth instar *Rhodinus* 1 day after feeding underwent parabiosis with a decapitated old one (8 days after feeding) or was injected with β-ecdysone, the median neurosecretory cells of the younger insect were induced to switch their normal sequence of cytological changes to those characteristic of the older insect. From these findings, Steel concluded that a feedback relationship exists between prothoracotropic "brain hormone" and ecdysone. The behavior of median neurosecretory cells produced "brain hormone" changes with the increased ecydsone titer; hence the effect of ecdysone on the median neurosecretory cells is a negative feedback on release of neurosecretory materials coupled with a stimulation of synthesis.

As mentioned earlier in this chapter (Section III,D), when the brain from diapausing *Mamestra* pupae was treated with β-ecdysone (0.3 μg/ml) *in vitro* for 2 days, the stainability of neurosecretory cells increased. This phenomenon indicates that neurosecretory material synthesis was enhanced as a direct result of ecdysone treatment *in vitro*. The brains treated with ecdysone were then transferred to ecdysone-free medium for 2 days in order to allow the release of the neurohormone into the medium and the conditioned medium from which the brains were removed was assayed. This conditioned medium contained highly active prothoracotropic hormone. This result suggests that the activated brain treated with ecdysone release neurosecretory materials into the medium if it was transferred from the medium with a high ecdysone titer to one without ecdysone. This *in vitro* evidence clearly indicated the feedback relation between the prothoracotropic hormone and ecdysone as suggested by Steel's *in vivo* experiment.

Marks *et al.* (1972) reported that the cultivated brain of cockroach *L. maderae* synthesized and stored hindgut-stimulating neurohormone in the median nuerosecretory cells, but if β-ecdysone (2.5 μg/ml) was applied to the cultured brain *in vitro*, the stored materials were released into the medium. Ecdysone thus acted as the releasing factor for the hindgut-stimulating neurohormone. Subsequently, the interesting fact was reported that the hindgut-stimulating neurohormone acted directly on the brain *in vitro*, and then it stimulated the release of the accumulated neurohormone. The corpora cardiaca were able to sequester this neurohormone from the medium when its concentration was above threshold level. Thus, Marks and Holman (1974) suggested that the brain–corpus cardiacum system may provide a homeostatic system for maintaining physiological levels of neurohormone.

In 1952, Williams suggested that the molting hormone can stimulate prothoracic glands. In addition, the fact that the prothoracic gland is stimulated and maintained by juvenile hormone was suggested by Gilbert and Schneiderman (1959) and Gilbert (1962). One is now faced with the question: What factors induce these glands to cease secreting? In an attempt to answer this question, Siew and Gilbert (1971) examined the effects of the molting hormone and juvenile hormone on nuclear RNA synthesis of the corpora allata and prothoracic gland of Saturniid pupae by autoradiography *in vivo*. When β-ecdysone (5 μg/gm) was injected into the diapausing cecropia pupae, β-ecdysone stimulated nuclear RNA synthesis in the prothoracic glands within 3 hours, followed by activation of the corpora allata 3 hours later. Injection of juvenile hormone (15 μg/gm) resulted first in corpora allata activation followed by prothoracic gland stimulation. When *H. cecropia*, which was fully chilled and in which adult development was initiated, was injected with B-ecdysone, the activity of RNA synthesis in the prothoracic gland decreased. This *in vivo* experiment suggests that the prothoracic gland activited by normal means (brain hormone) was controlled by a negative feedback mechanism induced by its own product, ecdysone.

The positive feedback mechanism of the prothoracic gland was ascertained by Agui and Fukaya (1973) using organ cultures of *Mamestra* prothoracic glands. Prothoracic glands (2, 5, 10, or 20) were cultured in Grace's medium which was renewed every 48 hours for 6 days and then harvested. The harvested medium was assayed in order to examine the successive changes of activity of ecdysone secreted from the cultured prothoracic glands. In the media precultured with 10 to 20 prothoracic glands the highest activity occurred within 48 hours, whereas in the media precultured for the last 48 hours, the activity was as low as that obtained with two prothoracic glands. In the culture with two prothoracic glands, the

differences among the activities of the hormone in the media conditioned
with the glands during the three successive stages were scarcely observed.
The fact that there was more prothoracic glands and that their output
was great during the earlier culture periods seems to suggest the presence
of a positive feedback system among prothoracic glands *in vitro*. Though
it has been demonstrated that prothoracic glands are stimulated by their
own secreted hormone, the question still remains of how the glands cease
to secrete. It may be caused by cessation of stimulation of prothoracotropic hormone secreted from the brain.

Another neuroendocrine tissue such as the subesophageal ganglion may
be able to reduce prothoracic gland activity. I used one pair of prothoracic glands and five subesophageal ganglia taken from 6-day-old
Mamestra larvae cocultured in Grace's medium. The ecdysone titer in
the conditioned medium was assayed by means of wing disc development.
The results revealed that the wing discs cocultured with both prothoracic
glands and subesophageal ganglia scarcely developed, but the wing discs
which were cocultured with prothoracic glands only developed to some
extent (N. Agui, unpublished). These *in vitro* experiments suggest that
the subesophageal ganglion released some factor which inhibits the function of the prothoracic gland. Moverover, the cessation of endocrine gland
activity seems to influence not only the humoral factor but also nervous
system control. The feedback relation clarified by the *in vitro* experiments is summarized in Fig. 8.

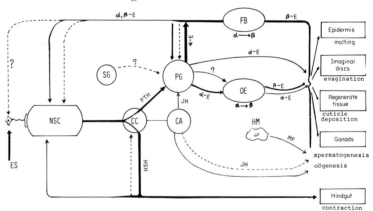

Fig. 8. Scheme of the relationship between secretory organs and target organs mentioned in this paper. ←, stimulation or releasing; ←--, inhibition; α-E, α-ecdysone;
β-E, β-ecdysone; CA, corpus allatum; CC, corpus cardiacum; ES, environment stimuli; FB, fat body; HM, hemocyte; HSN, hindgut-stimulating neurohormone; JH,
juvenile hormone; MF, macromolecular factor; NSC, neurosecretory cell; OE, oenocyte; PG, prothoracic gland; PTH, prothoracotropic hormone; SG, subesophageal
ganglion.

V. Conclusions

The author has attempted to answer the following questions: How does the interaction between cultured target organs and endocrine glands occur? What hormones act on cultured target organs? How is the secretory ability of endocrine glands controlled *in vitro*? The pertinent information available at this time is summarized in Fig. 8.

There are considerable advantages in using organ cultures of target organs. First, the sensitivity of target organs to hormones changes in relation to the aging of the organs: The younger target organs display refractoriness. On the other hand, when the target organs grow to a stage sensitive to hormones, the effects of the hormones are cumulative. Thus when organ culture methodology is used to study the effects of hormones on target organs, the age of the organs has to be taken into consideration. Second, it was ascertained that cultured target organs treated with hormones cyclically responded to the hormones *in vitro*. If hormone concentration in the culture medium could be controlled consecutively during cultivation as it is *in vivo*, it might be possible to clarify target organ response to hormones *in vivo*. However, further improvements in medium circulation in the culture vessel are needed to refine this technique. The mechanism by which the hormone is transported to the target organs in culture is worthy of future study.

Many problems have been encountered with endocrine glands cultured *in vitro*: The secreting ability of the endocrine glands changes simultaneously with the age of the gland in the same manner in which the hormone sensitivity of the target organs changes. This seems to be affected by internal factors that accompany aging or by external factors such as the metobolic rate or the speed with which other hormones are excreted. As for the latter, the feedback mechanism regulating the hormone titer in the blood must be considered. Thus, for the culture of endocrine glands such as prothoracic gland or corpus allatum which are activated by neurohormones, it is necessary to establish a culture system in which the hormone titer in the medium can be easily controlled because the glands are subject to negative or positive feedback regulation as a result of the hormones secreted within the glands themselves. It is also necessary to investigate further the manner in which the hormone precursors are dissolved in the medium in order to establish culture conditions suitable for precursor incorporation into endocrine glands and for hormone synthesis.

One of the most interesting studies is organ culture of neurosecretory glands which is directly controlled by the nervous system. As mentioned

in connection with culture of the brain, the synthesis or release of neurosecretory materials in the neurosecretory cells of the cultured brain are
stimulated by β-ecdysone *in vitro*. But, the mechanism is not clearly
understood: How are the external or internal neurostimuli transmitted
to the neurosecretory cells and how are the neurosecretory materials
released from the neurosecretory cells?

The accumulation of data concerning the interaction between target
organs and endocrine glands *in vitro* will facilitate the analysis of the
complicated phenomena of the insect endocrine system *in vivo* and possibly enable its study *in vitro*.

Acknowledgments

I am particularly indebted to the late Professor M. Fukaya and to Dr. S. Yagi,
Laboratory of Applied Zoology, Tokyo University of Education, for their help in
the preparation of this chapter and for their many suggestions for its improvement. I am also grateful to Professor K. Maramorosch for editing the English text.

References

Adams, T. S., and Eide, P. E. (1972). *Gen. Comp. Endocrinol.* **18**, 12–21.
Agui, N. (1973). *Appl. Entomol. Zool.* **8**, 236–239.
Agui, N. (1974). *Appl. Entomol. Zool.* **9**, 256–260.
Agui, N. (1975). *J. Insect Physiol.* **21**, 903–913.
Agui, N., and Fukaya, M. (1973). *Appl. Entomol. Zool.* **8**, 73–82.
Agui, N., and Yagi, S. (1973). *Appl. Entomol. Zool.* **8**, 239–240.
Agui, N., Yagi, S., and Fukaya, M. (1969a). *Appl. Entomol. Zool.* **4**, 156–157.
Agui, N., Yagi, S., and Fukaya, M. (1969b). *Appl. Entomol. Zool.* **4**, 158–159.
Agui, N., Kimura, Y., and Fukaya, M. (1972). *Appl. Entomol. Zool.* **7**, 71–78.
Berreur-Bonnenfant, J. (1972). *In* "Invertebrate Tissue Culture" (C. Vago, ed.), Vol.
    2, pp. 181–210. Academic Press, New York.
Bodenstein, D. (1954). *Recent Progr. Horm. Res.* **10**, 157–182.
Borst, D. W., Bollenbacher, W. E., O'Connor, J. D., King, D. S., and Fristrom
    J. W. (1974). *Develop. Biol.* **39**, 308–316.
Brooks, M. A., and Kurtii, T. J. (1971). *Annu. Rev. Entomol.* **16**, 27–52.
Burdette, W. J., Hanley, E. W., and Grosch, H. (1968). *Tex. Rep. Biol. Med.* **26**,
    173–180.
Chihara, C. J., and Fristrom, J. W. (1973). *Develop. Biol.* **35**, 36–46.
Chihara, C. J., Petri, W. H., Fristrom, J. W., and King, D. S. (1972). *J. Insect Physiol.*
    18, 1115–1123.
Chino, H., Sakurai, S., Ohtaki, T., Ikekawa, N., Miyazaki, H., Ishibashi, M., and
    Abuki, H. (1974). *Science* **183**, 529–530.
Courgeon, A. M. (1969). *C.R. Acad. Sci.* **268**, 950–952.
Demal, J., and Leloup, A. M. (1972). *In* "Invertebrate Tissue Culture" (C. Vago,
    ed.), Vol. 2, pp. 3–39. Academic Press, New York.
Diehl, P. A. (1973). *Nature (London)* **243**, 468–470.

Fujio, Y. (1962). *Jap. J. Genet.* **37**, 110–117.
Fukushima, T., and Yagi, S. (1975). *Appl. Entomol. Zool.* **10**, 220–225.
Gianfelici, E. (1968). *Ann. Endocrinol.* **29**, 496–500.
Gilbert, L. I. (1962). *Nature (London)* **193**, 1205–1207.
Gilbert, L. I., and Schneiderman, H. A. (1959). *Nature (London)* **184**, 171–173.
Goldschmidt, R. (1915). *Proc. Nat. Acad. Sci. U.S.* **1**, 220–222.
Gottschewski, G. (1958). *Naturwissenschaften* **45**, 400.
Gottschewski, G. (1960). *Wilhelm Roux' Arch. Entwicklungsmech. Organismen* **152**, 204–229.
Horikawa, M. (1958). *Cytologia* **23**, 468–477.
Horikawa, M. (1960). *Jap. J. Genet.* **35**, 76–83.
Horikawa, M., and Sugihara, M. (1960). *Radiat. Res.* **12**, 266–275.
Ittycheriah, P. I., and Stephanos, S. (1969). *Indian J. Exp. Biol.* **7**, 17–19.
Judy, K. J., Schooley, D. A., Dunham, L. L., Hall, M. S., Bergot, B. J., and Siddall, J. B. (1973). *Proc. Nat. Acad. Sci. U.S.* **70**, 1509–1513.
Kambysellis, M. P., and Williams, C. M. (1971a). *Biol. Bull.* **141**, 527–540.
Kambysellis, M. P., and Williams, C. M. (1971b). *Biol. Bull.* **141**, 541–552.
Karlson, P., and Bode, C. (1969). *J. Insect Physiol.* **15**, 111–118.
King, D. S. (1972). *Gen. Comp. Endocrinol. Suppl.* **3**, 221–227.
King, D. S., and Marks, E. P. (1974). *Life Sci.* **15**, 147–154.
King, D. S., and Siddall, J. B. (1969). *Nature (London)* **221**, 955–956.
King, D. S., Bollenbacher, W. E., Borst, D. W., Vedeckis, W. V., O'Connor, J. D., Ittycheriah, P. I., and Gilbert, L. I. (1974). *Proc. Nat. Acad. Sci. U.S.* **71**, 793–796.
Kuroda, Y. (1969). *Jap. J. Genet.* **44**, Suppl. 1,42–50.
Kuroda, Y. (1974). *J. Insect Physiol.* **20**, 637–640.
Kuroda, Y., and Yamaguchi, K. (1956). *Jap. J. Genet.* **31**, 98–103.
Laverdure, A. M. (1969). *C.R. Acad. Sci.* **269**, 82–85.
Leloup, A. M., and Demal J. (1968). *Proc. Int. Colloq. Invertebr. Tissue Cult., 2nd. 1967* pp. 126–137.
Leloup, A. M., and Gianfelici, E. (1966). *Ann. Endocrinol.* **17**, 506–508.
Lender, T., and Duveau-Hagége, J. (1962). *C.R. Acad. Sci.* **254**, 2825–2827.
Lender, T., and Duveau-Hagége, J. (1963a). *Ann. Epiphyt.* [2] **14**, 18–89.
Lender, T., and Duveau-Hagége, J. (1963b). *Develop. Biol.* **6**, 1–22.
Lender, T., and Laverdure, A. M. (1967). *C.R. Acad. Sci.* **265**, 451–454.
Locke, M. (1969). *Tissue & Cell* **1**, 103–154.
Mandaron, P. (1973). *Develop. Biol.* **31**, 101–113.
Marks, E. P. (1968). *Gen. Comp. Endocrinol.* **11**, 31–42.
Marks, E. P. (1969). *Biol. Bull.* **137**, 181–188.
Marks, E. P. (1970). *Gen. Comp. Endocrinol.* **15**, 289–302.
Marks, E. P. (1971). *Curr. Top. Microbiol. Immunol.* **55**, 75–85.
Marks, E. P. (1972). *Biol. Bull.* **142**, 293–301.
Marks, E. P. (1973). *Biol. Bull.* **145**, 171–179.
Marks, E. P., and Holman, C. M. (1974). *J. Insect Physiol.* **20**, 2087–2093.
Marks, E. P., and Leopold, R. A. (1970). *Science* **167**, 61–62.
Marks, E. P., and Reinecke, J. P. (1965). *Gen. Comp. Endocrinol.* **5**, 241–247.
Marks, E. P., Ittycheriah, P. I., and Leloup, A. M. (1972). *J. Insect Physiol.* **18**, 847–850.
Miciarelli, A., Sbrenna, G., and Colombo, G. (1967). *Experientia* **23**, 64–66.
Mitsuhashi, J. (1965). *Jap. J. Appl. Entomol. Zool.* **9**, 217–224.
Mitsuhashi, J. (1968). *Appl. Entomol. Zool.* **3**, 1–4.

160   NORIAKI AGUI

Moriyama, H., Nakanishi, K., King, D. S., Okauchi, T., Siddall, J. B., and Hafferl, W. (1970). *Gen. Comp. Endocrinol.* **15**, 80–87.
Oberlander, H. (1969). *J. Insect Physiol.* **15**, 297–304.
Oberlander, H. (1972a). In "Biology of Imaginal Discs" (H. Ursprung and R. Nöthinger, eds.), pp. 155–172. Springer-Verlag, Berlin and New York.
Oberlander, H. (1972b). *J. Insect Physiol.* **18**, 223–228.
Oberlander, H., and Fulco, L. (1967). *Nature (London)* **216**, 1140–1141.
Ohmori, K. (1974). *J. Insect Physiol.* **20**, 1697–1706.
Ohmori, K., and Ohtaki, T. (1973). *J. Insect Physiol.* **19**, 1199–1210.
Röller, H., and Dahm, K. (1970). *Naturwissenschaften* **57**, 31–32.
Romer, F. (1971). *Naturwissenschaften* **58**, 324–325.
Romer, F., Emmerich, H., and Nowock, J. (1974). *J. Insect Physiol.* **20**, 1975–1987.
Schaller, F., and Meunier, J. (1967). *C. R. Acad. Sci.* **264**, 1441–1444.
Schmidt, E. L., and Williams, C. M. (1953). *Biol. Bull.* **105**, 174–187.
Schneider, I. (1964). *J. Exp. Zool.* **156**, 91–104.
Schooley, D. A., Judy, K. J., Bergot, B. J., Hall, M. S., and Siddall, J. B. (1973). *Proc. Nat. Acad. Sci. U.S.* **70**, 2921–2925.
Sengel, P., and Mandaron, P. (1969). *C. R. Acad. Sci.* **268**, 405–407.
Shibuya, I., and Yagi, S. (1972). *Appl. Entomol. Zool.* **7**, 97–98.
Siew, Y. C., and Gilbert, L. I. (1971). *J. Insect Physiol.* **17**, 2095–2104.
Steel, C. G. H. (1973). *J. Exp. Biol.* **58**, 177–187.
Steel, C. G. H. (1975). *Nature (London)* **253**, 267–269.
Takeda, N. (1972a). *Appl. Entomol. Zool.* **7**, 37–39.
Takeda, N. (1972b). *J. Insect Physiol.* **18**, 571–580.
Williams, C. M. (1950). *Sci. Amer.* **182**, 24–28.
Williams, C. M. (1952). *Biol. Bull.* **103**, 120–138.
Willig, A., Rees, H. H., and Goodwin, T. W. (1971). *J. Insect Physiol.* **17**, 2317–2326.
Yagi, S., and Fukushima, T. (1975). *Appl. Entomol. Zool.* **10**, 77–83.
Yagi, S., Kondo, E., and Fukaya, M. (1969). *Appl. Entomol. Zool.* **4**, 70–78.

10

Characteristics of the Action of Ecdysones on *Drosophila* Imaginal Discs Cultured *in Vitro*

JAMES W. FRISTROM AND
MARY ALICE YUND

I. Introduction

The mechanisms by which hormones elicit responses from target tissues continue to be a subject of intense research. Investigations of the mode of action of vertebrate steroid hormones have been particularly successful. The results indicate that these hormones interact directly with the chromatin of target cells via a complex with a cytoplasmic receptor (for review, see Jensen and DeSombre, 1972). By contrast, studies on the modes of action of invertebrate hormones generally lag far behind those with vertebrate material, often for nontechnical reasons. Nevertheless, insect tissues offer unique properties unavailable in vertebrate systems for studying the action of hormones. This is particularly true in insects having a complete metamorphosis, where the dramatic destruction of larval tissues and the development of the imaginal discs into the form of the adult insect results from stimuli provided by the insect steroid molting hormone β-ecdysone.* An understanding of the mechanism of the hormone-induced development of imaginal discs may be useful for comparative endocrinology and lead toward a general understanding of action of steroid hormones so far heavily weighted by studies on vertebrates.

* The term ecdysone is used here to denote any member of a class of compounds having molting hormone activity.

For insects *Drosophila melanogaster* is particularly suited for studies on hormone action. The reasons are fourfold: (1) Responses to hormones, particularly ecdysones, occur *in vitro* under defined conditions and involve the complete set of normal responses to the hormone, e.g. complete differentiation of imaginal discs (Mandaron, 1973; Fristrom *et al.*, 1973; Milner and Sang, 1974); induction of the normal puffing pattern in polytene chromosomes of salivary glands (Ashburner, 1972); (2) the presence of "puffs" in polytene chromosomes allows the apparent direct visualization of gene activity, as has been particularly exploited by Ashburner and his colleagues (Ashburner *et al.*, 1973); (3) the *Drosophila* genome has less DNA with a simpler sequence complexity than that of vertebrates (see Laird, 1973, for review); and (4) *D. melanogaster* is susceptible to sophisticated genetic manipulations that may potentially be used for the study of hormone action. This last point can be briefly documented by a single example (unpublished observation). Genetic mosaics involving a recessive, nonpupariating X chromosome lethal mutant recovered by Stewart *et al.* (1972) were produced by mating females heterozygous for the lethal gene with ring X males carrying a wild-type allele of the lethal (Table I). Loss of the ring X chromosome in some cells during early embryonic cell division results in mosaics containing mutant cells with a single lethal carrying X chromosome and wild-type cells carrying two X chromosomes. When these mosaics attempt to pupariate, the result is an organism containing both tanned prepupal cuticle (wild-type) and untanned larval cuticle (mutant) (Fig. 1). This demonstrates that the failure to pupariate is an autonomous characteristic of the epidermal tissue and does not result from a failure in the production of molting hormone. This is particularly evident in the example shown in Fig. 1, where the anterior of the organism containing the endocrine glands is mutant. Furthermore, on the basis of preliminary genetic analysis, it

TABLE I
Production of a Mosaic Gynander Involving an X Chromosome Recessive Lethal

Parents: Ring X males × II-39 (X chromosome lethal)/+ heterozygous females
Progeny: Males: II-39/− [X chromosome hemizygous lethals (die without puparium formation)]
　　　　　+ viable males
　　Females: Ring X/+ heterozygote
　　　　　　Ring X/II-39 (X chromosome lethal) heterozygote
Formation of mosaic: Loss of Ring X during mitosis creates some cells that are II-39/−. These cells are male (X/O) and hemizygous for the lethal gene. They are found in the same organism with Ring X/II-39 heterozygous female cells.

Fig. 1. Larval pupal gynander. The anterior is mutant male tissue (identified by an independent genetic marker), and the posterior is wild-type female tissue (identified by the presence of ovaries). This individual lived about 4 days after the posterior end pupariated. Apparently the posterior also underwent the pupal molt.

seems likely that all such mutants of this type can be recovered and that one will have in hand a series of mutants preventing a particular target tissue from undergoing its characteristic response to a hormone. Thus, the genetic and, perhaps subsequently, the precise biochemical dissection of the response of a target tissue to its hormone can be carried out.

However, the main thrust of our work during the past 2–3 years has been an investigation of the parameters involved in the activity, uptake, binding, and distribution of ecdysones in imaginal disc cells. For our studies we have utilzed discs isolated *en masse* from mid to late third-instar larvae cultured in either the medium developed several years ago by James Robb (1969) or a minimal medium for evagination (MME) (Fristrom *et al.*, 1973). We routinely isolate 800,000 discs/day from 0.5–1kg of larvae. Such discs undergo evagination *in vitro* when exposed to ecdysones (Fig. 2) to produce wing and leg structures. Discs dissected from axenically grown larvae, differentiate adult structures (bristles, hairs, nerves, and muscles) in Robb's medium *in vitro* in response to

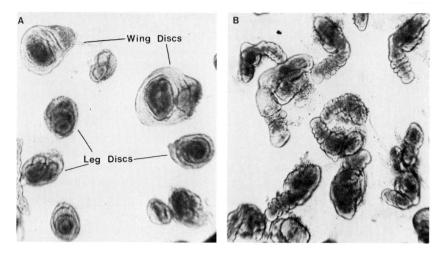

Fig. 2. (A) Unevaginated and (B) evaginated leg discs. Discs were incubated in Robb's culture medium with 0.1 μg/ml of β-ecdysone (B) or without hormone (A) for 20 hours at 25°C. The evagination process is quantitated using an Evagination Index (Chihara *et al.*, 1972) in which unevaginated discs (A) are scored as 0 and fully evaginated discs (B) are scored as 10.

β-ecdysone (Fristrom *et al.*, 1973). However, only evagination and the apolysis of the pupal cuticle occur using mass-isolated discs. Further development using mass-isolated discs has not been achieved, since despite the exhaustive washing of the discs during their isolation from nonaxenically grown larvae, contamination with yeast inevitably results. Thus, this system falls somewhat awkwardly between the biochemists "tissue slice" technique and the genuine organ culture described elsewhere in this volume by the real practitioners of the art. However, the system is, as we hope to demonstrate, quite suitable for studies of the action of ecdysones on target tissues.

II. Activity of Ecdysone Analogues

The *in vitro* induction of evagination has been used to determine the activity of different ecdysone analogues (Chihara *et al.*, 1972; Mandaron, 1973; Agui and Fukaya, 1973; Ohmori and Ohtaki, 1973). The assay using leg or wing discs of *D. melanogaster* is complete in about 20 hours and conveniently conducted overnight. The disc system is particularly attractive because discs do not metabolize ecdysones to a significant extent (Chihara *et al.*, 1972; Ohmori and Ohtaki, 1973). Therefore, in contrast to *in vivo* systems, the inherent activities of different ecdysone analogues can be ascertained. The earlier observations from our laboratory have been extended to include 24 different analogues. The results are presented in Table II and some appropriate structural formulas are shown in Fig. 3. As can be seen the range of activity among the active compounds varies about 10 millionfold from ponasterone A with the highest activity to rubrosterone with the lowest activity. Because of the wide range of activity, the purity of the compounds used in this test is particularly critical. This point is illustrated by assays on three different samples of 2-deoxyecdysone generously supplied by Denis Horn and his associates. Samples 1 and 2 are natural and 3 is synthetic. The three samples have different activities in the disc assay. Sample 2 is much less active than 1, and 3 is inactive up to 50 μg/ml. With the development

TABLE II

Activity of Ecdysone Analogues in Inducing Evagination of
Imaginal Discs *in Vitro*[a]

Ecdysone analogue	Concentration producing half-evagination (μg/ml)
Ponasterone A	0.0008
Polypodin B	0.007
β-Ecdysone	0.015
Cyasterone	0.015
2-Deoxy-β-ecdysone	0.15
Inokosterone	0.2
$2\beta,3\beta,5\beta,14\alpha$-tetrahydroxy-5$\beta$-cholest-7-en-6-one	0.5
α-Ecdysone	7
2,3-Diacetate of β-ecdysone	15
Iso-α-ecdysone	15
2,3-Acetonide of β-ecdysone	50
3,22-Diacetate of β-ecdysone	150
Rubrosterone	5000
2-Deoxy-α-ecdysone	
Sample 1, natural	0.005
Sample 2, natural	0.6
Sample 3, synthetic	Inactive

[a] Inactive compounds (to limit of solubility): 2,3,22-triacetate of β-ecdysone; "ketodiol;" $3\beta,5\beta,14\alpha$-trihydroxy-5β-cholest-7-en-6-one; $2\beta,3\beta,14\alpha$-trihydroxy-5β-cholest-7-en-6-one 2,3-dimethanesulfonate; $2\beta,3\beta,14\alpha$-trihydroxy(24R)5β-stigmast-7-en-6-one; $14\epsilon,15\epsilon$-epoxy-$2\beta,3\beta$-dihydroxy-5β-cholest-7-en-6-one; $2\beta,3\beta,14\alpha$-trihydroxy(24R)5β-ergost-7-en-6-one; $2\beta,3\beta,25$-trihydroxy-5β-cholestan-6-one; $2\beta,3\beta,14\alpha$-trihydroxy-5β-cholest-7-en-6-one; $3\beta,14\alpha,22$-trihydroxy-5β-cholest-7-en-6-one.

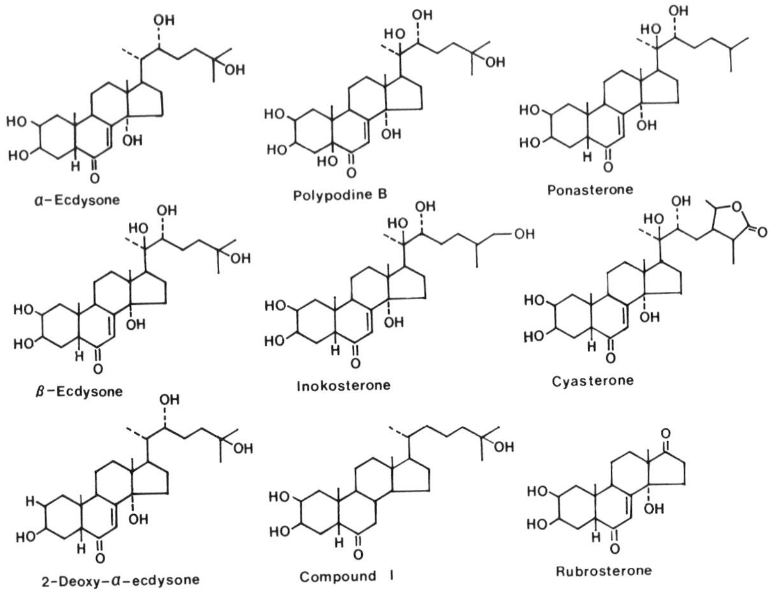

Fig. 3. Structures of ecdysones. Structures of some ecdysones listed in Table II and discussed in the text are shown.

of high resolution liquid chromatographic systems, D. Horn and his co-workers (personal communication) discovered that sample 1 contained substantial quantities of ponasterone A, and therefore the high activity of this material is trivially a result of a contaminant. The difference in activity of samples 2 and 3 is probably due to a slight contamination of the natural sample with ponasterone A (about 0.1%). It is possible that a minor contaminant in our sample of rubrosterone might be responsible for the activity of this compound, or that activities of other compounds are augmented by the presence of contaminants. Nevertheless, it seems clear that compounds containing the 20,22-dihydroxy configuration are the most active ones. Of particular note in this regard is the fact that of the two most commonly found insect ecdysones, β-ecdysone is 400 to 500 times more active than α-ecdysone. Our results on this point are not in accord with those of Mandaron (1973). However, a recent report by Milner and Sang (1974) confirms that β-ecdysone is substantially more active than α-ecdysone in inducing morphogenesis and differentiation of discs of *D. melanogaster*. Analysis of extracts of metamorphosing *D. melanogaster* shows that β-ecdysone is the predominant ecdysone recovered, with little or no α-ecdysone being detected (Borst *et al.*, 1974). Furthermore, the amount of β-ecdysone found *in situ* (0.1–0.4 μg/gm wet weight) agrees closely with the amount necessary to induce disc morphogenesis and differentiation *in vitro* (0.1 μg/ml) (Fristrom *et al.*, 1973; Milner and Sang, 1974). Therefore, there seems to be little doubt that β-ecdysone is responsible for inducing metamorphosis in *Drosophila*. As noted in Chapter 7, α-ecdysone is produced by the prothoracic glands and then subsequently metabolized to β-ecdysone.

In addition to detecting ecdysones active in inducing evagination *in vitro* we identified several compounds which at the limits of solubility are inactive in inducing evagination (Table II). We have tested some of these compounds for the possibility that they are antiecdysones, i.e., inhibit the action of β-ecdysone. At least two of the compounds, 2β,3β,25-trihydroxy-5β-cholestan-6-one (compound I, see Fig. 3) and 3β,5β,14α-trihydroxy-5β-cholest-7-en-6-one (compound II), appear to be anti-ecdysones. When these compounds are added to the culture medium (concentration: 1 μg/ml) they reduce the degree of evagination achieved by a minimally effective concentration of β-ecdysone. However, as seen in Fig. 4, the addition of increased amounts of β-ecdysone in the presence

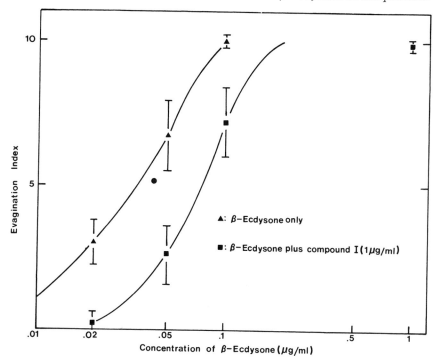

Fig. 4. Effects of compound I on β-ecdysone-induced evagination. Discs were incubated with different concentrations of β-ecdysone in the presence or absence of compound I as indicated. The Evagination Index is described briefly in the legend to Fig. 2.

of compound I results in complete evagination. Similar results are obtained with compound II. These observations demonstrate that these antiecdysones competitively inhibit the induction of evagination by β-ecdysone, presumably by preventing the binding of β-ecdysone to its specific receptors.

The activity of a given hormone analogue is a function of its equilibrium binding constant for specific receptors. The equilibrium constant in turn is a result of an affinity constant and a dissociation constant. Thus, the activity of an analogue may change because of a change in affinity for its receptors or because of a change in the rate of dissociation from the receptors. We have been interested in identifying an ecdysone with a low dissociation constant because such a compound will be particularly useful in studies aimed at purification of receptors and may be effective in the regulation of insect populations, since irreversible binding to receptor sites might adversely affect molting cycles.

Our approach to this problem has been indirect and utilizes an experiment in which an ecdysone is added to the culture medium for a short period of time and is then removed. The degree of evagination is then determined after 20 hours. Compounds with relatively low rates of dissociation might be expected to produce greater evagination than compounds with high rates of dissociation because they would remain attached to the receptors for longer periods of time. In an early experiment we dem-

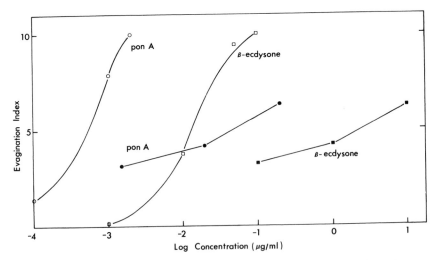

Fig. 5. Evagination produced by continuous or intermittent exposure to ecdysones. Discs were incubated with different concentrations of β-ecdysone and ponasterone A as indicated. Continual exposure (○, □); exposure for 2.5 hours (●, ■). The Evagination Index is described in the legend to Fig. 2. Pon A, ponasterone A.

onstrated that discs could undergo complete evagination after only a 6–9 hour exposure to β-ecdysone (Fristrom *et al.*, 1973). We have now conducted a similar study in which discs were exposed for 2.5 hours to different concentrations of ponasterone A. The results along with equivalent data for β-ecdysone are presented in Fig. 5. There is approximately a fiftyfold difference in the concentrations of the hormones which produce the same degree of evagination. Since this same fiftyfold difference in effective concentration is found in continuous exposure experiments, there is no idication that the rate of dissociation of ponasterone A from receptors is appreciably slower than that of β-ecdysone.

III. Characteristics of Uptake and Binding of β-Ecdysone

Another approach in the investigation of the action of ecdysones on imaginal discs has involved the use of ^{3}H-β-ecdysone. It should be emphasized from the start that ^{3}H-β-ecdysone is far more suitable than ^{3}H-α-ecdysone for studies on discs because of its higher activity. However, the availability of ^{3}H-β-ecdysone with high specific activity is extremely limited and therefore there are restrictions on the types of experiments which can be performed. We have chosen to limit our studies to experiments designed to elucidate some simple properties regarding the uptake and binding of β-ecdysone by discs. First, we have investigated the existence and location of specific, high affinity binding sites for β-ecdysone in discs. Second, we have compared the kinetics of uptake and release of β-ecdysone by discs with the kinetics of respective increases and decreases in RNA synthesis following addition and removal of β-ecdysone from the culture medium. Third, we have made an estimate of the equilibrium constant for specific binding of β-ecdysone to disc cells. The details of these experiments are published elsewhere (Yund and Fristrom, 1975a).

The specificity of binding of ^{3}H-β-ecdysone has been investigated by incubating discs with different concentrations of ^{3}H-β-ecdysone in the presence and absence of excess ^{1}H-β-ecdysone. Specific binding is demonstrated by the competition of the nonradioactive hormone for binding of the ^{3}H-β-ecdysone. In Fig. 6 the difference in the amount of β-ecdysone bound in the presence and absence of a fiftyfold excess of nonradioactive β-ecdysone is presented. The concentration at which the difference in binding becomes constant is that at which the ^{1}H-β-ecdysone has displaced all the ^{3}H-β-ecdysone from specific binding sites. The fact that this point is reached indicates that there is a limited number of specific sites per disc (about 500/disc cell based on these results). Saturation

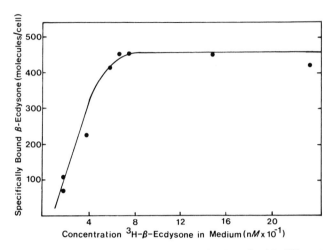

Fig. 6. Specific binding of β-ecdysone. Discs were incubated with different concentrations of ^{3}H-β-ecdysone as indicated in the presence or absence of a fiftyfold excess of [1]H-β-ecdysone in Robb's medium for 90 minutes at 25°C. The difference in binding of ^{3}H-β-ecdysone in the presence and absence of the unlabeled competitor is expressed in molecules per cell. From Yund and Fristrom (1975).

of specific binding sites in this experiment occurs at a β-ecdysone concentration of about 0.8×10^{-7} M (0.04 μg/ml) which is similar to the concentration of β-ecdysone effecting complete evagination (0.08 μg/ml).

The morphogenetic significance of these ecdysone specific sites has been investigated by incubating a constant amount of ^{3}H-β-ecdysone with varying amounts of ponasterone A, β-ecdysone, and α-ecdysone—compounds with widely varying activities in inducing evagination. As can be seen from the results shown in Table III the ability of these three analogues to compete for binding of ^{3}H-β-ecdysone differs greatly and is a function of their biological activity with ponasterone A being most effective and α-ecdysone being least effective. In addition the concentration at which competition is first detected is roughly equivalent to the concentration of the compound which produces evagination. This result indicates that these specific binding sites are involved in the morphogenetic response of the discs to the ecdysones. Under the experimental conditions used, about 50% of the binding of ^{3}H-β-ecdysone to discs is not competed by the nonradioactive analogues. Thus, there is substantial nonspecific binding of ^{3}H-β-ecdysone to disc cells that is not involved in the biological activity of the hormone. The data in Table III allow an estimation of about 1000 specific receptor sites for β-ecdysone per cell. This value is in good agreement with that derived from saturation experiments.

TABLE III

Inhibition of Uptake and Retention of ^{3}H-β-Ecdysone by Unlabeled
β-Ecdysone, α-Ecdysone, and Ponasterone A[a,b]

Competitor (μg)	Ponasterone A (% control)	β-Ecdysone (% control)	α-Ecdysone (% control)
0.00001	101 (3)[c]		
0.0001	101 (3)		
0.001	101 (3)[d]	98 (1)	
0.01	81 (4)	94 (4)[d]	95 (1)
0.05		67 (2)	
0.1	57 (3)	63 (9)	96 (1)
1.0	64 (1)	52 (5)	101 (4)
10.0	63 (1)	47 (5)	83 (3)[d]
100.0		61 (2)	

[a] From Yund and Fristrom (1975).

[b] Retention of ^{3}H-β-ecdysone was measured when 20,000 discs/ml Robb's medium were incubated for 90 minutes at 25°C with 8 ng ^{3}H-β-ecdysone and varying concentrations of other steroids. Competition is expressed as percent of retention of ^{3}H-β-ecdysone by discs incubated in the absence of added steroid.

[c] Numbers in parentheses equal the number of independent determinations in duplicate for each concentration.

[d] Concentration which gives half evagination (from Table II).

It is important to determine the location of the specific binding sites. The fact that there is appreciable nonspecific binding of the hormone to disc cells requires that distinctions be made between specifically and nonspecifically bound hormone. This is accomplished by comparing binding in the presence and absence of excess cold β-ecdysone.

In Table IV we present data on the distribution of ^{3}H-β-ecdysone into nuclear and cytoplasmic fractions in the presence and absence of excess nonradioactive β-ecdysone. As can be seen from the data, after a 2-hour exposure to hormone there is substantial competition for nuclear binding of ^{3}H-β-ecdysone, but no detectable competition for cytoplasmic binding. This result demonstrates that almost all the binding of ^{3}H-β-ecdysone to nuclei is specific. Experiments similar to those described for whole cells were conducted on nuclear binding of ^{3}H-β-ecdysone using ponasterone A, β-ecdysone, or α-ecdysone as competitors (Yund and Fristrom, 1975b). Virtually all the nuclear binding was eliminated. Again, the competitors acted in order of their activity in inducing evagination. Thus, the nuclear binding appears to be morphogenetically significant as well. It should be emphasized that these experiments were done when the amount of

172   JAMES W. FRISTROM AND MARY ALICE YUND

TABLE IV
Specificity of Binding of ^{3}H-β-Ecdysone to Subcellular Fractions of Imaginal Discs[a]

Incubation condition	Cytosol (cpm)	Nuclei (cpm)	Nuclei (% cpm)	Binding in intact discs (%)
^{3}H-β-Ecdysone, 2 hours	1458	1100	43	100
^{3}H-β-Ecdysone, 2 hours ⎱ ^{1}H-β-Ecdysone, 2 hours ⎰	2479	184	7	55
^{3}H-β-Ecdysone, 2 hours ⎱ ^{1}H-β-Ecdysone, 0.5 hours ⎰	3718	213	5	57

[a] Discs were incubated in Robb's medium with ^{3}H-β-ecdysone (8 ng/ml) for 2 hours at 25°C. In separate experiments 100 ng/ml of ^{1}H-β-ecdysone were added to the culture for the entire 2-hour incubation period or only during the last 0.5 hours of incubation. The discs were then homogenized and nuclear and cytoplasmic fractions were prepared. About 750,000 discs were divided into three groups for the fractionation experiments and 20,000 discs were independently used to determine binding to intact discs for each of the experimental conditions. The results are the average of two experiments.

^{3}H-β-ecdysone bound reached steady-state conditions (see below). Thus, they relate only to the final disposition of the specifically bound hormone in the cell and not to possible intermediate states, for example, involving a cytoplasmic receptor carrying the hormone to the nucleus.

To briefly summarize these results, discs are found to contain an average of about 500–1000 specific ecdysone receptors per cell. These receptors are apparently involved in the morphogenetic response of discs to ecdysones, and are located ultimately in nuclei.

One of the earliest responses of discs to β-ecdysone is an increase in RNA synthesis, mainly of ribosomal RNA (rRNA) (Petri et al., 1971, and unpublished observations). We have followed the kinetics of ^{3}H-β-ecdysone uptake and binding by discs and its release following removal of exogenous hormone (Yund and Fristrom, 1975a). These results are compared in Figs. 7A and B with the kinetics of increase and decrease in RNA synthesis resulting from the addition and removal of the hormone from the medium. As can be seen in the figure, the uptake and release of hormone from disc cells closely precedes the respective increase and decrease in the rate of RNA synthesis. There is a particularly close relationship during release of ^{3}H-β-ecdysone when one considers that release of hormone from nuclei is slower than the rate for total cells by about 15–30 minutes (unpublished observation). Thus, based on the temporal

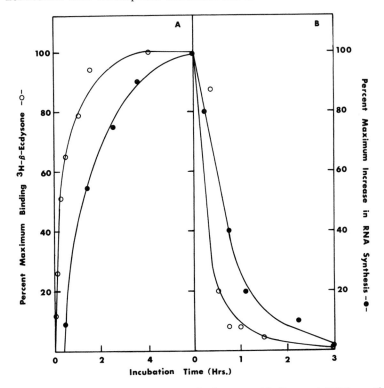

Fig. 7. Kinetics of increase and decrease in hormone binding and RNA synthesis. (A) The kinetics of uptake of ^{3}H-β-ecdysone (○) and the ecdysone-induced increase in RNA synthesis (●). (B) Kinetics of release of ^{3}H-β-ecdysone (○) and the decrease in ecdysone-induced RNA synthesis (●). From Yund and Fristrom (1975).

relationships, these data are consistent with the view that specific nuclear binding of ecdysones is directly responsible for an increase in RNA synthesis.

An estimate has been made for the intrinsic association constant (K) for binding of β-ecdysone to disc receptors (Yund and Fristrom, 1975a). K is computed by dividing the rate constant for association (binding) (k_A) of β-ecdysone with its receptors by the rate constant (k_D) for the dissociation (release) of the ecdysone–receptor complex. The association constant k_A is derived from the kinetics of specific binding of ^{3}H-β-ecdysone to discs. The kinetics are second-order with the estimate that $k_A = 1.5 \times 10^5$ M^{-1} min^{-1} (Fig. 8). The dissociation constant k_D is derived from data on the rate of exchange of ^{3}H-β-ecdysone bound to discs with excess ^{1}H-β-ecdysone in the culture medium. The kinetics of dissociation are pseudo first-order with an average value of k_D from two

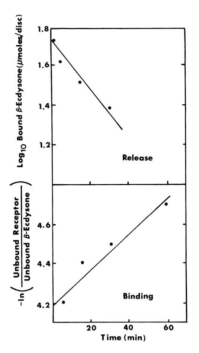

Fig. 8. Kinetics of specific binding and release of ^{3}H-β-ecdysone. The kinetics of specific binding are plotted in terms of second-order reaction kinetics. The kinetics of ecdysone release are plotted in terms of first-order reaction kinetics. From Yund and Fristrom (1975).

experiments being 3×10^{-2} min^{-1} (Fig. 8). Thus, $K = k_A/k_D = 0.5 \times 10^7$ M^{-1}. At equilibrium $k_A[R_f][E_f] = k_D[RE]$ and $[R_f][E_f]/[RE] = k_D/k_A$; where $[R_f]$ = the concentration of unbound receptor, $[E_f]$ = the concentration of unbound ecdysone, and $[RE]$ = the concentration of ecdysone–receptor complex. For the condition when 50% of the receptor is occupied, i.e., $[R_f] = [RE]$, then $[E_f] = 2 \times 10^{-7}$ M (about 0.1 μg/ ml). Since the concentration of receptor is much smaller than that of ecdysone, $[E_f]$ is essentially identical to the total concentration of ecdysone in the culture medium and is indistinguishable from the concentration of β-ecdysone needed to induce complete evagination. The calculated percent of receptor occupied at different concentrations of β-ecdysone is presented in Table V and compared with the increase in the synthesis of rRNA induced by the hormone. As can be seen from the data there is a good correlation between the percent of receptor bound with β-ecdysone and the percent of the maximum increase in rRNA synthesis.

TABLE V

Comparison of Amount of Ecdysone–Receptor Complex Present and Stimulation in RNA Synthesis as a Function of β-Ecdysone Concentration

Ecdysone concentration [E$_f$]	Estimated percent of receptor in complex with β-Ecdysone[a]	Incorporation into RNA (cpm/μg RNA)[b]	Increase over control	Percent of maximum increase
2×10^{-5} M (10 μg/ml)	99	66	56	100
2×10^{-6} M (1 μg/ml)	91	60	50	89
2×10^{-7} M (0.1 μg/ml)	50	37	27	48
2×10^{-8} M (0.01 μg/ml)	9	22	12	21
None	0	10	0	0

[a] Calculated from the expression $k_D/k_A = [E_f][R_f]/[ER] = 2 \times 10^{-7}$ M where k_D = dissociation constant; k_A = association constant; [E$_f$] = concentration of unbound β-ecdysone; [R$_f$] = concentration of unbound receptor and [ER] = concentration of ecdysone–receptor complex. It is assumed that [E$_f$] is essentially identical to the concentration of total β-ecdysone.

[b] Based on the data for Fig. 3 of Fristrom *et al.* (1973).

IV. Discussion

There are several observations in this paper which we feel merit further discussion. First, there are enormous differences in the intrinsic activities of different ecdysones and ecdysone analogues. In general the presence of hydroxyl groups on the 20 and 22 positions is associated with high activity. However, of particular note is the observation that some analogues of ecdysone that are incapable of eliciting a response from a target tissue, in this case imaginal discs, interfere with the response induced by β-ecdysone. The two compounds identified have been studied in the laboratory of Dr. William Robbins (Robbins *et al.*, 1970; Thompson *et al.*, 1971). Both compounds I and II were found to have little activity in inhibiting insect development. Compound II had significant activity in the housefly molting hormone assay, but compound I had none. The results with the molting hormone assay are not inconsistent with the data reported here where neither compound is active, since the *in vivo* activity of compound II may result from metabolism and not be an inherent property of this analogue. The failure of these "antiecdysones" to interfere with insect development is also not surprising since the antiecdysone activity of the analogues, although easily detected, is not great. Hopefully, however, the observations presented here will encourage the development and synthesis of compounds with greater antiecdysone activity.

Second, we have found that under the conditions employed in this study there is substantial nonspecific binding of β-ecdysone to disc cells (see Table III). This observation indicates that studies in which the distribution of isotopically labeled ecdysone in cells is investigated must be carefully controlled. For example, in the work presented here we find ^3H-β-ecdysone in both the nuclear and the cytoplasmic fractions (Table IV). On the basis of these data alone we can make no judgment as to the site(s) of action of the hormone. It is only when we determine, using competition experiments, that the specifically bound β-ecdysone is found in the nuclear fraction that we have an indication of the site of action of the hormone. These data do not eliminate the possibility that a small fraction of the hormone found in the cytoplasmic fraction is also specifically bound. They simply indicate that at maximal binding (equilibrium) it is a small fraction. If the results from our studies are generalizable to other systems, they suggest that autoradiographic studies already done without the use of competition experiments tell us little about the site of action of ecdysones in target cells.

Third, the evidence presented strongly indicates that the mechanism of action of β-ecdysone involves binding to nuclear components. The binding to nuclei is specific and the majority of specific receptors are ultimately found in the nuclei. The kinetics of binding and release from the nuclear fraction are similar to those for the β-ecdysone-stimulated increase and decrease in RNA synthesis in discs. Also the increase in RNA synthesis correlates closely with the calculated amount of ecdysone–receptor complex present. These results would hardly be predicted from a model of ecdysone action which purely involved effects on the cell membrane and changes in ion permeability (e.g., Kroeger and Lezzi, 1966). That is not to suggest that there are no direct effects on the cell membrane. Other work from our laboratory has demonstrated that β-ecdysone increases the rate of transport of uridine (Chihara and Fristrom, 1973) and thymidine (Logan et al., 1975) into disc cells. This action of ecdysone is not unique and is a common property of many steroid hormones (Munns and Katzman, 1971; Jensen and DeSombre, 1972).

Fourth and last, it is interesting to compare the intrinsic association constant of disc receptors with association constants of other steroid receptors. We find that the value estimated for discs (about $10^7 \, M^{-1}$) is several orders of magnitude lower than those for other steroids (ca. 10^9–$10^{11} \, M^{-1}$). A closer look reveals that the k_A for the binding of β-ecdysone is on the same order of magnitude as those for other hormones, but the k_D (ca. 10^{-2} min^{-1}) is several orders of magnitude higher than those found

for other steroids (ca. 10^{-6} min^{-1}). The significance of such a high dissociation constant remains to be determined, but we suspect that it is necessary because of the nature of insect development. In *Drosophila* in particular, changes in ecdysone titer occur rapidly, as short as every 20 hours
during larval development and perhaps within 5–10 hours during the prepupal period. Thus, in order for the target tissue to effectively change
from high to low titer physiology, there may be a real need to release
hormone rapidly following a decline in hemolymph titer. Such a rapid
release would be characterized by a high dissociation constant.

Other comparisons can be made between the results with insect and
vertebrate steroid hormone systems. In both there is evidence for nuclear
binding of hormone and that an early site of action involves changes
in nuclear function. In vertebrates these are changes in RNA synthesis
and the production of new mRNA's (Rosenfeld *et al.*, 1972; Schwartz,
1972). In insects increases in RNA synthesis also occur and the induction
of specific puffs by ecdysones is an indication of the possible production
of new messenger RNA's. In each system the concentration of hormone
which is physiologically active and the hormone concentration resulting
in half of the specific receptor sites being occupied are similar. The number of specific binding sites per cell has been estimated in both systems
and is 500–1000 for *Drosophila* imaginal discs and about 10 times as much
in vertebrates (Jensen and DeSombre, 1972). That there is also a tenfold
difference in DNA content between these two groups of organisms is
intriguing.

Acknowledgments

The authors are indebted to Drs. Denis Horn and William Robbins for generously providing the ecdysones and ecdysone analogues used in this study. We
especially wish to acknowledge the expert assistance of Mrs. Odessa Eugene for
patiently and routinely isolating the large numbers of imaginal discs used in these
experiments. Previously unpublished work presented in this paper was supported
by grant number GM-19937 from the USPHS. We wish to thank Dr. Eva Fekete
for taking the photograph shown in Fig. 1 and Mr. Andrew Kuniyuki for making
the prints. Some of the research reported here was done in the Genetics Institute,
Biological Research Center, Szeged, Hungary.

References

Agui, N., and Fukaya, M. (1973). *Appl. Entomol. Zool.* **8**, 73–82.
Ashburner, M. (1972). *Chromosoma* **38**, 255–281.
Ashburner, M., Chihara, C., Meltzer, P., and Richards, G. (1973). *Cold Spring Harbor Symp. Quant. Biol.* **38**, 655–662.

Borst, D., Bollenbacher, W., O'Connor, J., King, D., and Fristrom, J. (1974). *Develop. Biol.* **39**, 308–316.

Chihara, C. J., and Fristrom, J. W. (1973). *Develop. Biol.* **35**, 36–46.

Chihara, C. J., Petri, W. H., Fristrom, J. W., and King, D. S. (1972). *J. Insect Physiol.* **18**, 1115–1123.

Fristrom, J. W., Logan, W. R., and Murphy, C. (1973). *Develop. Biol.* **33**, 441–456.

Jensen, E. V., and DeSombre, E. R. (1972). *Annu. Rev. Biochem.* **41**, 203–230.

Kroeger, H., and Lezzi, M. (1966). *Annu. Rev. Entomol.* **11**, 1–22.

Laird, C. (1973). *Annu. Rev. Genet.* **7**, 177–204.

Logan, W. R., Fristrom, D., and Fristrom, J. W. (1975). *J. Insect Physiol.* **21**, 1343–1354.

Mandaron, P. (1973). *Develop. Biol.* **31**, 101–113.

Milner, M., and Sang, J. (1974). *Cell* **3**, 141–143.

Munns, T. W., and Katzman, P. A. (1971). *Biochemistry* **10**, 4941–4948.

Ohmori, K., and Ohtaki, T. (1973). *J. Insect Physiol.* **19**, 1199–1210.

Petri, W. H., Fristrom, J. W., Stewart, D. J., and Hanly, E. W. (1971). *Molec. Gen. Genetics* **110**, 245–262.

Robb, J. A. (1969). *J. Cell Biol.* **41**, 876–884.

Robbins, W. E., Kaplanis, J. N., Thompson, M. J., Shortino, T. J., and Joyner, S. C. (1970). *Steroids* **16**, 105–125.

Rosenfeld, G. C., Comstock, J. P., Means, A. R., and O'Malley, B. W. (1972). *Biochem. Biophys. Res. Commun.* **46**, 1695–1703.

Schwartz, R. J. (1972). *Nature (London), New Biol.* **237**, 121–125.

Stewart, M., Murphy, C., and Fristrom, J. W. (1972). *Develop. Biol.* **27**, 71–83.

Thompson, M. J., Robbins, W. E., Cohen, C. F., Kaplanis, J. N., Dutky, S. R., and Hutchins, R. F. N. (1971). *Steroids* **17**, 399–409.

Yund, M. A., and Fristrom, J. W. (1975a). *Develop. Biol.* **43**, 287–298.

Yund, M. A., and Fristrom, J. W. (1975b). "Squaw Valley Conference on Developmental Biology," W. A. Benjamin, New York, in press.

Viruses and
Microorganisms
in Invertebrate Cells

11

Applications of Insect Tissue Culture to an Elucidation of Parasite Transmission

MARION A. BROOKS

I. Introduction

The growth of parasitic protozoa in tissue cultures, both vertebrate and invertebrate, was reviewed by Trager and Krassner (1967). Chao (1973) continued the review of invertebrate tissue culture, listing those parasites that have been cultured. The tissues were predominantly from insects but included a few from ticks and mollusks. The purpose of writing this article was to point out examples of insect cell culture in which penetration of host cells by intracellular parasites was thought to be comparable to natural infections; and the examples were sought primarily from studies involving parasitic protozoa transmitted by insect vectors. However, little work of this kind has been done with cultured insect cells; therefore, occasional reference is made to vertebrate cell culture. No consideration is given here to viruses and rickettsiae.

Defining the metabolism and nourishment of intracellular parasites is an important goal of parasitologists and is a valid subject of tissue culture work. Presumably, this is because, the behavior and development of the parasites *in vitro* reflect the same processes occurring *in vivo*. However, *in vitro* conditions enabling parasites to actually enter cultured insect cells have been practically unattained so far. The following remarks are directed to factors that, in my opinion, may be worth further investigation.

Not many different kinds of cells from invertebrates have been cultured. The development of different cell types and lines from more species is urgently needed. If a greater variety of cell types were available, the chances of finding one susceptible to parasite invasion might be increased.

II. Parasitic Protozoa

Aikawa and Sterling (1974) present a splendid collection of studies on the fine structure of intracellular protozoa, but these do not show protozoa cultured in insect cells. The electron microscopic examination of infected cells, even in short-term primary cultures, is to be desired. Morphological alterations in cultured parasites, compared to those found *in vivo*, may be a guide to needed changes in the medium.

The parasitic protozoa, in the discussion below, are taken up by taxonomic groups following the terminology in Aikawa and Sterling (1974).

A. THE PARASITOPHOROUS VACUOLE OF SINGLE-HOST PARASITES

In the invertebrate parasites, such as *Toxoplasma, Eimeria,* and other coccidia which are not known to be transmitted by insect vectors, the parasite lies in a parasitophorous vacuole which is lined by a host cell membrane (Hammond, 1971). In some species of *Eucoccidia,* the membrane of the vertebrate host cell is thrown into fine villuslike protuberances projecting into the vacuole; these projections disintegrate into fine particles which are ingested by the parasite through micropores, or by pinocytosis. Parasitized cells indicate an increased rate of metabolism by hypertrophy and an increased number of mitochondria, evidently associated with the production of materials used by the parasite for nourishment and growth (Aikawa and Sterling, 1974). Penetration of the cultured vertebrate cells by the parasite, as seen by direct microscopic observation, is accomplished within a few seconds, gliding sporozoites invaginating or passing through the host cell membranes (Vickerman, 1972). Coccidia grow readily in a very great number of mammalian and avian cells *in vitro* (Doran, 1973), but not in mosquito cell lines (Buckley, 1973).

B. THE PARASITOPHOROUS VACUOLE OF PARASITES TRANSMITTED BY INSECT VECTORS

A different situation with respect to the parasitophorous vacuole is seen in the insect-transmitted malaria organism. In vertebrates, the erythrocytic states of *Plasmodium* species and related genera in the Haemosporina are bound by two closely apposed unit membranes, the outer one of which has been derived from the erythrocyte, the cytoplasm of the host cell being immediately adjacent to the outer membrane (Trager, 1974).

1. Nourishment in Vertebrate Hosts

The erythrocytic *Plasmodium* ingests nourishment by pinocytosis and through cytosomes as do the coccidia (Rudzinska, 1969). Required nutrients not furnished by the host cell seem to be taken up from the external milieu as if the parasite were an organelle in the host cell. Extracellular, *in vitro* cultures point to a deficiency in the parasite's membrane absorptive capabilities (Siddiqui *et al.*, 1969; Trager, 1971; Trigg and Gutteridge, 1972). It is difficult to culture the vertebrate stages of *Plasmodium* and related genera compared to *Toxoplasma* and *Eimeria*. The difficulties encountered in culturing the Haemosporina are undoubtedly associated with the increased host dependency developed in an organism with two hosts. Increased complexities in morphological variations in the life cycle parallel a corresponding degree of metabolic specializations at each stage (Levine, 1972).

2. Relationship to Insect Hosts

A high degree of intimacy of the parasite with the insect host cell has not been attained. The oocyst normally develops on the outside, that is, the hemocoel side, of the mosquito stomach (Vanderberg *et al.*, 1967). The solid oocyst bounded by a capsule rests tenuously upon the basal lamina of the midgut epithelium (Terzakis, 1971). The growth of the parasite in the insect is thus not intracellular.

a. ZYGOTE FORMATION IN THE MOSQUITO STOMACH. Haemosporina parasites in erythrocytes commence maturation shortly after malarial blood has been taken into the stomach of a mosquito. Gametocytes emerge from the erythrocyte and within 10–15 minutes become extracellular gametes. The extracellular gamete stage seems to be a delicate period in the life cycle because it seldom has been seen, is of short duration, and has not been sustained *in vitro*. The fine structure of the gametes is somewhat simpler than that of the other stages, particularly in the lack of a mitochondrion (Aikawa *et al.*, 1970; Sterling and Aikawa, 1973). The microgamete and the macrogamete quickly fuse to form the zygote (or ookinete) which in 4–6 hours *in vivo* becomes motile and is the tissue penetrating stage. Ball (1964) saw zygotes which had formed *in vitro* in blood removed from the stomach of a mosquito. Schneider (1972) failed to detect exflagellation or zygote formation from monkey erythrocytes which had been separated from the other blood elements by centrifugation in bovine serum albumin gradients. Zygotes failed to form when infected erythrocytes were suspended in insect tissue culture medium or in mosquito cell cultures. Speer *et al.* (1975) succeeded in obtaining zygote formation by injecting rodent erythrocytes infected with gametocytes into mosquito cell cultures.

Ookinetes are much more durable than gametes, surviving longer both *in vivo* and *in vitro*. Ookinetes penetrate the peritrophic membrane of the mosquito gut, pass through the midgut epithelial cells, and attach to the hemocoel side of the gut where they transform into oocysts (Freyvogel, 1966; Garnham *et al.*, 1969). Ball (1964) observed this in stomachs isolated from mosquitoes. Freyvogel (1966) used phase-contrast microscopy to study dissected mosquito stomachs cultured in TC199 supplemented with salts and sugars, in which the ookinetes were motile for up to 21 hours. Ookinetes remained attached to the inside of the peritrophic membrane after it was removed from the blood coagulum. If the stomach wall was laid out flat, the ookinetes passed through the peritrophic membrane in 12 minutes, glided over the surface microvilli of the midgut epithelial cells for 8 minutes, and then abruptly inserted their anterior ends into the midgut cells and disappeared into the cells. A chemical attractant of the midgut cells for the ookinetes could not be demonstrated. The ookinetes would go into the midgut cells only when the peritrophic membrane was laid directly onto the midgut epithelium and not when the midgut epithelium was simply in the "immediate vicinity" of the blood coagulum. Having gone through the epithelium, the young oocysts on the outside of the gut wall enlarge, differentiate internally, and in 10–12 days release mature sporozoites which are infective for the vertebrate host.

b. FINE STRUCTURE OF ZYGOTES. The fine structure features of ookinetes include one or more paracrystalline bodies. These bodies have been reported for *Plasmodium gallinaceum, P. cynomolgi, Leucocytozoon simondi,* and *Parahaemoproteus velans* (Rudzinska, 1969; Desser and Trefiak, 1971; Desser, 1972). The crystalloids lie near the nucleus, have a patterned appearance, are not bounded by a membrane, and often make intimate contact with the dilated cisternae of the endoplasmic reticulum. The nature and function of the crystalloids is unknown, but Desser (1972) suggested that they constitute reserve energy for the organism which must survive extracellularly for some time. Based on histochemical tests, he found that the crystalloids are of a lipoprotein composition in which the lipid is neutral. He proposed that the macrogamete synthesizes the precursor which is converted to the crystalloid by the addition of a second component from the microgamete during fertilization. Ookinetes in cell culture systems should be examined for the integrity of such fundamental features as the crystalloids. If they are depleted before the ookinetes transform into oocysts, this might be an indication of a nutrient deficiency.

Crystalloids are also known for an intracellular prokaryotic symbiote of a leafhopper which invades the embryo (Körner and Feldhege, 1970). These bodies can be digested with pronase and chymotrypsin. It is tempting to speculate that the crystalloids perform some function in the penetration of insect cells, but such bodies have also been seen in free-living bacteria and in blue-green algae (Shively, 1974).

C. *In Vitro* CULTURE OF OOCYSTS

For many years, oocysts of *Plasmodium relictum* isolated from *Culex tarsalis* stomach walls, or attached to fragments of stomach walls, were cultured in cell culture media (Ball, 1964, 1972; Chao and Ball, 1964). The stomach wall was never found to be beneficial in maintaining the cultures. In early experiments, only brief phases of development occurred in each culture, with the younger oocysts being particularly short-lived. Numerous changes in the medium, the methods of obtaining the oocysts, and in the physical conditions imposed on the cultures led to only disappointingly small increments in the duration of the cultures and therefore in the development of the sporozoites. Eventually (as noted below), when gametocytes were used as inoculum, oocysts and sporozoites did develop.

This indicates that unrecognized factors or manipulations were more important than the medium.

Plasmodium gallinaceum was studied in a modification of Grace's cell culture fluid by Schneider (1968a). She found that 9-day-old oocysts from *Aedes aegypti* could be maintained for 24 hours and then would release infective sporozoites. All younger stages of oocysts which did not naturally contain nearly mature sporozoites did not develop *in vitro*.

1. The Use of Insect Cell Cultures to Maintain Oocysts

Culturing the oocysts with actively growing insect cells evoked a slight improvement in the growth and differentiation of sporozoites from 8-day-old oocysts (Schneider, 1968b). In a concentration of 10^5 cells/ml, 7-day-old oocysts continued to develop for the first 24 hours after being placed in culture. The replacement of heterologous insect hemolymph by *A. aegypti* hemolymph had no observable beneficial effect. In all cultures, the insect cells grew so well that they overgrew the cultures and seemed to cause degeneration of the parasites. While it was thought that the cells were *A. aegypti* (Grace's 1966 line), it was shown later by Greene and Charney (1971) and Greene *et al.* (1972) that this line consisted of cells of *Antheraea*, a moth.

Ball and Chao (1971), employing a similar cell system to culture oocysts of *P. relictum*, obtained slightly better results than they had earlier, and even found that Grace's (supposed *A. aegypti*) line supported considerably better growth than did Singh's line of *A. aegypti*. This anomaly was discussed with much insight by Schneider (1972), who was unable to get mature sporozoites from *P. cynomolgi* oocysts younger than 8 days. She inoculated the stomach fragments into an established cell line of *Anopheles stephensi*, and into 4- to 7-day-old primary cultures of larvae or adults of the same species. Oocyst development was the same in all cell culture systems, but invariably better than in medium alone. This was a definite indication that something from the mosquito cells was beneficial, and it points up the need for establishing a line from mosquito midgut epithelium. There is still the question of whether the ookinetes attach to stomach cells because these cells are conveniently located or because the cells offer a particular and necessary substrate or nutrient. Perhaps this question could be answered by culturing oocysts separated from cells by a membrane. Such an experiment could also address the problem of whether the stomach cells *in vitro* are antagonistic.

Working with hemogregarines of a snake, Chao and Ball (1972) cultured *Hepatozoon* oocysts with Grace's and with Singh's cell lines. The results were the same as with *Plasmodium*, i.e., development to mature sporozoites from 6-day-old oocysts in Grace's line but not in Singh's. At the conclusion of the experiment, the authors learned of the moth identity of the Grace line.

2. The Use of Gametocytes to Develop Oocysts in
Insect Cell Lines

The culmination of this work with blood parasites was reached by Ball and Chao (1973) with the complete development of *Hepatozoon rarefaciens* in a *Culex pipiens* cell line they had initiated. The line was heterogeneous, having been derived from embryos. The cultures were inoculated with infected blood from the snake, rather than with oocysts from the mosquito. Only four drops (0.2 ml) of blood were adequate to inoculate 3 ml of cell culture, but the cell densities of blood and of cultures were not stated. Oocysts formed as free-floating bodies, i.e., attachment to cells or substrate evidently was not necessary, questioning again the significance of the role played by the stomach wall cells *in vivo*. If particular cell types are not needed, an interaction between heterogeneous types may be stimulatory. Yet, Schneider (1972) found no advantage in primary cultures, which are presumably more heterogeneous than lines, while working with *A. stephensi*. Ball and Chao (1973) observed that "Many oocysts developed in cell cultures 2 weeks or more of age, fresh medium having been added 4 days before the inoculation of infected blood. On the other hand, fewer oocysts developed in newly transferred subcultures." This effect of new subcultures may be responsible for much of the failure of gametes to survive. Perhaps the physical traumas inflicted on cells by handling and transferring breaks them, releasing proteases from them. The destructive capabilities of the enzymes may be lost after some days have elapsed.

A primary culture of *A. stephensi*, started by Rosales-Ronquillo *et al.* (1972), involved hatching larvae from surface-sterilized eggs directly into sterile trypsin–saline solution, avoiding the tedious and risky task of cleaning up contaminated larvae. Cell cultures 10–28 days old were inoculated with blood cells containing gametocytes of *Plasmodium berghei* from a rodent (Rosales-Ronquillo and Silverman, 1974). Exflagellation was observed within 7–10 minutes, and mature ookinetes were present in the cultures in 24 hours. Ookinetes which had been in culture as long as 40 hours were examined with scanning electron microscopy, and they gave the impression that they had been motile, but none was ever arrested in the process of penetrating the insect cells (Speer *et al.*, 1974). With cinephotomicrography of living cells, the ookinetes were definitely proven to be motile for over 30 hours (Speer *et al.*, 1975). Occasional ookinetes were observed indenting the surface of the erythrocytes, or probing and pushing with an anterior projection. Some parasites penetrated the erythrocytes in this manner, whereupon the blood cells immediately lysed. There were no observations of penetration of the mosquito cells by the ookinetes.

To summarize, in culturing the mosquito phases of malaria and related blood parasites, as long as the inoculum was oocysts, the developing sporozoites within them grew only slightly and degenerated in a short time. The addition of actively growing insect cell cultures (not necessarily from the homologous mosquito host, since even moth cells were helpful) conferred an advantage. But no one has obtained sporozoites in cultures started with oocysts. Growth was promoted best by using gameotocyte-infected vertebrate blood cells as the inoculum, in mosquito cell cultures which had been allowed to stabilize for at least 4 days. Gametes released in the culture medium proceeded to develop into ookinetes and in one case proceeded on to oocysts and ultimately to sporozoites. Tissue attachment was not necessary for the formation of oocysts. It is not known whether these oocysts were comparable on the ultrastructural level to oocysts formed *in vivo*, since there are no electron micrographs of cultured oocysts. Also, it is not known if the resulting sporozoites were infective for the vertebrate host, because there is no indication that susceptible animals were challenged with them.

Using blood cells as the inoculum results in a culture of great complexity: heterogeneous insect cell types plus diverse vertebrate blood cells. This may seem to be a step backward, since one of the aims of cell culture is the identification of factors involved in growth. Yet, this mixed cell system does at last permit growth and development of the test organism, so that it should be possible to start simplifying the components and progress to a more nearly defined system.

D. HOST CELL MEMBRANES IN TICKS

The genus *Babesia* in the class Piroplasmea is an important cause of disease in cattle and is transmitted by ixodid ticks. The organisms infect erythrocytes of ruminants, horses, dogs, and rodents (Aikawa and Sterling, 1974). The life cycle, especially in the tick, has not been verified, so the terminology is still somewhat flexible. Isogametes are released in the blood of the vertebrate host, and even though it is thought that the gametes unite after ingestion by the tick, the process has not been demonstrated. Zygotes may pass directly to the salivary gland cells of the tick, where sporozoites develop. In some cases the zygotes enter the ovaries, where sporozoites develop, infect the eggs, and, ultimately, the salivary glands of the offspring. Little work has been done with electron microscopy of the stages in the tick vectors. The spherical form (trophozoite) of the organism in tick ovary cells possesses a pellicle composed of two membranes, an outer plasmalemma, and an inner, thick, continuous membrane (Friedhoff and Scholtyseck, 1969). The pellicle is thrown into folds and invaginations. The parasite lies within a parasitophorous vacuole surrounded by a host cell membrane of the tick, a situation comparable to the stages of coccidia in the vertebrate host. But when it is in the vertebrate host erythrocyte, *Babesia* is separated only by its own membrane. In tick ovary or salivary gland cells, the trophozoite is transformed to an elongate form, called the merozoite, which has a pellicle like that of the spherical form except that it is smoothed out. Schizogony in the salivary gland cells and ovaries results in many merozoites which are infectious for vertebrates.

Tick tissues and cells have been studied and cultured for the propagation of viruses and rickettsiae (Reháček, 1971, 1972; Weiss, 1975). Tick tissue culture for the study of *Babesia bigemina* was recently initiated by Hoffmann (1972). Various tissues of infected and noninfected *Boophilus annulatus* were maintained and compared in culture medium. Cyclical development of the microorganisms was observed in the infected tissues for as long as 12 days, while the noninfected tissues survived somewhat longer. Hoffmann found that the concentrations of vitamins and amino acids in the culture medium affected survival of the tissues, and that the addition of 0.01% ribose prolonged survival 15–30%, except for hemocytes. Multiplication of the *Babesia* organism and penetration of new cells was not achieved.

E. PARASITES WITHOUT ALTERNATE HOSTS

Parasites belonging to subphylum Microspora (Microsporida), like the Eimeriina which were discussed above, have no alternate hosts.

Members of the Nosematidae, especially in the genus *Nosema*, are important disease agents in both beneficial and harmful insects, other arthropods, fish, and a few mammals, including man (Weidner, 1972). The Haplosporidiidae parasitize aquatic invertebrates, oyster pathogens being very important (Aikawa and Sterling, 1974). Microspora differ, however, from the Eimeriina in that the latter are surrounded by a parasitophorous membrane in the vertebrate host cell while a parasitophorous vacuole is not seen around the Microspora organism in the invertebrate host cell. Weidner (1972) studied *Nosema michaelis*, which normally parasitizes blue crabs. He found that it has only one membrane after it is injected into cells of the host; but if it is caused to enter cells of a vertebrate, or is ejected into a culture medium, it has two membranes.

1. Germination of Spores

Nosema spores are passed into the environment where they infect other hosts; but in the case of insects, transovarial transmission is also possible. Ingested spores do not germinate in the sense of germination of bacterial spores. Nosema spores explosively evert a coiled tubule, called the polar filament, and then the sporoplasm flows out through the polar filament. If the end of the polar filament contacts a host cell, the sporoplasm enters the cell, is transformed to a binucleate trophozoite lacking a mitochondrion, and enters vegetative growth.

Hemolymph and tissues of insects which had been fed Nosema spores were placed in culture conditions; 24 hours later the microorganism had multiplied and it eventually filled the tissues with spores (Trager, 1937; Gupta, 1964; Petri, 1966). Employing a pure suspension of spores to infect cultured cells is complicated because of the necessity of disinfecting the spores if they have been collected from diseased insects or if they have been contaminated in handling or storage. However, after feeding spores to insects, tissues containing mature spores may be harvested aseptically after the passage of an appropriate length of time, and they will yield a large number of clean spores which can be concentrated by centrifugation. These spores may be used to inoculate cell cultures, on a quantitative basis, after they have been treated, i.e., activated, with alkali.

It has long been known that spores will extrude their filaments if suspended in a drop of the digestive juice of the caterpillar host, or in other fluids of alkaline pH (Trager, 1937). They apparently do not evert if injected directly into the hemocoel. Ohshima (1937) found that prior incubation of spores in a 0.1 M KOH solution for 40 minutes makes them infectious upon injection into the hemocoel.

2. Culturing Microsporida in Insect Cells

Applying Ohshima's method of alkaline treatment of spores, Ishihara and Sohi (1966) succeeded in infecting ovarian cells of Bombyx mori with N. bombycis, and Kurtti and Brooks (1971) infected cells of Malacosoma americanum with a microsporidan from Malacosoma disstria. Ishihara and Sohi (1966) identified vegetative stages and found that sporogony commenced 72 hours after inoculation, with host cells becoming filled with spores 21 days after inoculation. Sporoplasms which failed to invade host cells degenerated in the tissue culture medium. Weidner and Trager (1973) found that sporoplasms may be kept in culture medium containing, among other nutrients, an extract of duck erythrocytes, and supplemented with adenosine triphosphate (ATP), pyruvate and coenzyme A. Adenosine triphosphate is seemingly the most important supplement. The extracellular sporoplasms in this medium maintained their morphological integrity as seen in electron microscopy for up to 4 hours.

Although there have been arguments proposed for the existence of an extracellular primary infectious form, intermediate between the ejected sporoplasm and the trophozoite, there is no morphological evidence for it. Ishihara (1969) has reported what he believes to be a "secondary infective form" of *N. bombycis* in *B. mori* cells. It is binucleate, like the trophozoite, but is larger and more variable in shape although usually round, has a stronger affinity for Giemsa's stain, and is found outside of the host cells. Photomicrographs of stained cells were presented as evidence for this migratory form, leaving one cell and invading another in disseminating the protozoan from cell to cell *in vitro*. Evidently no electron microscopy work has been done to ascertain the exact method of cell penetration by these secondary forms, whether it is by motile invasion or phagocytosis by the host cell.

3. Phagocytosis of Spores by Insect Cells

Gilliam and Shimanuki (1967) observed phagocytosis of large numbers of spores of *Nosema apis* by hemocytes of *Apis mellifera in vitro*. So many spores were phagocytosed that the cells ruptured. Rupturing of engorged cells was proposed to account for the lowered number of hemocytes in diseased bees. However, infection *in vivo* is not accomplished through hemocytic phagocytosis of viable, noneverted spores. A number of factors involved in preparing and washing spores and the length of storage affect viability of spores. The principle of activation with alkali is that when the spores are transferred to a lower pH, as they are when inoculated into culture media, the sudden change causes the sporoplasms to evert. With the exception noted above (Weidner and Trager, 1973), sporoplasms everted into extracellular fluids soon perish. Nonviable, empty spore shells are nonrefractile with phase-contrast microscopy and can be readily distinguished from refractile, viable spores (Ishihara, 1967). Dead spores, and the empty spore walls of everted spores, are probably phagocytosed as foreign objects by hemocytes. Kurtti and Brooks (1971) noted that spores which failed to evert in cell cultures were phagocytized by hemocytes.

4. Host Specificity in Cell Culture Systems

Host specificity of *Nosema* and other microsporida may be examined by cell culture methods. Ishihara (1968) infected cells of mammals and birds with *N. bombycis*, and the parasite was able to complete its life cycle in rat and chick embryo cells at 28°C. At 37°C, the normal temperature for homeotherms, the insect parasite failed to grow.

F. THE CHANGING MORPHOLOGY OF FLAGELLATES

The last group of parasitic protozoa to be considered is the family Trypanosomatidae in the subphylum Sarcomastigophora. Two genera, *Trypanosoma* and *Leishmania*, known as hemoflagellates, are important from the standpoint of human health, and both are transmitted by insect vectors. The known intermediate hosts of mammalian hemoflagellates are represented by hemotophagous Hemiptera, Diptera, and Siphonaptera. The trypanosomes are a large and varied group for which it is impossible to write a generalized account of their life cycles and vector relationships; therefore, for any particular species, an authoritative source such as Hoare (1972) should be consulted.

The African trypanosomes (Salivarian) multiply in the gut lumen and salivary glands (*Trypanosoma brucei*, etc.), or proboscis (*T. vivax*), of various species of *Glossina*, the tsetse flies, and to some extent in tabanids. The trypanosomes are transmitted by the bite of the fly to man and animals. The American species (Stercorarian) such as *T. cruzi* and *T. rangeli* multiply in the gut lumen of at least 20 species of triatomids and reduviids, but especially in *Triatoma infestans*, *Rhodnius prolixus*, and *Panstrongylus megistus* (Faust *et al.*, 1970). *Trypanosoma cruzi* is transmitted by mechanical abrasion of fecal material from the bug into the skin, conjunctiva, and mucous membranes, while *T. rangeli* is most commonly transmitted by inoculation ("biting").

Multiplication of all the species of the two genera occurs in both vertebrate and invertebrate hosts. The site of multiplication of the African trypanosomes (*T. brucei* and its related species and subspecies) in the vertebrate is primarily in the bloodstream (although amastigote forms have been discovered in some tissues) (Hoare, 1972). *Trypanosoma cruzi* multiples in various tissues of the vertebrate host and not in the blood at all.

All the trypanosomes can be cultured, with varying degrees of success, in vertebrate cell culture or in several kinds of cell-free media, blood agar usually serving the best. However, the cultured organism always assumes the morphological characteristics of the stages in the insect gut, and in young cultures loses its infectivity for vertebrates. Infectivity is usually regained with age and is optimum on about the sixteenth–eighteenth day (Amrein *et al.*, 1965). The factors responsible for the change in morphology, and the relationship between form and infectivity, are challenging questions.

1. The Surface Coat of Flagellates

Perhaps the fact that the trypanosomes grow so readily in artificial media has diverted attention from culturing them in insect cell cultures. But information is needed regarding what it is in the insect that confers infectivibility on the parasite (Zeledón, 1971; Vickerman, 1972). Some knowledge regarding this might be obtained from *in vitro* work. The African trypanosomes, while in the blood of vertebrates, undergo a series of antigenic changes, paralleled by changes in morphology (Vickerman, 1971, 1974). The entire body and flagellum is enveloped by a 12–15 nm thick surface coat when in the bloodstream, but this is lost in the gut of the fly, and is also absent from organisms cultured on blood agar. In this respect, cultured forms always resemble the insect gut forms. After multiplication in the gut of the fly, and passage to the salivary glands, the trypanosomes at first are shorter epimastigote forms with kinetoplast anterior to the nucleus. They multiply more and transform to the longer, metacyclic form with posterior kinetoplast, and they reacquire the surface coat. In this stage they are infective again for the vertebrate host. It is thought that the surface coat is a highly immunogenic surface which is replaced by antigenic variant coats in successive populations in the vertebrate. The use of tsetse fly cultures, possibly derived from midgut and salivary gland cells, would be a highly interesting way to explore the loss and restoration of the surface coat.

2. Culturing Flagellates in Insect Cell Lines

There are no cell lines of *Glossina* (Dolfini, 1971; Hink, 1972). Attempts to establish tsetse fly cultures have been done with organs, which sometimes send out migrating cells or tissue extensions, but these do not result in a population of cells capable of being subcultured. Trager (1959), Nicoli and Vattier (1964), Amrein *et al.* (1965), and Cunningham (1973) have described cultures of several species of *Trypanosoma* (*brucei, congolense, rhodesiense, vivax*) in primary tissue cultures of various organs of *G. brevipalpis, G. fuscipes, G. morsitans, G. pallidipes,* and *G. palpalis.* In some experiments, the flagellates exhibited tissue tropism, moving to

194   MARION A. BROOKS

the salivary glands (Amrein *et al.*, 1965), concentrating on a membrane overlying the gut (Cunningham, 1973), or congregating in and around the alimentary tract, hypopharynx, and proboscis (Trager, 1959). Trager was able to infect sheep with metacyclic forms from some of his cultures of *T. vivax*, grown with alimentary tract and salivary glands. *Trypanosoma vivax* normally multiplies only in the proboscis of the fly, and this is the only successful case of culturing infective forms of it.

In the absence of cell lines of trypanosome vectors, some attempt has been made to use cells of nonvector insects to culture flagellates. *Trypanosoma rangeli*, a neotropical parasite of man and laboratory mammals, is most commonly transmitted by *R. prolixus*. The protozoan multiplies in the gut and can be transmitted by fecal contamination but the more effective method is by inoculation. The organism is also capable of invading the hemocoel, where it is sometimes phagocytosed by hemocytes. Pipkin *et al.* (1968) cultured *T. rangeli* in Grace's 1966 *A. aegypti* line and found that by the sixteenth day the parasites had invaded some of the cells and 4 days later had destroyed them. But in the culture medium alone, which contained fetal bovine serum (FBS) and insect hemolymph, the trypanosomes grew much faster, doubling every 10 hours, and reaching the trypomastigote form on the fifth day. Wood and Pipkin (1969) found that although *T. cruzi* could be cultured with Grace's 1962 line of *An. eucalypti*, the organisms did as well in the medium alone. The authors studied varying proportions of FBS and *Philosamia cynthia* hemolymph in the medium for effect on growth and transformation of the trypanosome. They found that 20% FBS + 0.5% hemolymph gave the maximum number of metacyclic forms in 16 days. The metacyclic forms were infective and virulent for mice. FBS above 20% was inhibitory, and 0.25% hemolymph was not beneficial. Hemolymph without FBS failed to support any growth.

Chao (1971) found that Grace's medium supplemented with 10% FBS could support *T. scelopori* from a lizard. When cultured with Grace's 1966 line of *A. aegypti* or with a hemocyte line of *Samia cynthia*, the trypanosomes overgrew the insect cells, but in Singh's *A. aegypti* line the trypanosomes failed to grow well enough to be serially transferred. It is unlikely that these differences are attributable to the cells per se, since the media were different. Chao found that the trypanosomes behaved the same in cell-free media as they did in the presence of cells in the corresponding media. He never saw parasites inside of the cells although extracellular amastigotes were observed in Grace's medium. The parasites were still infective more than a year after culturing in Grace's medium.

G. *Leishmania* NOT CULTURED WITH INSECT CELLS

There are many fascinating problems regarding the life cycle and transmission of the *Leishmania*, which are found in tropical and subtropical regions around the world (Lainson and Shaw, 1971). However, there are no known records of cell culture studies with the sandfly vectors, species of *Phlebotomus* and *Lutzomyia*.

III. Culturing Microfilariae with Insect Cells

Little information is available for using cell cultures to study parasites other than the parasitic protozoa and rickettsiae. Wood and Suitor (1966) cultured microfilaria *Macacanema formosana*, from a Taiwan monkey *Macaca cyclopsis*. Grace's 1966 line of *A. aegypti* cells in medium supplemented with FBS + *P. cynthia* hemolymph formed the substrate. The larvae grew in the complete medium without insect cells for 6 days, then died. But in the presence of growing insect cells, the larvae passed through the second stage and possibly underwent some development into the third stage. Since no interaction between the microfilariae and the insect cells was noted, it was suggested that the cultures simply provided growth factors for both biological entities.

Microfilariae of *Dirofilaria immitis* matured to the "sausage form" in continuous cell lines of *A. aegypti*, *Culiseta inornata*, and *Aedes vexans* (Cupp, 1972). This was said to be comparable to development in an unnatural intermediate host.

IV. Conclusions

A. SUMMARY

Insect cell culture has not elucidated the means of entry of parasites into cells. In most reported cases of attempts to study parasites in cell culture, the organisms have failed to develop, or have developed extracellularly in the medium. There are not enough cell lines available as yet from insect vectors, or the lines do not represent a great enough variety of cell types, to insure the right combination of factors for parasite growth. There are over 50 lines at present (Hink, 1972), comprising moths, mosquitoes, fruit flies, cockroaches, and leafhoppers. Selection of cell types capable of supporting parasites might be helpful, and cloning techniques should be advanced (McIntosh and Rechtoris, 1974). The only parasites which have been cultured within insect cells *in vitro* are the Microsporida. Years of work on *Plasmodium* in insect cell culture systems have resulted in the development of sporozoites, but neither oocysts nor sporozoites penetrated or developed inside of the insect cells. *Babesia* has been maintained in infected tissues of ticks *in vitro*, but penetration of new tissues did not occur.

Possession of an organelle for active penetration, the evertible polar filament, distinguishes the Microsporida. Parasites lacking such an organelle enter vertebrate cells by a gliding and pressing motion which seems to induce phagocytosis by the host cell. This has not been observed in insect cells. Phagocytosis of the spores of *Nosema* by honey bee hemocytes *in vitro* has been reported (Gilliam and Shimanuki, 1967). Phagocytosis of bacteria by cockroach hemocytes *in vitro* has been reported (Anderson *et al.*, 1973). In neither instance was it definitely known if the microorganisms were viable at the time they were phagocytosed. *Nosema* spores which failed to evert in culture medium, and which therefore were presumed to be dead, were phagocytosed by hemocytes (Kurtti and Brooks, 1971). The ingestion of bacteria by hemocytes *in vitro* was shown to be accomplished by pseudopod formation, but again it was not known if the phagocytized bacteria were killed and degraded by the hemocytes (Ratcliffe and Rowley, 1974). The need to know if a phagocytized parasite remains viable is obvious.

The presence or absence of a parasitophorous vacuole was emphasized in this treatment of various groups of parasitic protozoa because its formation has been suggested as a factor of possible significance in the penetration by the parasites into host cells and, therefore, in the ability of the protozoa to be maintained *in vitro*. Yet as Trager (1974) pointed out, "There is, of course, no actual or even theoretical necessity for a host cell membrane around a parasite." The extent to which an intracellular parasite is sequestered by host membranes or vacuoles seems to be immaterial with respect to its ability to survive *in vitro*. Likewise, it has been impossible to draw a correlation between the presence of these membranes and the extent to which parasitized insect cells have been cultured.

B. PROSPECTUS

This review has brought out the lack of studies on the ultrastructure of surviving parasites *in vitro*. Since much is known of the ultrastructure of parasites while they are within the host cell (Aikawa and Sterling, 1974), periodic examination of the organisms subjected to culture conditions might reveal many guideposts to the suitability of the culture conditions, as was done by Weidner and Trager (1973). Some of the unique features of insect-inhabiting stages, for example, the crystalloids of *Plasmodium* and related genera (Desser, 1972), should be sought to determine if they are lost in culture.

The flagellates evidently never have been studied in insect cell culture systems because they grow so readily on cell-free media and in vertebrate cell cultures. Yet, critical questions regarding loss of infectivity by these organisms in culture have not been answered in available systems. Further development of cultures from reduviid and triatomid bugs, tsetse flies, tabanid flies, and fleas needs to be undertaken.

There is no evidence that insect-transmitted protozoa penetrate cells of the insect host *in vitro*. In most cases, the parasites grew as well in cell-free insect culture medium as they did in the presence of insect cells. When unnatural host cells were used, if the parasites invaded them, the "host" cells were destroyed.

Fetal bovine serum, essential for all insect cell lines, was found to be beneficial for at least one trypanosome. The essential constituents in serum need to be identified.

Is it thinkable that the insect cell constitutes a hostile environment for parasites having a definitive vertebrate host? During their sojourn in insect vectors, the parasites live extracellularly. They are taken up by mouth, inhabit the gut lumen, the hemocoel, or the salivary gland lumen, where they undergo further development before returning to the vertebrate host. Only in the tissues of vector ticks do we find such organisms as *Babesia* developing intracellularly.

The idea has been suggested from time to time that perhaps a growth factor, excreted by cultured cells or tissues, should be sought to promote development of the parasites. On the contrary, there is strong evidence that excretions of cultured insect cells may be harmful to microorganisms (Bernier *et al.*, 1974). Spent culture media should be examined for metabolites having a detrimental effect on structure or viability of the parasites.

Acknowledgments

This work was supported in part by Research Grant No. AI 09914 from the National Institute of Allergy and Infectious Diseases, USPHS. This is Paper No. 9032, Scientific Journal Series, Minnesota Agricultural Experiment Station.

References

Aikawa, M., and Sterling, C. R. (1974). "Intracellular Parasitic Protozoa." Academic Press, New York.
Aikawa, M., Huff, C. G., and Strome, C. P. A. (1970). *J. Ultrastruct. Res.* **32**, 43–68.

198 MARION A. BROOKS

Amrein, Y. U., Geigy, R., and Kaufmann, M. (1965). *Acta Trop.* **22**, 193–203.
Anderson, R. S., Holmes, B., and Good, R. A. (1973). *J. Invertebr. Pathol.* **22**, 127–135.
Ball, G. H. (1964). *J. Parasitol.* **50**, 3–10.
Ball, G. H. (1972). In "Invertebrate Tissue Culture" (C. Vago, ed.), Vol. 2, pp. 321–342. Academic Press, New York.
Ball, G. H., and Chao, J. (1971). *J. Parasitol.* **57**, 391–395.
Ball, G. H., and Chao, J. (1973). *J. Parasitol.* **59**, 513–515.
Bernier, I., Landureau, J. C., Grellet, P., and Jollés, P. (1974). *Comp. Biochem Physiol. B* **47**, 41–44.
Buckley, S. M. (1973). *Exp. Parasitol.* **33**, 23–26.
Chao, J. (1971). *46th Annu. Meet., Amer. Soc. Parasitol.* Abstract No. 131, p. 53.
Chao, J. (1973). *Curr. Top. Comp. Pathobiol.* **2**, 107–144.
Chao, J., and Ball, G. H. (1964). *Amer. J. Trop. Med. Hyg.* **13**, 181–192.
Chao, J., and Ball, G. H. (1972). *J. Parasitol.* **58**, 148–152.
Cunningham, I. (1973). *Exp. Parasitol.* **33**, 34–45.
Cupp, E. W. (1972). In "Workshop on Development of Filariae in Mosquitoes," U.S.-Japan Coop. Med. Sci. Program, U.S. Panel Parasitic Dis., pp. 7–8. University of California, Los Angeles.
Desser, S. S. (1972). *Can. J. Zool.* **50**, 477–480.
Desser, S. S., and Trefiak, W. D. (1971). *Can. J. Zool.* **49**, 134–135.
Dolfini, S. (1971). In "Invertebrate Tissue Culture" (C. Vago, ed.), Vol. 1, pp. 247–265. Academic Press, New York.
Doran, D. J. (1973). In "The Coccidia. Eimeria, Isospora, Toxoplasma, and Related Genera" (D. M. Hammond and P. L. Long, eds.), pp. 183–252. Univ. Park Press, Baltimore, Maryland.
Faust, E. C., Russell, P. F., and Jung, R. C. (1970). "Clinical Parasitology," 8th ed. Lea & Febiger, Philadelphia, Pennsylvania.
Freyvogel, T. A. (1966). *Acta Trop.* **23**, 201–222.
Friedhoff, K., and Scholtyseck, E. (1969). *Z. Parasitenk.* **32**, 266–283.
Garnham. P. C. C., Bird, R. G., Baker. J. R., Desser, S. S., and El-Nahal, H. M. S. (1969). *Trans. Roy. Soc. Trop. Med. Hyg.* **63**, 187–194.
Gilliam, M., and Shimanuki, H. (1967). *J. Invertebr. Pathol.* **9**, 387–389.
Greene, A. E., and Charney, J. (1971). *Curr. Top. Microbiol. Immunol.* **55**, 51–61.
Greene, A. E., Charney, J., Nichols, W. W., and Coriell, L. L. (1972). *In Vitro* **7**, 313–322.
Gupta, K. S. (1964). *Curr. Sci.* **33**, 407–408.
Hammond, D. M. (1971). In "Ecology and Physiology of Parasites" (A. M. Fallis, ed.), pp. 3–20. Univ. of Toronto Press, Toronto.
Hink, W. F. (1972). *Advan. Appl. Microbiol.* **15**, 157–214.
Hoare, C. A. (1972). "The Trypanosomes of Mammals." Blackwell, Oxford.
Hoffmann, G. (1972). *Z. Angew. Entomol.* **71**, 26–34.
Ishihara, R. (1967). *Can. J. Microbiol.* **13**, 1321–1332.
Ishihara, R. (1968). *J. Invertebr. Pathol.* **11**, 328–329.
Ishihara, R. (1969). *J. Invertebr. Pathol.* **14**, 316–320.
Ishihara, R., and Sohi, S. S. (1966). *J. Invertebr. Pathol.* **8**, 538–540.
Körner, H. K., and Feldhege, A. (1970). *Cytobiologie* **1**, 203–207.
Kurtti, T. J., and Brooks, M. A. (1971). *Curr. Top. Microbiol. Immunol.* **55**, 204–208.
Lainson, R., and Shaw, J. J. (1971). In "Ecology and Physiology of Parasites" (A. M. Fallis, ed.), pp. 21–57. Univ. of Toronto Press, Toronto.

11. PARASITE TRANSMISSION

199

Levine, N. D. (1972). *Res. Protozool.* **4**, 291–350.
McIntosh, A. H., and Rechtoris, C. (1974). *In Vitro* **10**, 1–5.
Nicoli, J., and Vattier, G. (1964). *Bull. Soc. Pathol. Exot.* **57**, 213–219.
Ohshima, K. (1937). *Parasitology* **29**, 220–224.
Petri, M. (1966). *Acta Pathol. Microbiol. Scand.* **66**, 13–30.
Pipkin, A. C., Wood, D. E., and Suitor, E. C., Jr. (1968). *43rd Annu. Meet., Amer. Soc. Parasitol.* Abstract No. 64, p. 39.
Ratcliffe, N. A., and Rowley, A. F. (1974). *Nature (London)* **252**, 391–392.
Reháček, J. (1971). *Ann. Parasitol.* **46**, Suppl., 197–231.
Reháček, J. (1972). *In* "Invertebrate Tissue Culture" (C. Vago, ed.), Vol. 2, pp. 279–320. Academic Press, New York.
Rosales-Ronquillo, M. C., and Silverman, P. H. (1974). *J. Parasitol.* **60**, 819–824.
Rosales-Ronquillo, M. C., Simons, R. W., and Silverman, P. H. (1972). *Ann. Entomol. Soc. Amer.* **65**, 721–729.
Rudzinska, M. A. (1969). *Int. Rev. Cytol.* **25**, 161–199.
Schneider, I. (1968a). *Exp. Parasitol.* **22**, 178–186.
Schneider, I. (1968b). *Proc. Int. Collog. Invertebr. Tissue Cult., 2nd, 1967* pp. 247–253.
Schneider, I. (1972). *Proc. Helminthol. Soc. Wash. Spec. Issue* **39**, 438–444.
Shively, J. M. (1974). *Annu. Rev. Microbiol.* **28**, 167–187.
Siddiqui, W. A., Schnell, J. V., and Geiman, Q. M. (1969). *Mil. Med.* **134**, 929–938.
Speer, C. A., Rosales-Ronquillo, M. C., and Silverman, P. H. (1974). *J. Invertebr. Pathol.* **24**, 179–183.
Speer, C. A., Rosales-Ronquillo, M. C., and Silverman, P. H. (1975). *J. Invertebr. Pathol.* **25**, 73–78.
Sterling, C. R., and Aikawa, M. (1973). *J. Protozool.* **20**, 81–92.
Terzakis, J. A. (1971). *J. Protozool.* **18**, 62–73.
Trager, W. (1937). *J. Parasitol.* **23**, 226–227.
Trager, W. (1959). *Ann. Trop. Med. Parasitol.* **53**, 473–491.
Trager, W. (1971). *J. Protozool.* **18**, 392–399.
Trager, W. (1974). *Science* **183**, 269–273.
Trager, W., and Krassner, S. M. (1967). *Res. Protozool.* **2**, 357–382.
Trigg, P. I., and Gutteridge, W. E. (1972). *Parasitology* **65**, 265–271.
Vanderberg, J., Rhodin, J., and Yoeli, M. (1967). *J. Protozool.* **14**, 82–103.
Vickerman, K. (1971). *In* "Ecology and Physiology of Parasites" (A. M. Fallis, ed.), pp. 58–91. Univ. of Toronto Press, Toronto.
Vickerman, K. (1972). Functional aspects of parasite surfaces, *Brit. Soc. Parasitol., Symp.* **10**, 71–91.
Vickerman, K. (1974). Trypanosomiasis and leishmaniasis with special reference to Chagas' disease, *Ciba Found. Symp.* [N.S.] **20**, 171–198.
Weidner, E. (1972). *Z. Parasitenk.* **40**, 227–242.
Weidner, E., and Trager, W. (1973). *J. Cell Biol.* **57**, 586–591.
Weiss, E. (1975). *Bacteriol. Rev.* **37**, 259–283.
Wood, D. E., and Pipkin, A. C., Sr. (1969). *Exp. Parasitol.* **24**, 176–183.
Wood, D. E., and Suitor, E. C., Jr. (1966). *Nature (London)* **211**, 868–870.
Zeledón, R. (1971). *Rev. Biol. Trop.* **19**, 197–210.

12

Arboviruses and *Toxoplasma gondii* in Diptera Cell Lines

SONJA M. BUCKLEY

I. Introduction

At present, more than 300 arboviruses (for "arthropod-borne viruses") are recognized. These heterogeneous viruses, incorporated into a general system of virus classification (Casals, 1971), can be arranged for the most part into groups of related, but distinct agents. Four taxons have already been well characterized, namely, alphaviruses, flaviviruses, orbiviruses, and rhabdoviruses; a fifth subset, the bunyaviruses, has been proposed recently (Porterfield *et al.*, 1973, 1974). In addition, there is a sixth taxon, the iridoviruses, discussed with some of the other taxons in a recent symposium on comparative virology (Brown and Tinsley, 1973).

By definition, "arboviruses are viruses which are maintained in nature principally or to an important extent, through biological transmission between susceptible vertebrate hosts by hematophagous arthropods" (World Health Organization Scientific Group, 1967). While much has been known for many decades about anopheline mosquitoes and *Aedes aegypti* and *Culex quinquefasciatus*, during the past 20 years attention has been focused on dozens of species of other *Aedes* and *Culex* and also on *Mansonia* and other mosquitoes suspected or proved to be actively involved in arbovirus transmission cycles (Theiler and Downs, 1973). Generally, the recognized range of natural vectors extends from mosquitoes to ticks, phlebotomines, *Culicoides*, and, possibly, mites (Casals, 1971).

"Vector tissue culture research began in 1966" (Yunker, 1971) with the establishment of a stable cell line from larval *A. aegypti* (Grace, 1966). According to Hink (1972), Diptera cell lines are represented by 14 species with primary explants derived from embryos, larvae, imaginal discs, adult ovaries, or adult species. Since 1966, *in vitro* studies of viruses in invertebrate tissue culture have burgeoned. Recent and comprehensive reviews have covered arthropod cell cultures and their application to the study of arboviruses (Singh, 1971, 1972; Yunker, 1971; Reháček, 1972; Dalgarno and Davey, 1973). A detailed survey of the literature is, therefore, not attempted in this communication. Examples indicative of the current direction of research are presented. In addition to arboviruses, *Toxoplasma gondii* and its survival in mosquito cell lines will be discussed.

II. "Growth or Nongrowth" of Arboviruses in Diptera Cell Lines

In this review of the susceptibility of Diptera cell lines to members of the different taxons of arboviruses, the following determinants are recognized: (1) marked innate differences between individual mosquito cell lines and (2) heterogeneity of viruses with regard to vector and sensitivity to sodium deoxycholate (SDC) (Theiler, 1957; Borden *et al.*, 1971; Casals, 1971). Singh's *Aedes albopictus* cell line (Singh, 1967) has excelled, not only in terms of overall viral growth performance, but also in terms of ease of handling under routine laboratory conditions, i.e., cellular dispersion by mechanical means and growth in Mitsuhashi–Maramorosch (1964) medium. Other mosquito cell lines used with varying frequency have been established from *Aedes aegypti* (Grace, 1966; Singh, 1967; Peleg, 1968), *Aedes vittatus* (Bhat and Singh, 1970), *Aedes w-albus* (Singh and Bhat, 1971), *Anopheles stephensi* (Schneider, 1969), *Anopheles gambiae* (M. G. R. Varma and M. Pudney, unpublished data), and *Culex quinquesfasciatus* (Hsu *et al.*, 1970). Growth media, as well as characteristics of these cell lines, including population doubling time and ploidy, have been described in detail by Hink (1972).

In the alphavirus taxon, at least eight viruses (chikungunya, eastern equine encephalitis, O'nyong-nyong, Ross River, Semliki Forest, Sindbis, Venezuelan equine encephalitis, and western equine encephalitis) have multiplied in one or more of the abovementioned mosquito cell lines (Singh, 1971, 1972; Yunker, 1971; Reháček, 1972; Dalgarno and Davey, 1973). Few investigators, however, have carried out comparative alphavirus growth studies simultaneously in a number of Diptera cell lines. Figure 1 shows such growth patterns of small- and large-plaque (SP and LP) variants of Chikungunya virus (Ross strain S27, high mouse passage level) (Buckley, 1973a) in two vertebrate and seven invertebrate cell lines (Buckley *et al.*, 1975). In both the *Aedes* cell lines (incubated at 29 ± 1°C) and the vertebrate cell lines (incubated at 36°C), infectivity titers of extracellular virus reached a peak rapidly; cytopathic effect (CPE) occurred only in the vertebrate cells and is credited with the subsequent decrease in infectivity titers. In *Anopheles* cells (incubated at 29 ± 1°C), the infectivity titers increased slowly to a peak at 10 days postinoculation; in *Culex* cells (incubated in one experiment at 29 ± 1°C and in the other at room temperature of 20–25°C), persistence of virus only or no multiplication was observed. Growth differences found *in vitro* parallel, to some extent, *in vivo* differences found in experimental transmission; thus, under laboratory conditions, *Aedes aegypti* (Ross, 1956), *Aedes calcatus* (Taylor, 1967), *Aedes furcifer/taylori* group (Paterson and McIntosh, 1964), and *Aedes albopictus* (Taylor, 1967) transmitted Chikungunya virus efficiently, while *Anopheles stephensi*, maintaining the virus for 10 days, failed to transmit (Rao, 1964). *Culex pipiens* (Taylor, 1967) similarly did not transmit. Another alphavirus, O'nyong-nyong, characterized biologically as an anopheline isolate and serologically as closely related yet distinguishable from Chikungunya, multiplies in established cell lines from larvae of *Anopheles stephensi* but not in *Aedes aegypti* cell lines (Varma and Pudney, 1971; Buckley, 1971a). Singh's *A. albopictus* cell lines are susceptible to infection with both Chikungunya and O'nyong-nyong viruses (Buckley, 1971b).

Little is known about the behavior of bunyaviruses (including the Bunyamwera supergroup viruses) in mosquito cell lines. California encephalitis virus multiplies in Singh's *Aedes* cell lines (Whitney and Deibel, 1971). Tahyna virus propagates in the *A. albopictus* cell line (Adamcova-Otova and Marhoul, 1974) and also in an *Anopheles gambiae* cell line (Marhoul, 1973). Both viruses are mosquito isolates and SDC-sensitive.

The susceptibility of mosquito cell lines to "other possible members"

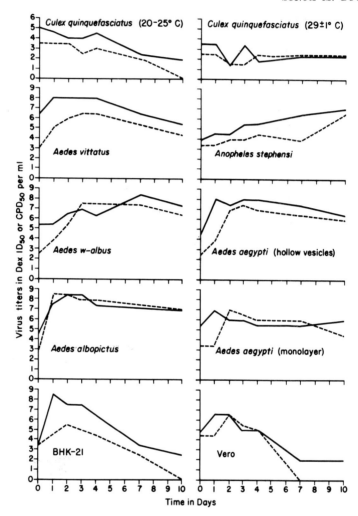

Fig. 1. Growth curves of small plaque (- - - - -) and large plaque (——) variants of Chikungunya virus in vertebrate and invertebrate cell lines. Dex, decimal exponent. (Buckley *et al.*, 1975). Reproduced with permission from *Acta Virol. (Prague)* **19,** 10–18.

of the bunyaviruses (morphologically similar, but serologically unrelated) (Porterfield *et al.*, 1973/1974) has been studied more extensively. Thus, Ganjam virus, isolated from both mosquitoes and ticks (Theiler and Downs, 1973) and SDC-sensitive, multiplies in the *Aedes albopictus* but not in the *A. aegypti* cell line (Singh, 1971). The following phlebotomine isolates have failed to grow in Singh's *Aedes* cell lines: Pacui (Aitken *et al.*, 1975; S. M. Buckley, unpublished data), and Sandfly

Fever Neapolitan and Sandfly Fever Sicilian (Buckley, 1969). The following tick isolates have failed to propagate in Singh's *Aedes* cell lines: Congo, Kaisodi, and Silverwater (Buckley, 1971b). To the extent tested, these agents are SDC-sensitive.

The flaviviruses can be subdivided into three sets according to whether the vector is a mosquito, a tick, or an unknown. Flaviviruses are SDC-sensitive. As far as examined, the mosquito-borne flaviviruses multiply in one or more of the Diptera cell lines; these flaviviruses include dengue 1, 2, 3, and 4, Ilheus (Geneviève Panon, personal communication), Japanese B encephalitis, Kunjin, Murray Valley encephalitis, St. Louis encephalitis, West Nile, and yellow fever (Singh, 1971, 1972; Yunker, 1971; Reháček, 1972; Dalgarno and Davey, 1973). The tick-borne flaviviruses, such as Kyasanur Forest disease (Singh, 1971), Langat (Buckley, 1969), and Kadam (Mugo and Shope, 1972), fail to grow. Of the flaviviruses without adequate indication of an arthropod vector, those tested (i.e., Modoc and Cowbone Ridge) do not infect Singh's *Aedes* cell lines (Buckley, 1969).

The iridovirus taxon, so far, contains only one arbovirus: African swine fever. This icosahedral cytoplasmic deoxyribovirus propagates in vertebrate cell cultures (Hess, 1971) but has not yet been studied in invertebrate cell cultures. Since the virus is sensitive to lipid solvents (DeTray, 1963) and, furthermore, has been transmitted by both adult and nymphoid ticks (*Ornithodoros moubata porcinus*) fed sequentially on infected and susceptible swine (Heuschele and Coggins, 1965; Plowright *et al.*, 1970), the likelihood that the virus will propagate in mosquito cell lines seems remote.

In the orbivirus taxon, vectors extend from mosquitoes to ticks, phlebotomines, and *Culicoides*. Viruses are resistant or relatively SDC-resistant (Borden *et al.*, 1971). The following agents with suspected or proved mosquito vectors have been propagated in Singh's *A. albopictus* cell cultures: Lebombo and Palyam (Buckley, 1972) and epizootic hemorrhagic disease of deer (Willis and Campbell, 1973), the latter also growing in the *A. aegypti* cell line. African horse-sickness virus, a *Culicoides* isolate, multiplies in Singh's *A. albopictus* cells (Mirchamsy *et al.*, 1970). Irituia virus, isolated from *Oryzomys* sp. and serologically related by complement-fixation test to Changuinola virus, grows in Singh's *Aedes* cell lines (Buckley, 1972) and the same is true for additional Changuinola subgroup viruses, isolated from vertebrates or phlebotomines, such as Changuinola, BeAr 35646, Bt 766, Bt 2365, Bt 104, BeAr 41067, Bt 2164, and CO Ar 2837 (S. M. Buckley, unpublished). Comparative serum neutralization titers of various Changuinola group viruses (Baker, 1973) are shown in Table I.

TABLE I

Comparative Serum Neutralization Titer of Various Changuinola Sero Group Viruses (Vero Plaque Reduction Test)

Virus	Serum[a]								
	Chan-guinola	BeAr 35646	Bt 766	Bt 2365	Bt 104	BeAr 41067	Bt 2164	CO Ar 2837	Irituia
Changuinola (BT 436)	>256	0	>256	0	0	0	0	0	0
BeAr 35646	0	>256	0	4	0	8	0	0	0
Bt 766	>256	0	256	4	4	0	2	0	0
Bt 2365	0	0	0	64	0	0	128	2	0
Bt 104	0	0	0	2	32	0	0	0	0
BeAr 41067	0	0	0	0	0	>256	0	0	0
Bt 2164	0	0	0	32	0	0	32	2	0
CO Ar 2837	0	2	0	2	0	0	0	>256	
Irituia (BeAn 28873)	0	0	0	0	0	0	0	0	>256

[a] The serum neutralization titer is the reciprocal serum dilution which gave a 50% reduction in plaques when reacted with 100 plaque forming units. Italic numbers indicate homologous and cross-reactions.

Of the tick-borne orbiviruses, Colorado tick fever grows only in the *Aedes albopictus* cell line (Yunker and Corey, 1969; Buckley, 1969) and Chenuda, Kemerovo, Lipovnik, and Tribec grow in both Singh's *Aedes* cell lines (Buckley, 1971c; Libikova and Buckley, 1971).

In the rhabdovirus taxon, all viruses tested are SDC-sensitive. Those belonging to the vesicular stomatitis (VS) subgroup with proved mosquito susceptibility, such as Chandipura, Cocal, and VSV-Indiana type, multiply in both Singh's *Aedes* cell lines, with the *A. aegypti* cell line apparently being the more sensitive (Buckley, 1969; Singh, 1971; Artsob and Spence, 1974a). In the rabies serogroup (Shope *et al.*, 1970) ; Shope, 1975), Obodhiang (isolated from *Mansonia uniformis*) and kotonkan (isolated from *Culicoides*) (Kemp *et al.*, 1973) multiply in Singh's *A. albopictus* cell line but not in *A. aegypti* cells (Buckley, 1973b). A third member, Mokola virus (isolated from *Crocidura* sp.), behaves similarly (S. M. Buckley, unpublished data) ; at present, there is no other evidence indicative of an arthropod cycle in the maintenance of Mokola virus.

Of the still unclassified tick-borne viruses that have been tested, Hughes, Johnston Atoll, Qalyub, Quaranfil, and Soldado are SDC-sensitive and do not grow in Singh's *Aedes* cell lines (Buckley, 1971b).

TABLE II

Growth of Arboviruses in Singh's *Aedes albopictus* Cell Cultures

Taxon	No. of arboviruses tested	Proved or suspected vector[a]	SDC-sensitive or relatively resistant	Growth	
Alphavirus	8	Mosquito	Sensitive	Yes	(8/8)
Bunyavirus	3	Mosquito	Sensitive	Yes	(3/3)
Flavivirus	11	Mosquito	Sensitive	Yes	(11/11)
Orbivirus	3	Mosquito	Relatively resistant	Yes	(3/3)
Rhabdovirus	4	Mosquito	Sensitive	Yes	(4/4)
Bunyavirus	3	Tick	Sensitive	No	(0/3)
Flavivirus	3	Tick	Sensitive	No	(0/3)
Orbivirus	5	Tick	Relatively resistant	Yes	(5/5)
Not classified	5	Tick	Sensitive	No	(0/5)
Bunyavirus	3	Phlebotomine	Sensitive	No	(0/3)
Orbivirus	9	Phlebotomine	Relatively resistant	Yes	(9/9)
Rhabdovirus	1	*Culicoides*	Sensitive	Yes	(1/1)
Orbivirus	1	*Culicoides*	Relatively resistant	Yes	(1/1)

[a] In addition to having been isolated from mosquitoes, bunyavirus Ganjam and some rhabdoviruses (VSV subgroup) have also been isolated from ticks and phlebotomine flies, respectively.

In conclusion and in an attempt to instill order into this "tale about growth and nongrowth" (Wagner, 1973), it is emphasized that the selective characteristics of vector, sensitivity or relative resistance to SDC, and individual invertebrate cell line correlate with the ability of a given arbovirus to multiply or not in Diptera cell lines. The presently known relationships as they pertain to Singh's *A. albopictus* cell line are summarized in Table II.

III. Quantitative Aspects of Arbovirus Multiplication

Invertebrate cell lines maintained by individual investigators as well as heterogeneity of arboviruses, viral strain, viral passage history, multiplicity of infection (MOI), and temperature of incubation following viral inoculation are factors determining the specifics of kinetics of viral growth *in vitro*. Examples pertaining to various taxons are listed below.

A. ALPHAVIRUS

A comprehensive review pertaining to kinetics, biochemistry, and ultra-structural studies in vertebrate and invertebrate cell cultures with special reference to such viruses as Semliki Forest (SF), Sindbis, eastern equine encephalitis, western equine encephalitis, and Venezuelan equine enceph-alitis has recently been published by Dalgarno and Davey (1973). In their hands, the short-term growth kinetics of SF virus in *Aedes albopic-tus* cells were similar in all respects to those in three vertebrate cell cul-tures (porcine kidney, Vero, and chick embryo fibroblast). The percent-age of *A. albopictus* cells infected was dependent on the conditions of SF virus adsorption and also on cell concentration during adsorption. The temperature giving the shortest latent period of SF growth was close to the optimal temperature for cell growth (*A. albopictus*, 28°C; vertebrate cell cultures, 37°C). In a long-term experiment in *A. aegypti* cells (Peleg, 1968), SF virus at 1–5 MOI infected productively 2.8–8% of cells at 28°C; 6–13 days following inoculation, only 0.2–0.9% of cells produced virus. The percentage dropped to less than 0.1% within 20–140 days post-inoculation, with an estimated mean virus yield of 1–6 plaque-forming units (PFU) per infected cell throughout the experiment (Peleg, 1969a). Sindbis virus, under Indian conditions, produced peak virus titers 72–96 hours after inoculation of *A. albopictus* cell cultures (Singh and Paul, 1968), thus indicating low multiplicity infections; at Rutgers University Medical School, the same alphavirus showed maximum virus titers 25 hours postinoculation (Stevens, 1970). The *A. albopictus* cell line had been subcultured at least 50 times prior to Stevens' experiment; Sindbis virus, grown initially in chick embryo fibroblasts, was derived from the eighth passage in BHK-21 cells and had a titer of 10^8 PFU/ml.

B. BUNYAVIRUS (Including Bunyamwera Supergroup
   Viruses)

In quantitative studies with California encephalitis (CE) virus in Singh's *Aedes* cell lines (Singh, 1967), Whitney and Deibel (1971) found a 4-hour duration in the eclipse phase in *A. aegypti* cells and 8 hours in *A. albopictus* cells. The latent period was followed by a steep increase in cell-associated virus. In the first growth cycle, 300–1500 mouse lethal doses (LD_{50}) of virus were produced per infected cell. A relatively higher virus production was observed in the second growth cycle. Virus was readily released from the cells. Immunofluorescence studies indicated that only a small fraction of the cells in a culture were infected; these cells showed tiny perinuclear granules, bright cytoplasmic bodies, or diffuse cytoplasmic staining. Propagation of CE virus in Singh's *Aedes* cell lines was of a magnitude to allow comparative studies between mosquito cell-adapted strains and mammalian cell-adapted strains. Tahyna virus was produced in *Anopheles gambiae* cells during an observation period of 28 days postinoculation with peak virus titers occurring on the fifth day (Marhoul, 1973). In an immunofluorescence study of the same virus in *Aedes albopictus* cells, specific fluorescence was first detected 6 hours postinoculation. The maximal number (33%) of fluorescent cells was observed 2 days following inoculation; then there was a rapid decrease and stabilization at a 2% level. The specific fluorescence was cytoplasmic in localization and diffuse in nature (Adamcova-Otova and Marhoul, 1974).

C. FLAVIVIRUS

Dalgarno and Davey (1973) have extensively reviewed contributions pertaining to kinetics, biochemistry, and ultrastructural studies of flaviviruses in vertebrate and invertebrate cell cultures. As with alphaviruses, optimal flavivirus replication in invertebrate and vertebrate cell systems was apparently related to the optimum temperature of cell growth. Thus, Kunjin virus had a latent period of 12–16 hours in *A. albopictus* cells and latent period of 18–20 and 10–12 hours respectively, in Vero and porcine kidney cells (Davey *et al.*, 1973). The importance of strain differences for the kinetics of flaviviruses has been stressed by Igarashi *et al.* (1973). These Japanese investigators studied the growth of three strains of Japanese encephalitis (JE) virus with different passage histories in three different mosquito cell lines, i.e., Singh's *A. albopictus* and *A. aegypti*, as well as Peleg's *A. aegypti*. The virus yield at 28°C was highest in Singh's *A. albopictus*, followed by that in Peleg's *A. aegypti* and then in Singh's *A. aegypti* cells. Virus strains with *in vitro* passages in monkey kidney or BHK-21 cell cultures appeared to produce higher virus titers in Singh's *A. albopictus* cells than did strains passed only in infant mice. The BHK-21 plaque-purified progenies of the JaOH-0566 strain of JE virus, a strain that had been passaged 3 times in infant mice and 62 times in primary cultures of monkey kidney cells, were outstanding. Following several passages in BHK-21 cells, the strain was plaque-passaged 3 times in BHK-21 cells under nutrient agar overlay. Plaque-purified viral progenies were inoculated into *A. albopictus* cells at a MOI of 10 PFU/cell. An "alphaviruslike" latent period of only 6 hours was observed in *A. albopictus* cells incubated at 28°C. The virus titer increased between 6 and 36 hours, the maximum yield being more than 10^9 PFU/ml in the fluid phase (Fig. 2). Singh and Paul (1968)

SONJA M. BUCKLEY

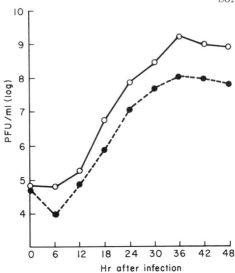

Fig. 2. Growth of Japanese encephalitis virus, Ja OH-0566 strain, in Singh's *Aedes albopictus* cells at 28°C; O——O, fluid virus; ●----●, cell-associated virus (Igarashi *et al.*, 1973). Reproduced with permission from *Biken J.* **16**, 17–23.

reported LD_{50} or tissue culture dose (TCD_{50}) titers of $10^{-7.5}$/ml in *A. albopictus* cells inoculated with an Indian topotype strain of JE virus.

D. ORBIVIRUS

A strain of African horse-sickness (AHS) virus, type 9, adapted to mouse brain and monkey kidney cells, readily propagated in *A. albopictus* cells (Mirchamsy *et al.*, 1970). Samples collected at different post-inoculation intervals revealed a latent period of approximately 12 hours; there was no difference between titers of cell-released and cell-associated virus. Maximal titers were observed 96–120 hours following inoculation. A virus strain of epizootic hemorrhagic disease of deer (EHD), isolated from infected deer spleen by passage in infant mice, was further adapted to grow in BHK-21 cells and then given five successive passages in *A. albopictus* cells incubated at 27°C; by the fifth passage the infectious virus titer produced was comparable to that obtained in BHK-21 cells incubated at 37°C (Willis and Campbell, 1973). In studies with a tick-borne orbivirus, Kemerovo, strain L75, was used following isolation in chick embryo cell cultures, seven yolk sac passages in chick embryos, and three passages in infant mice. After inoculation of *A. albopictus* cell cultures at a MOI of 10 TCD_{50} of virus per cell, the intracellular and extracellular virus titers increased slightly by 72 hours postinoculation.

The yield of virus during primary infection was very low and consisted approximately of 1 PFU/50 cells (Libikova and Buckley, 1971).

E. RHABDOVIRUS

Cocal virus (subgroup vesicular stomatitis) multiplied in Singh's *A. albopictus* and *A. aegypti* cell lines incubated at 30°C, producing a low-titered infection involving only a small proportion of the cell population (Artsob and Spence, 1974a). Prior to the preparation of virus stock in primary chick monolayers, the Cocal virus strain had been passed in infant mice and once in BHK-21 cell cultures. In *A. aegypti* cells, irrespective of initial input multiplicity of virus, the PFU/cell ratio remained fairly constant; in contrast, this ratio at a MOI of 10 PFU/*A. albopictus* cell was consistently higher than at a MOI of 1 PFU cell. In attempts to increase the percentage of infected *A. albopictus* cells as well as the PFU/cell ratio, it was found that a MOI of 10 PFU/cell at an incubation temperature of 37°C resulted in 100% infection of cells 24, 36, and 60 hours postinoculation (Table III); the PFU/cell ratio increased correspondingly from 1.93 to 10.43 and 17.11, respectively. These conditions allowed a subsequent electron microscopy study (see below). Shope *et*

TABLE III

Effect of Temperature on Cocal Virus (Subgroup Vesicular Stomatitis) Production in *Aedes albopictus* with an Input Multiplicity of 10[a]

Time (hours)	% Infected cells (PFU/cell)	
	30°C	37°C
6	0.54 (0.0032)	0.67 (0.0024)
12	1.20 (0.0157)	2.39 (0.0136)
18	0.97 (0.0067)	16.18 (0.0856)
24	0.65 (0.0048)	100 (1.93)
36	0.68 (0.0114)	100 (10.43)
60	5.0 (0.1900)	100 (17.11)

[a] From Artsob and Spence (1974a). Reproduced with permission of the National Research Council of Canada from the *Can. J. Microbiol.* **20,** 329–336,

TABLE IV

Percentage of Infective Centers and Production of Intracellular and Extracellular Virus Obtained with Obodhiang and Kotonkan Viruses on Primary Infection of *Aedes albopictus* Cells, as Determined by Subinoculation into Vero Cells

Virus	Days after subinoculation into Vero cells	No. A. *albopictus* cells plated on Vero cells (×100)	No. plaques counted	Infective center (%)	Amount of virus detected[a]	
					Intracellular	Extracellular
Obodhiang, SudAr 1154-64	8	660	0	0	0	0
	14	1280	0	0	0	0
	22	50	3	0.06	1585	0
	36	40	22	0.55	7080	100
Kotonkan, IbAr 23380	8	1320	0	0	0	0
	14	1300	0	0	0	0
	22	2600	0	0	0	0
	36	60	3	0.05	2510	80

[a] Expressed as plaque-forming units per milliliter. From Buckley (1973b). Reproduced with permission from *Appl. Microbiol.* **25,** 695–696.

al. (1970) established the existence within the rhabdoviruses of a rabies serogroup composed of rabies, Lagos bat, and Mokola viruses. Subsequent studies (Shope, 1975) have added two other rhabdoviruses to this serogroup: Obodhiang, strain Sud Ar 1154–64, isolated from unengorged *Mansonia uniformis* in the Sudan (Schmidt *et al.*, 1965) and kotonkan, strain IbAr 23380, isolated from *Culicoides* sp. in Ibadan, Nigeria (Kemp *et al.*, 1973). In preliminary experiments in this laboratory, mouse brain preparations of Obodhiang and kotonkan were tested for their ability to infect cultures of invertebrate and vertebrate cell lines. Both viruses readily infected monolayer cultures of *A. albopictus* cells incubated at 30°C; neither virus multiplied in *A. aegypti* cells incubated at 30°C or in cultures of the LLC-MK$_2$, Vero, and BHK-21 cell lines incubated at 36°C. Initially, the presence of virus in the mosquito cells was determined by subinoculation of either undiluted fluid phase or combined fluid and cell phases into infant mice by the intracerebral route. Subsequently, it was observed that subinoculation from *A. albopictus* cells infected with either virus repeatedly and consistently induced plaque formation and CPE in Vero cell cultures. These *A. albopictus* "helper cells" were then studied with regard to percentage of infected cells and production of intracellular and extracellular virus by methods described in detail elsewhere (Libikova and Buckley, 1971). In Vero cell cultures inoculated with Obodhiang- or kotonkan-infected *A. albopictus* cells at a MOI of 1 LD$_{50}$/50 mosquito cells, plaques were formed only when at least 0.06% (Obodhiang) or 0.05% (kotonkan) of the *A. albopictus* cells were infected. As shown in Table IV, amounts of intracellular virus exceeded amounts of extracellular virus, with plaque-forming intracellular virus demonstrated on postinoculation days 22 (Obodhiang) and 36 (kotonkan) and plaque-forming extracellular virus (both agents) demonstrated on postinoculation day 36. In a subsequent experiment, inoculation of cell phase (containing at least 0.05–0.06% infective centers) of Obodhiang- and kotonkan-infected *A. albopictus* cell cultures regularly induced CPE in both Vero and BHK-21 cell cultures; plaque formation in Vero cell cultures and the development of CPE in both Vero and BHK-21 cell cultures under fluid medium were specifically inhibited by Obodhiang and kotonkan hyperimmune mouse ascitic fluids (Buckley, 1973b).

IV. Ultrastructural Studies

Electron microscopic studies of arboviruses in vertebrate cells have recently been joined by some ultrastructural studies in invertebrate cells (Dalgarno and Davey, 1973).

A. ALPHAVIRUS

Raghow *et al.*, (1973b) have examined Ross River (RR) virus replication in cultured mosquito and mammalian cells with particular emphasis on short-term virus kinetics and correlated ultrastructural changes. In both cell systems, latent periods (5–6 hours) and maximum yields were similar; temperatures of incubation were 30° and 37°C, respectively. In *Aedes albopictus* cells, RR virus induced an inapparent persistent infection with essentially no difference in cell division rate between uninoculated control and inoculated cultures. Virus matured cytoplasmically at the cell membrane or within large, electron-dense inclusions where it increased in quantity; nucleocapsids failed to accumulate (18 hours). At that time, at least 50% of the cells were infected. Subsequently and concomitant with a 1-log unit decrease in cell-associated virus titer, the inclusions lost their electron-dense material (28 hours) and were transformed into microvesiculated vacuoles. In parallel, virus disappeared from these structures and also from the cell membrane. The authors suggest that during the establishment of the persistent infection, digestion of the contents of the inclusions occurs as a result of fusion with lysosomal microvesicles. In Vero cells, RR virus was found in small cytoplasmic vesicles early in the infection (8 hours). Type 1 cytopathic vacuoles (Grimley *et al.*, 1968) were present. Accumulation of nucleocapsids, at times in the form of paracrystalline arrays, was marked by late in infection. Ultimately, cell death occurred.

Somewhat different results were obtained by Raghow *et al.* (1973a) when they studied the ultrastructural morphology of *A. albopictus* cells infected with Semliki Forest (SF) virus. The maximum cell-associated virus titer, although similar to that found in RR virus-infected *A. albopictus* cells, occurred at 12 hours and thus considerably earlier. At that time, 100% of the mosquito cells were infected as determined by both infective center and immunofluorescence assay. Intracellular virus-specific structures were absent; however, substantial quantities of extracellular virus were occasionally observed. Free nucleocapsids were not detected in infected cells at this or later times. At 24 hours, less than 5% of cells showed numerous electron-dense cytoplasmic inclusions, some of them containing enveloped virus as well as microvesicles. Type 1 cytopathic vacuoles, described in SF virus-infected mammalian cells (Grimley *et al.*, 1968), were observed occasionally in few cells. The authors presume relative rapidity and high efficiency with regard to SF virus maturation in *A. albopictus* cells and most likely budding from the cell membrane. Consistent with such an assumption is the high ratio of extracellular virus to cell-associated virus in SF virus-infected cells as compared with RR virus-infected cells. In salivary glands from eastern equine encephalitis virus-infected *Aedes triseriatus* (Whitfield *et al.*, 1971), nucleocapsids were first found in the cytoplasm. *In vivo*, nucleocapsids apparently were enveloped immediately when formed, i.e., while passing through intracytoplasmic or plasma membranes. In contrast to these findings pertaining to short-term experiments and primary infection with alphaviruses, S. R. Webb (personal communication, 1974) discovered viruslike, typically paracrystalline structures (Fig. 3) localized within an inclusion body as well as occasional budding of particles from cell membranes of uninoculated *A. albopictus* cells. The cytoplasmic, viruslike contaminant (icosahedral in shape) measured about 60 nm in diameter in thin sections and 63 nm in negatively stained preparations. Light microscopy revealed presence of polykaryocytes as well as massive syncytia formation in the uninoculated mosquito cells. Serologically (see Section VI), the contaminant was identified as alphavirus Chikungunya (Cunningham *et al.*, 1975). Thus, under conditions of persistence, paracrystalline structures may also be found in mosquito cells. *Aedes albopictus* cells infected with morphologically homogeneous Sindbis virus were found to produce progeny virions which could be divided into three classes based on size. However, these morphological variants of Sindbis virus were not serologically identified (Brown and Gliedman, 1973).

B. BUNYAVIRUS

The aspects of the developmental morphology of California encephalitis (CE) virus, LaCrosse strain, were studied by Lyons and Heyduk (1973), both *in vitro* (Vero cell and *A. albopictus* cell cultures) and *in vivo* (infant mice). Replication of CE virus in mice has been studied previously by Murphy *et al.* (1968). Ultrastructurally, similar virus particles (approximately 95 nm diam.) were visualized in all three host systems; there appeared to be a common mode of virus assembly and maturation. Thus, the authors showed that virus assembly was limited to the internal cytomembrane interfaces, with the Golgi complex apparently representing the initial assembly site. Concomitant with progress of infection, there was proliferation of Golgi smooth membranes and also dilation of cisternae and vesicles. Virions became enveloped while budding into cisternae and vesicular lumina. In infected *A. albopictus* cells, at times of maximal virus production, virions were detected within dilated cisternae and vacuoles and extracellularly as shown in Fig. 4. In the mosquito cells, cellular changes such as vacuolization and smooth membrane proliferation were similar to those described for infected Vero cells. Although the mosquito cell line displayed the three major morphological

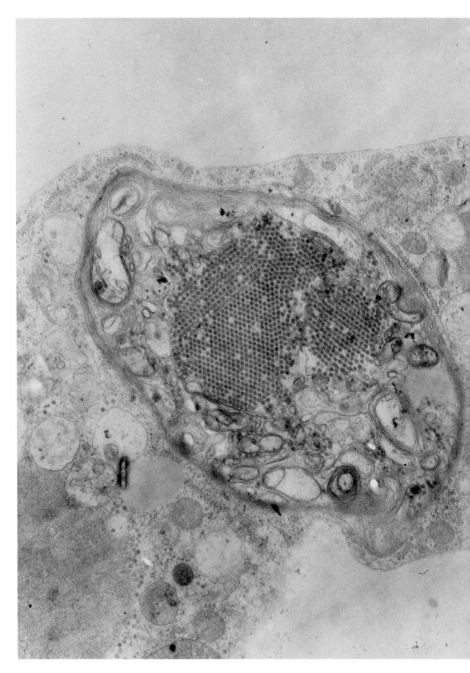

Fig. 3. Cytoplasmic inclusion containing paracrystalline structures. Uninoculated
Aedes albopictus cells ("Webb" cell line). ×19,000.

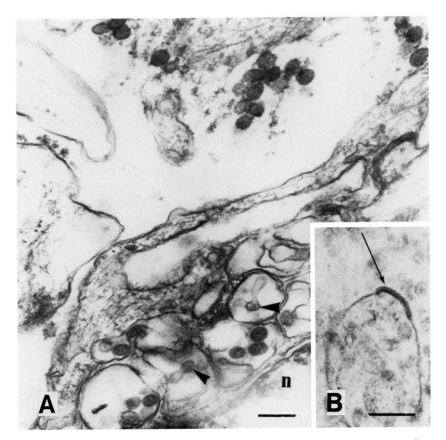

Fig. 4. (A), Part of the vacuolated cytoplasm and the nucleus (n) of an *A. albopictus* cell, 24 hours postinfection, occupy the lower part of the field. Typical viral particles are present within the vacuoles. Possible viral budding forms are indicated by arrow heads. The viral particles in the upper part of the figure appear to have been shed into an extracellular space. Scale, 200 nm. ×42,500. (B) A small section of a degenerating mosquito cell, 48 hours postinfection, in which a cytoplasmic process is visualized as extending into a relatively vacuolated space. There is a slight outpouching of a short segment of the cytoplasmic membrane and the alignment underneath, on the cytoplasmic side, of a closely apposed dense layer (arrow). Scale, 200 nm. ×59,500. (From Lyons and Heyduk, 1973).

cell types described originally by Singh (1967), with the small round cell type (6–20 μm diam.) most prevalent (80–95% of the total), in the experiments carried out by Lyons and Heyduk the ultrastructural changes indicative of CE infection were found predominantly in the small round cell type. Several hundred sections were examined, but the relative susceptibility of the spindle-shaped and binucleated cells could not be ascertained. The end results of CE infection in mice and Vero cells was species death or cell death. In *A. albopictus* cells, an inapparent persistent infection developed. While less than 10% of mosquito cells became productively infected as determined by electron microscopy, the infection caused degeneration of some of the infected cells.

C. FLAVIVIRUS

In vertebrate cells infected with flaviviruses, characteristically there is a massive proliferation of intracytoplasmic membrane. Maturation of virus particles takes place when nucleocapsids bud through the cytoplasmic reticulum. Enveloped particles accumulate within distended cisternae of the endoplasmic reticulum and few free nucleocapsids are seen in the cytoplasm. Virions may be released from the cell when the vesicle migrates to the cell surface or when the cell ruptures (Ota, 1956; Yasuzumi and Tsubo, 1965). Studies of the morphology of Murray Valley encephalitis and Japanese encephalitis viruses growing in a line of *A. aegypti* cells have been reported by Filshie and Reháček (1968). Final maturation was observed at the internal membranes of mosquito cells; incomplete particles and nucleoids were not detected.

D. RHABDOVIRUS

With animal rhabdoviruses, nucleocapsid formation occurs in the cytoplasm and is frequently located near plasma membranes or intracytoplasmic membranes. Once formed, the structures bud through and acquire their outer coat (Knudson, 1973). In studies with regard to development of Cocal virus, vesicular stomatitis subgroup, in *A. albopictus* cells, Artsob and Spence (1974a) found cytoplasmic virus assembly and maturation of virus at intracellular vesicles. Owing to technical difficulties in obtaining sections of intact mosquito cells with well-defined plasma membranes, possible virus maturation at plasma membranes of some cells could not be excluded. While long (B) and short (T) particles (Schaeffer *et al.*, 1969) of Cocal virus could not be ascertained in cell sections of *A. albopictus*, B and T particles were demonstrated by the cell-spread method.

V. Persistence, Attenuation, and Interferon

Mosquitoes, once infected with an arbovirus, presumably remain infected throughout their life span. With regard to arboviruses, their origin and evolution, "it seems almost certain that one is dealing either with

insect-parasites which have become secondarily adapted to living in vertebrates or the other way round. The first alternative seems preferable"
(Andrews, 1973). Unfortunately, results of studies with arboviruses in
invertebrate cell lines, while consistent with those observed in the intact
insects, have not yet rendered this state of parasitism intelligible.

Inapparent persistent infection of mosquito cell cultures may be induced with any arbovirus capable of replication in a given cell system,
such as *A. albopictus* and *A. aegypti* (monolayer; hollow vesicles)
(Singh, 1967), *A. aegypti* (Peleg, 1968), and *A. w-albus* (Singh and Bhat,
1971). As discussed in Section III, individual cell lines, viral strains,
viral passage history, multiplicity of infection (MOI), split ratios, and
temperature of incubation are factors determining the specifics of persistence. Thus, differences in techniques may account for different results
obtained by laboratories in different countries.

Stevens (1970) studied the kinetics with regard to release of the alphavirus Sindbis from persistently infected *A. albopictus* cells. During eight
transfers carried out at weekly intervals, the virus titers were constant,
ranging between 1×10^7 and 10×10^7 PFU/ml. Stevens obtained the
A. albopictus cell line from the Yale Arbovirus Research Unit, New
Haven. The source of Enzmann's (1973) *A. albopictus* cell line is not
stated; the latter investigator found a drop in Sindbis virus titer from
1.0×10^7 PFU 24 hours after inoculation to 1.0×10^1 PFU/0.2 ml at
both 11 and 28 days postinoculation. In the YARU cell line, complete
homologous interference was demonstrated by Stollar and Shenk (1973)
in cultures chronically infected with Sindbis virus; interferon apparently
was ruled out as playing any role in viral interference in mosquito cells.
In contrast, Enzmann (1973) reported that persistently infected *A. albopictus* cells were able to manufacture a small percentage of the amounts
of Sindbis virus produced in *A. albopictus* cells inoculated for the first
time; he also commented that the maintenance of cell–virus equilibrium
in persistently infected cells appeared to be due to the presence of interferonlike antiviral substances. In studies with the alphavirus Semliki
Forest, Davey and Dalgarno (1974) confirmed results reported by Peleg
(1969a,b, 1972) which indicate that persistence is not established or
maintained by interfering substances released into the medium. Davey
and Dalgarno stress the fact that significant amounts of virus were released from about 2% of cells in persistently infected cultures. In studies
with the alphavirus Chikungunya, Ross S27 strain (Buckley *et al.*, 1975),
small-plaque (SP) and large-plaque (LP) variants were obtained by
methods described elsewhere (Buckley, 1973a). Virus stocks of both variants were prepared in cultures of *A. albopictus* cells and used for primary
infection and establishment of persistence. As seen in Fig. 5, the majority

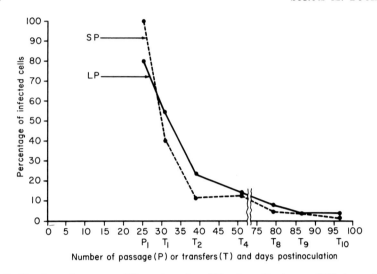

Fig. 5. Number of passage (P) or transfers (T) of small plaque (SP) (-----) and large plaque (LP) (———) variants of Chikungunya virus (Ross S27 strain) and days postinoculation) (Buckley *et al.,* 1975). Reprinted with permission from *Acta Virol. (Prague)* **19,** 10–18.

of cells were infected in the first passage, as determined by infectious center assay (Peleg, 1969a; Libikova and Buckley, 1971). The percentage of persistently infected cells in the carrier cultures decreased rapidly during the initial transfers and then apparently stabilized. Table V shows comparative percentages of cells persistently infected with SP and LP variants in transfer 10 of *A. aegypti* (monolayer), *A. aegypti* (hollow vesicles), *A. w-albus,* and *A. albopictus.* The lowest percentages, 0.6 and 0.8%, were observed in the *A. aegypti* (monolayer) cell line. It should

TABLE V

Comparative Percentages of Cells Persistently Infected with Small Plaque (SP) and Large Plaque (LP) Variants in Transfer 10 of *A. aegypti* **(Monolayer),** *A. aegypti* **(Hollow Vesicles),** *A. w-albus,* **and** *A. albopictus*

Carrier culture	Ratio of infected to uninfected cells		% of infected cells	
	SP	LP	SP	LP
A. aegypti (monolayer)	1/125	1/167	0.8	0.6
A. aegypti (hollow vesicles)	1/50	1/28	2.0	3.6
A. w-albus	1/31	1/40	3.2	2.5
A. albopictus	1/71	1.31	1.4	3.2

be noted here that, in comparison with the *A. albopictus* cell line, the *A. aegypti* (monolayer) cell line is less susceptible to arboviruses. Biochemically, the lipid profile of the *A. aegypti* (monolayer) cell line (Townsend *et al.*, 1972) is distinctly different from the lipid profiles of the *A. aegypti* (hollow vesicles) and *A. albopictus* cell lines (Luukkonen *et al.*, 1973), which are similar. While O'nyong-nyong virus fails to multiply in *A. aegypti* (monolayer) cells, it multiplies in both *A. aegypti* (hollow vesicles) and *A. albopictus* cells (S. M. Buckley, unpublished data). K. Banerjee (personal communication, 1971) found 0.2–2.0% of *A. albopictus* cells persistently infected with Chikungunya virus, strain I 634029, Indian topotype by the infectious center assay; by the fluorescent antibody method, however, 10–15% of the carrier culture cells showed specific cytoplasmic fluorescence. This difference in the numbers of persistently infected cells found may indicate (1) presence of incomplete virus in persistently infected cultures, as recently suggested for Dugbe virus (David-West and Porterfield, 1974), or (2) contamination of the cell line with Chikungunya virus that does not readily induce plaques in Vero cells (see Section VI).

Attenuation during persistence has been reported for alphaviruses Chikungunya (Banerjee and Singh, 1969; Buckley, 1973a) and Semliki Forest (Peleg, 1971; Davey and Dalgarno, 1974); attenuated Chikungunya was characterized by loss of virulence for infant mice, and attenuated Semliki Forest by loss of virulence for adult mice. The attenuated viruses, insofar as tested, were good immunogens and serologically indistinguishable from the parent viruses.

Persistence with flaviviruses (Banerjee and Singh, 1968; Filshie and Reháček, 1968; Reháček, 1972) and the bunyavirus California encephalitis (Whitney and Deibel, 1971; Lyons and Heyduk, 1973) have been described in detail. The establishment of persistence could not be related to release of an interferonlike antiviral substance (Murray and Morahan, 1973; Kascsak and Lyons, 1974). On the basis of conclusive results, the latter authors go so far as to suggest absence of the interferon defense mechanism in arthropods as well as in arthropod cell cultures.

Persistent infection of *A. albopictus* cells has been established with the orbiviruses African horse-sickness (Mirchamsy *et al.*, 1970), Kemerovo (Libikova and Buckley, 1971), and Lebombo, Palyam, and Irituia (Buckley, 1972). Results obtained by infectious center assay and by immunofluorescence of cells persistently infected with Kemerovo virus, strain L75, showed agreement, as documented in Tables VI and VII. Only a small percentage of cells was infected (Fig. 6). Interferon production could not be conclusively demonstrated.

Artsob and Spence (1974b) established a persistent infection with

TABLE VI

Percentage of Cells Containing Infective Virus in *Aedes albopictus* **Cell Cultures Infected Persistently with Kemerovo Virus**[a]

Transfer level of cells	Cells per ml of tested cell suspensions (log)	log PFU (L75)/ml of tested cell suspension	Ratio of infective and noninfective cells	Percentage of infective cells
Transfer 7	7.0	5.6	1/26	3.8
Transfer 8	6.8	5.0	1/64	1.6
Transfer 21	7.7	5.8	1/80	1.7

[a] From Libikova and Buckley (1971). Reprinted with permission from *Acta Virol·* (Prague) **51**, 393–403.

vesicular stomatitis virus (VSV), rhabdovirus taxon, in Singh's *A. aegypti* and *A. albopictus* cell lines. No evidence was found for either interferon production or for antiviral factors in the medium. Actinomycin D increased VSV titers in persistently infected *A. albopictus* cells, but

TABLE VII

Proportion of Fluorescent to Nonfluorescent Cells in *Aedes albopictus* **Cell Line Infected Persistently with Kemerovo Virus, Transfer No. 35**

Microscopic field No. counted[a]	No. of fluorescent cells[b]	Total number of cells counted per microscopic field
1	6 adjacent cells + 1 single	162
2	None	244
3	10 adjacent cells + 2 single	179
4	None	200
5	None	210
6	None	190
7	None	195
8	None	205
9	6 single	217
10	3 adjacent cells	221
10	25 fluorescent cells (1.2%)	2023

[a] Microscopic fields selected at random from one preparation.
[b] Fluorescent cells adjacent to each other, suggesting spread of virus from cell to cell; in some instances single, fluorescent cells present. Most fluorescent cells characterized by granular fluorescence; only few cells with diffuse fluorescence. From Libikova and Buckley (1971). Reprinted with permission from *Acta Virol.* **15**, 393–403.

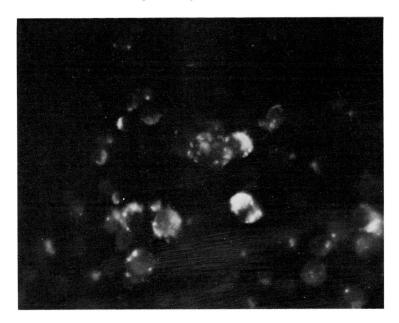

Fig. 6. *Aedes albopictus* cells persistently infected with Kemerovo virus, strain L75, and stained by indirect method of fluorescent antibody technique. Cells showing specific cytoplasmic staining in the form of bright cytoplasmic bodies or diffuse immunofluorescence. ×380.

not in cells freshly infected with VSV. Artsob and Spence state that "the enhancement of VSV titers by 1–2 logs after just 6 hours actinomycin treatment leads to the assumption that VSV is present in persistently infected cells in a partially replicated form." Persistent infection of *A. albopictus* cells with plaque-purified kotonkan, Obodhiang, and Mokola viruses, rhabdovirus taxon, has been established by S. M. Buckley (unpublished data). Table VIII shows percentage of cells containing infective virus in *A. albopictus* cell cultures at different transfer levels.

In conclusion, in persistence, only a small number of cells produce virus productively and in significant amounts. This can be demonstrated impressively by immunofluoresence, where brilliant specific cytoplasmic fluorescence is present in single cells or in foci of from 3 to 10 cells. Generally, with one exception (Enzmann, 1973), interferon is not responsible for the state of persistence, as assayed by methods presently available. Possibly, partial or incomplete virus assembly may be the explanation for the establishment of persistence. One might postulate also that some cell types are capable of productive infection and are cured (by a mechanism not yet known), while a very few others, perhaps of a different

TABLE VIII

Percentage of *Aedes albopictus* Cells Containing Infective Kotonkan, Obodhiang, and Mokola Viruses

Virus	Strain	Transfer level of infected cells	Cells/ml of tested cell suspension ($\times 10^6$)	PFU[a]/ml of tested cell suspension ($\times 10^3$)	Ratio of infective and non-infective cells	Percentage of infective cells
Kotonkan	Ib Ar 23380	4	8.0	1.5	1/5555	0.02
	Ib Ar 23380	9	10.4	225.0	1/46	2.60
Obodhiang	Sud Ar 1154-64	4	5.0	7.5	1/667	0.15
	Sud Ar 1154-64	9	5.2	125.0	1/42	2.41
Mokola	Ib An 27377	4	8.0	19.0	1/420	0.24
	Ib An 27377	7	5.8	550.0	1/11	9.48

[a] PFU = plaque forming units; plaque assay in Vero cells.

cell type, remain as carriers. In the mosquitoes, incidentally, the gut cells are initially infected and are cured; finally neurons and salivary glands remain persistently infected (Marshall, 1973).

VI. Cytopathic Effect and Cell Contamination

Vertebrate cell cultures commonly develop a cytopathic effect (CPE) following inoculation with arboviruses. Addition of specific antibody to cell cultures along with the viral inoculum prevents the development of CPE. Under fluid medium, CPE consists of moderate to marked cell destruction (Scherer and Syverton, 1954; Buckley, 1964; Karabatsos and Buckley, 1967), whereas under nutrient agar overlay, plaque formation is induced (Dulbecco, 1952; Stim 1969). A CPE has been described with arboviruses of six taxons (alphavirus, bunyavirus, flavivirus, iridovirus, orbivirus, and rhabdovirus) in a large number of vertebrate cell systems. With invertebrate cell cultures, CPE is seen only in Singh's (1967) *Aedes albopictus* cell line and predominantly with mosquito-borne viruses such as Japanese B encephalitis, West Nile, and dengue types 1, 2, 3, and 4 (Paul *et al.*, 1969). The observation of plaque formation with Japanese B encephalitis virus in *A. albopictus* cells by Suitor and Paul (1969) was later confirmed and extended to West Nile, dengue, and yellow fever flaviviruses and to vesicular stomatitis virus, type Indiana, a rhabdovirus (Corey and Yunker, 1972). What is puzzling with regard to the CPE of invertebrate cell cultures is that different *A. albopictus* sublines in combination with a variety of arbovirus strains appear to react differently in different laboratories. This suggests that arbovirus-infected *A. albopictus* cells only show CPE "when stressed in a particular way" (Dalgarno and Davey, 1973). Thus, dengue virus, type 2, failed to produce CPE in the hands of Sinarachatanant and Olsen (1973), whereas CPE with the same virus was reported by Paul *et al.* (1968) and confirmed by Suitor and Paul (1969), as well as by Sweet and Unthank (1971). That the CPE caused by the stress situation might be dependent on latent viral contamination came to light when two institutions, the Boyce Thompson Institute, Yonkers (H. Hirumi, personal communication, 1972) and Purdue University, West Lafayette (S. R. Webb, personal communication, 1974) reported spontaneous syncytia formation in two sublines of uninoculated *A. albopictus* cells, now called the "Hirumi" and "Webb" sublines. Arbovirus studies were not being carried out by Drs. Hirumi and Webb at the time the cell line was carried in their laboratories. When the cells were examined by electron microscopy, a cytoplasmic, viruslike contaminant was present in both sublines. It was icosahedral in shape and measured 60–70 nm in diameter in thin sections. Paracrystalline structures (Fig. 7) and occasional budding from cell membranes were observed (findings characteristic of an alphavirus). The Yale Arbovirus Research Unit (YARU), New Haven, was asked to aid in the identification of the contaminating agent by serological means. In the Hirumi cell line, complete extracellular virions were present. They were cytopathic for Vero and BHK-21 cell cultures and were identified serologically by complement fixation (J. Casals, personal communication, 1974) and plaque reduction neutralization tests as Chikungunya (an alphavirus). However, the syncytia formation was not prevented specifically by addition of a Chikungunya hyperimmune serum to cell cultures, YARU subline, along with Hirumi's syncytia-forming agent. In the Webb cell line, isolation of a virus cytopathic in Vero cells was effected only by combining sonication, concentration with Aquacide II, rate zonal centrifugation, and subsequent plating of fractions on Vero cells under nutrient agar overlay (Cunningham *et al.*, 1975). The virus at first produced hazy plaques in Vero cells and, subsequently, CPE in BHK-21 cell cultures; it was identified serologically as Chikungunya virus. Both strains of Chikungunya multiplied in *A. aegypti cells*, but caused neither death nor disease on inoculation into infant and adult mice, nor were they immunogenic in mice.

It has been known since 1968 that *A. albopictus* cells may support the growth of Chikungunya virus without causing CPE (Banerjee and Singh, 1968) and that the virus may become avirulent for mice through

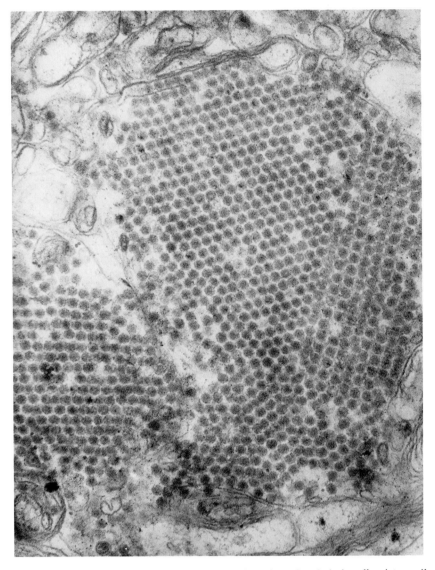

Fig. 7. Cytoplasmic, paracrystalline structures in uninoculated *Aedes albopictus* cells (Webb cell line). ×44,455.

transfers of persistently infected cells (Banerjee and Singh, 1969). In another instance, mouse-pathogenic Chikungunya virus was isolated as a contaminant of mosquito cells (B. H. Sweet, personal communication, 1974). It is assumed in the present cases that Chikungunya virus, representing either an extraneous virus or a virus inherent to the cultured larvae themselves (Sanders, 1973), became cytopathic under the conditions of routine transferring in the respective laboratories. An alternative explanation might be that Chikungunya activated a latent virus, such as the agent first described as "viruslike particles" in the nucleoli of *A. albopictus* cells by Suitor and Paul (1969). H. Hirumi (personal communication, 1974) has observed intranuclear, parvoviruslike structures by electron microscopy; in his opinion, the activation of these intranuclear viruslike particles results in production of polykaryocytes or syncytia formation. The YARU finding that syncytia formation could not be inhibited specifically with a Chikungunya hyperimmune mouse serum adds weight to Hirumi's hypothesis.

VII. Other Taxons and *Toxoplasma gondii* in Diptera Cell Lines

In the arenavirus taxon, Junin virus has been isolated from mites on a few occasions (Mettler, 1969) and Tacaribe virus once from a mixed pool of mosquitoes (Downs *et al.*, 1963). However, evidence to support an arthropod cycle in the maintenance of viruses of the arenavirus taxon is lacking (Casals, 1971). *In vitro*, Junin (Mettler and Buckley, 1973) and Lassa virus (Buckley and Casals, 1970) failed to infect Singh's (1967) *Aedes* cell lines.

Nariva virus, paramyxovirus taxon, is an ether-sensitive agent which has been isolated exclusively from forest rodents in Trinidad. "The demonstrated ether-sensitivity warrants consideration of this agent for inclusion as an arbovirus, but final resolution of this point must await further study" (Tikasingh *et al.*, 1966). Propagation of the virus was attempted in mosquito cell lines. Viral multiplication was not detected by subinoculation of culture fluid into infant mice and BHK-21 cell cultures (Karabatsos *et al.*, 1969).

Nodamura virus, picornavirus taxon, has been proposed as an arbovirus (Murphy *et al.*, 1970). While *in vitro* propagation of Nodamura virus in invertebrate cell cultures has not as yet been reported, a picornavirus has recently been isolated from *Aedes taeniorhynchus* (Wagner *et al.*, 1974).

Cotia virus, poxvirus taxon, is presumbaly transmitted by mosquitoes (Theiler and Downs, 1973). Viral multiplication in mosquito cell lines is, therefore, likely.

The possibility of arthropod transmission of *Toxoplasma gondii* was first suggested by Nicolle and Connor (1913). Their hypothesis gained support in 1937, when Sabin and Olitsky (1937) isolated *T. gondii* from the blood of infected animals. Since then, observations on ectoparasites of animals in nature, as well as experimental attempts to demonstrate arthropod transmission, have been numerous. In transmission studies involving mosquitoes in particular, Jacobs *et al.* (1950) reported negative

TABLE IX

Tests in Mice for Presence of *Toxoplasma gondii* **in Media without Cells in Washed Infected Cell Phases**

Test material	Temperature of incubation (°C)	No. mice with positive peritoneal exudate (Giemsa)	Ascites in mice on days	Tests in mice on days after inoculation
Aedes albopictus				
Medium without cells[a]	35	0/3	—	15
Passage 1	35	3/3	8,8,8	15
Passage 1	35	0/3	—	28
Transfer 1	35	0/3	—	57
Aedes aegypti				
Passage 1	35	3/3	6,6,6	15
Passage 1	35	1/3	9	28
Transfer 1	35	0/3	—	57
Aedes w-albus				
Passage 1	35	3/3	6,6,6	15
Passage 1	35	1/3	8	28
Transfer 1	35	0/3	—	57
Anopheles stephensi				
Medium without cells	29	0/3	—	15
Passage 1	29	0/3	—	15
Passage 1	29	0/3	—	28
Transfer 1	29	0/3	—	57
Culex quinquefasciatus				
Medium without cells	29	0/3	—	15
Passage 1	29	0/3	—	15
Passage 1	29	0/3	—	28
Transfer 1	29	0/3	—	57
Vero				
Medium without cells	35	0/3	—	15
Passage 1	35	3/3	7,7,7	160
Transfer 1	35	3/3	5,6,6	195
Transfer 2	35	3/3	7,7,7	225

[a] The same medium was used for cultures of all three *Aedes* cell lines. From Buckley (1973c).

results with *Culex quinquefasciatus*, as did Weyer (1951) with *Aedes aegypti* and *Anopheles stephensi*.

Attempts have been made (Buckley, 1973c) to propagate *T. gondii* (strain RH, kindly supplied by Professor David Weinmann, Yale University) in cultures of five different mosquito cell lines—*Aedes albopictus*, *A. aegypti* (hollow vesicles) (Singh, 1967), *A. w-albus* (Singh and Bhat, 1971), *Anopheles stephensi* (Schneider, 1969), and *C. quinquefasciatus* (Hsu *et al.*, 1970)—as well as in cultures of the Vero cell line which served as a vertebrate control. Under the experimental conditions, proliferative forms of *T. gondii* survived but did not multiply in three *Aedes* cell lines capable of growing at 35°C: *Aedes albopictus*, *A. aegypti* (hollow vesicles), and *A. w-albus*. Results were negative for *C. quinquefasciatus* and *Anopheles stephensi* cell cultures, which had to be maintained at 29°C. In Vero cells, a continuous infection was established during the first passage and maintained through two consecutive cell transfers of persistently infected cells. The results are summarized in Table IX.

In conclusion, isolation of an agent from an ectoparasite does not necessarily incriminate it as a vector.

Acknowledgments

I am grateful to Dr. J. Casals, Dr. R. E. Shope, and Anne Newbery for their helpful advice and critical review of the manuscript. Some of the samples for illustration were kindly provided by Dr. S. R. Webb and Dr. J. D. Paschke. The skillful technical assistance of Mr. C. Mullen, Mrs. E. Gilson, and Mrs. M. Malhoit is gratefully acknowledged. Immunofluorescence microphotographs were kindly supplied by G. Martine. I am indebted to Mrs. Carmel Bierwirth for her competent assistance with the typescript.

Previously unpublished studies of the author were supported by the U.S. Army Medical Research and Development Command (DADA-17-12-C-2170), National Institute of Allergy and Infectious Diseases (PHS-RO-1-AI 10984), The World Health Organization, and The Rockefeller Foundation.

References

Adamcova-Otova, B., and Marhoul, Z. (1974). *Acta Virol. (Prague)* **18**, 158–160.
Aitken, T. H. G., Woodall, J. P., Andrade, A. H. P. de Bensabath, G., and Shope, R. E. (1975). *Amer. J. Trop. Med. Hyg.* **24**, 358–368.
Andrews, C. (1973). *In* "Viruses and Invertebrates" (A. J. Gibbs, ed.), pp. 1–13. Amer. Elsevier, New York.
Artsob, H., and Spence, L. (1974a). *Can. J. Microbiol.* **20**, 329–336.
Artsob, H., and Spence, L. (1974b). *Acta Virol. (Prague)* **18**, 331–340.
Baker, S. T. (1973). Doctor of Medicine Thesis, Yale University School of Medicine, New Haven, Connecticut (unpublished).
Banerjee, K., and Singh, K. R. P. (1968). *Indian J. Med. Res.* **56**, 812–814.

Banerjee, K., and Singh, K. R. P. (1969). *Indian J. Med. Res.* **57**, 1003–1005.
Bhat, U. K. M., and Singh, K. R. P. (1970). *Curr. Sci.* **39**, 388.
Borden, E. C., Shope, R. E., and Murphy, F. A. (1971.) *J. Gen. Virol.* **13**, 261–271.
Brown, D. T., and Gliedman, J. B. (1973). *J. Virol.* **12**, 1534–1539.
Brown, F., and Tinsley, T. W. (1973). *J. Gen. Virol.* **20**, Suppl., 1–130.
Buckley, S. M. (1964). *Proc. Soc. Exp. Biol. Med.* **116**, 354–358.
Buckley, S. M. (1969). *Proc. Soc. Exp. Biol. Med.* **131**, 625–630.
Buckley, S. M., (1971a). *Trans. Roy. Soc. Trop. Med. Hyg.* **65**, 102.
Buckley, S. M. (1971b). *Curr. Top. Microbiol. Immunol.* **55**, 133–137.
Buckley, S. M. (1971c). *Proc. Int. Symp. Tick-borne Arboviruses (Excluding Group B), 1969* pp. 43–52.
Buckley, S. M. (1972). *J. Med. Entomol.* **9**, 168–170.
Buckley, S. M. (1973a). *Proc. Int. Colloq. Invertebr. Tissue Cult., 3rd, 1971* pp. 307–324.
Buckley, S. M. (1973b). *Appl. Microbiol.* **25**, 695–696.
Buckley, S. M. (1973c). *Exp. Parasitol.* **33**, 23–26.
Buckley, S. M., and Casals, J. (1970). *Amer. J. Trop. Med. Hyg.* **19**, 680–691.
Buckley, S. M., Singh, K. R. P., and Bhat, U. K. M. (1975). *Acta Virol. (Prague)* **19**, 10–18.
Casals, J. (1971). *In* "Comparative Virology" (K. Maramorosch and E. Kurstak, eds.), pp. 307–333. Academic Press, New York.
Corey, J., and Yunker, C. E. (1972). *Acta Virol. (Prague)* **16**, 90.
Cunningham, A., Webb, S. R., Buckley, S. M., and Casals, J. (1975). *J. Gen. Virol.* **27**, 97–100.
Dalgarno, K., and Davey, M. W. (1973). *In* "Viruses and Invertebrates" (A. J. Gibbs, ed.), pp. 245–270. Amer. Elsevier, New York.
Davey, M. W., and Dalgarno, L. (1974). *J. Gen. Virol.* **24**, 1–11.
Davey, M. W., Dennett, D. P., and Dalgarno, L. (1973). *J. Gen. Virol.* **20**, 225–232.
David-West, T. S., and Porterfield, J. S. (1974). *J. Gen. Virol.* **23**, 297–307.
DeTray, D. E. (1963). *Advan. Vet. Sci.* **8**, 299–333.
Downs, W. G., Anderson, C. R., Spence, L., Aitken, T. H. G., and Greenhall, A. H. (1963). *Amer. J. Trop. Med. Hyg.* **12**, 640–646.
Dulbecco, R. (1952). *Proc. Nat. Acad. Sci. U.S.* **38**, 747–752.
Enzmann, P. J. (1973). *Arch. Gesamte Virusforsch.* **41**, 382–389.
Filshie, B. K., and Reháček, J. (1968). *Virology* **34**, 435–443.
Grace, T. D. C. (1966). *Nature (London)* **211**, 366–367.
Grimley, P. M., Berczesky, I. M., and Friedman, R. M. (1968). *J. Gen. Virol.* **2**, 1326–1338.
Hess, W. R. (1971). *In* "Virology Monographs" (S. Gard, C. Hallauer, and K. F. Meyer, eds.), No. 9, pp. 1–33. Springer-Verlag, Berlin and New York.
Heuschele, W. P., and Coggins, L. (1965). *Proc. U.S. Livestock Sanit. Ass.* **69**, 94–100.
Hink, W. F. (1972). *In* "Invertebrate Tissue Culture" (C. Vago, ed.), Vol. 2, pp. 363–387. Academic Press, New York.
Hsu, S. H., Mao, W. H., and Cross, J. H. (1970). *J. Med. Entomol.* **7**, 703.
Igarashi, A., Sasao, F., Wungkobkiat, S., and Fukai, K. (1973). *Biken J.* **16**, 17–23.
Jacobs, L., Woke, P., and Jones, F. E. (1950). *J. Parasitol.* **36**, Suppl., 36–38.
Karabatsos, N., and Buckley, S. M. (1967). *Amer. J. Trop. Med. Hyg.* **16**, 99–105.
Karabatsos, N., Buckley, S. M., and Ardoin, P. (1969). *Proc. Soc. Exp. Biol. Med.* **130**, 888–892.
Kascsak, R. J., and Lyons, M. J. (1974). *Arch. Gesamte Virusforsch.* **44**, 1–6.

Kemp, G. E., Lee, V. H., Moore, D. L., Shope, R. E., Causey, O. R., and Murphy, F. A. (1973). *Amer. J. Epidemiol.* **98**, 43–49.

Knudson, D. L. (1973). *J. Gen. Virol.* **20**, Suppl., 105–130.

Libikova, H., and Buckley, S. M. (1971). *Acta Virol.* (*Prague*) **15**, 393–403.

Luukkonen, A., Brummer-Korvenkontio, M., and Renkonen, O. (1973). *Biochim. Biophys. Acta* **326**, 256–261.

Lyons, M. J., and Heyduk, J. (1973). *Virology* **54**, 37–52.

Marhoul, Z. (1973). *Acta Virol.* **17**, 507–509.

Marshall, I. D. (1973). *In* "Viruses and Invertebrates" (A. J. Gibbs, ed.), pp. 406–427. Amer. Elsevier, New York.

Mettler, N. (1969). *Pan. Amer. Health Org. Sci. Publ.* **183**, 1–55.

Mettler, N. E., and Buckley, S. M. (1973). *Proc. Int. Colloq. Invertebr. Tissue Cult., 3rd, 1971* pp. 255–265.

Mirchamsy, H., Hazrati, A., Bahrami, S., and Shafyi, A. (1970). *J. Vet. Res.* **31**, 1755–1761.

Mitsuhashi, J., and Maramorosch, K. (1964). *Contrib. Boyce Thompson Inst.* **22**, 435–460.

Mugo, W. N., and Shope, R. E. (1972). *Trans. Roy. Soc. Trop. Med. Hyg.* **66**, 300–304.

Murphy, F. A., Whitfield, S. G., Coleman, P. H., Calisher, C. H., Rabin, E. R., Jenson, A. B., Melnick, J. L., Edwards, M. R., and Whitney, E. (1968). *Exp. Mol. Pathol.* **9**, 44–56.

Murphy, F. A., Scherer, W. F., Harrison, A. K., Dunne, H. W., and Gary, G. W., Jr. (1970). *Virology* **40**, 1008–1021.

Murray, A. M., and Morahan, P. S. (1973). *Proc. Soc. Exp. Biol. Med.* **142**, 11–15.

Nicolle, C., and Connor, M. (1913). *Bull. Soc. Pathol. Exot.* **6**, 160–165.

Ota, Z. (1965). *Virology* **25**, 372–378.

Paterson, H. E., and McIntosh, B. M. (1964). *Ann. Trop. Med. Parasitol.* **58**, 52.

Paul, S. D., Singh, K. R. P., and Bhat, U. K. M. (1969). *Indian J. Med. Res.* **57**, 339–348.

Peleg, J. (1968). *Virology* **35**, 617–619.

Peleg, J. (1969a). *J. Gen. Virol.* **5**, 463–471.

Peleg, J. (1969b). *Nature* (*London*) **221**, 193–194.

Peleg, J. (1971). *Curr. Top. Microbiol. Immunol.* **55**, 155–161.

Peleg, J. (1972). *Arch. Gesamte Virusforsch.* **37**, 54–61.

Plowright, W. C. T., Perry, C. T., Pierce, M. A., and Parker, J. (1970). *Arch. Gesamte Virusforsch.* **31**, 33–50.

Porterfield, J. S., Casals, J., Chumakov, M. P., Gaidamovich, S. Y., Hannoun, C., Holmes, I. H. Horzinek, M., Mussgay, M., and Russell, P. K. (1973/1974). *Intervirology* **2**, 270–272.

Raghow, R. S., Davey, M. W., and Dalgarno, L. (1973a). *Arch. Gestamte Virusforsch.* **43**, 165–168.

Raghow, R. S., Grace, T. D. C., Filshie, B. K., Bartley, W., and Dalgarno, L. (1973b). *J. Gen. Virol.* **21**, 109–122.

Rao, T. R. (1964). *Indian J. Med. Res.* **52**, 719.

Reháček, J. (1972). *In* "Invertebrate Tissue Culture" (C. Vago, ed.), Vol. 2, pp. 279–320. Academic Press, New York.

Ross, R. W. (1965). *J. Hyg.* **54**, 177–191.

Sabin, A. B., and Olitsky, P. K. (1973). *Science* **85**, 336–338.

Sanders, F. K. (1973). *In* "Contamination in Tissue Culture" (J. Fogh, ed.), pp. 243–256. Academic Press, New York.

Schaeffer, F. L., Hackett, A. J., and Soergel, M. E. (1969). *Fed. Proc., Fed. Amer. Soc. Exp. Biol.* **28**, 1867–1874.
Scherer, W. F., and Syverton, J. T. (1954). *Amer. J. Pathol.* **30**, 1075.
Schmidt, J. R., Williams, M. C., Lulu, M., Mivule, A., and Mujombe, E. (1965). *East Afr. Virus Res. Inst. Rep.* **15**, 24.
Schneider, I. (1969). *J. Cell Biol.* **42**, 603–606.
Shope, R. E. (1975). *In* "The Natural History of Rabies" (G. M. Baer, ed.), Academic Press, New York.
Shope, R. E., Murphy, F. A., Harrison, A. K., Causey, O. R., Kemp, G. E., Simpson, D. I. H., and Moore, D. L. (1970). *J. Virol.* **6**, 690–692.
Sinarachatanant, P., and Olsen, L. C. (1973). *J. Virol.* **12**, 275–283.
Singh, K. R. P. (1967). *Curr. Sci.* **36**, 506–508.
Singh, K. R. P. (1971). *Curr. Top. Microbiol. Immunol.* **55**, 127–133.
Singh, K. R. P. (1972). *Advan. Virus Res.* **17**, 187–206.
Singh, K. R. P., and Bhat, U. K. M. (1971). *Experientia* **27**, 142–143.
Singh, K. R. P., and Paul, S. D. (1968). *Curr. Sci.* **37**, 65–67.
Stevens, T. M. (1970). *Proc. Soc. Exp. Biol. Med.* **134**, 356–361.
Stim, T. B. (1969). *J. Gen. Virol.* **5**, 329–338.
Stollar, V., and Shenk, T. E. (1973). *J. Virol.* **11**, 592–595.
Suitor, E. C., and Paul, F. J. (1969). *Virology* **38**, 482–485.
Sweet, B. H., and Unthank, H. D. (1971). *Curr. Top. Microbiol. Immunol.* **55**, 150–154.
Taylor, R. M. (1967). "Catalogue of Arthropod-borne Viruses of the World," Pub. Health Serv. Publ. No. 1760. US Govt. Printing Office, Washington, D.C.
Theiler, M. (1957). *Proc. Soc. Exp. Biol. Med.* **96**, 380–382.
Theiler, M., and Downs, W. G. (1973). "The Arthropod-borne Viruses of Vertebrates." Yale Univ. Press, New Haven, Connecticut.
Tikasingh, E. S., Jonkers, A. H., Spence, K., and Aitken, T. H. G. (1966). *Amer. J. Trop. Med. Hyg.* **15**, 235–238.
Townsend, D., Jenkin, H. M., and Yang, T. K. (1972). *Biochim. Biophys. Acta* **260**, 20–25.
Wagner, G. W., Webb, S. R., Paschke, J. D., and Campbell, W. R. (1974). *J. Invertebr. Pathol.* **24**, 380–382.
Wagner, R. R. (1973). *Amer. Soc. Microbiol. News* **39**, 355–356.
Weyer, F. (1951). *Z. Tropenmed. Parasitol.* **3**, 65–72.
Whitfield, S. G., Murphy, F. A., and Sudia, L. D. (1971). *Virology* **43**, 110–112.
Whitney, E., and Deibel, R. (1971). *Curr. Top. Microbiol. Immunol.* **55**, 138–139.
Willis, N. G., and Campbell, J. B. (1973). *Proc. Int. Colloq. Invertebr. Tissue Cult.*, pp. 347–366. Slovak Academy of Sciences, Bratislava.
World Health Organization, Scientific Group. (1967). *World Health Organ., Tech. Rep. Ser.* **369**.
Yasuzumi, G., and Tsubo, I. (1965). *U. Ultrastruct. Res.* **12**, 304–316.
Yunker, C. E. (1971). *Curr. Top. Microbiol. Immunol.* **55**, 113–126.
Yunker, C. E., and Corey, J. (1969). *J. Virol.* **3**, 631–632.

13

Viral, Microbial, and Extrinsic Cell Contamination of Insect Cell Cultures

HIROYUKI HIRUMI

I. Introduction

Various viruses, bacteria, mycoplasmas, protozoa, and extrinsic cells are known to be responsible for the contamination of cell and tissue cultures. Many bacteria and fungi cause turbidity in culture media and/or destructive effects on cultured cells. They can be readily detected by direct inspection (Fogh *et al.*, 1971). However, many viral and mycoplasmal contaminants, as well as antibiotic-suppressed bacterial and fungal contaminants, often cause chronic or latent infection and are difficult to detect in cultures. Certain foreign cells are also compatible with some cell cultures. In many instances, the latent contaminants alter the morphology and growth of cells to some extent; the contaminants also interact with experimentally introduced viruses. The existence of these contaminants has often accounted for incorrect interpretation of both diagnostic and experimental results. During the past two decades, extensive studies of the contamination in vertebrate tissue cultures have been made by a number of investigators. The reader is referred to recent review articles on this subject (Brown and Officer, 1968; Fogh *et al.*, 1971; Barile, 1973a; Coriell, 1973a; Fogh, 1973; Kenny, 1973; Ludovici and Holmgren, 1973; Sanders, 1973; Stulberg, 1973).

Invertebrate tissue culture techniques, particularly of insects, have
been rapidly developed during the last 15 years. Many successful primary
cultures have been obtained and cell lines from various species established. These culture systems are presently being employed in many laboratories in studying various disciplines of the biosciences. In the past,
most laboratories which reared insects under nonaseptic conditions for
tissue culture work encountered bacterial and fungal contaminations. The
contaminants generally originated from microbes which were present on
the surface and/or in the digestive organs of the insects. The problems,
however, were overcome by application of antibiotics, surface sterilization, and/or aseptic rearing of the insects (Schneider, 1967; Vago, 1967;
Brooks and Kurtti, 1971; Hirumi and Maramorosch, 1971).

Although some attention has been given to latent viral infection of
certain mosquito cell cultures, information concerning the contamination
of other insect tissue cultures is presently sparse. This chapter is not in-
tended to be a comprehensive review but emphasizes recent findings of
viral, microbial, and extrinsic cell contamination of insect cell cultures,
including unpublished data obtained in this laboratory. Similar incidences
may occur in many other invertebrate tissue cultures. It is hoped that
this chapter will make workers aware of the problems caused by latent
contaminants in tissue culture systems. The reader who is interested in
more details on technical procedures for detection, identification, and prevention of contaminants should consult recent technical articles on the
subject (Barile, 1973b; Coriell, 1973b; De Harven, 1973; Gartler and
Farber, 1973; Greene and Charney, 1973a; Hayflick, 1973; House, 1973;
Stulberg and Simpson, 1973).

II. Viral Contamination

A. POSSIBLE SOURCES

Many extraneous viruses in vertebrate cell cultures are known to originate from (a) medium ingredients such as trypsin and sera, (b) original
tissue explants used to initiate the cultures, or (c) accidental introduction
by investigators (Sanders, 1973). All these sources of viral contamination
can be expected in insect cell culture systems.

Trypsin, suggested to be a source of porcine parvovirus contamination
in human cell lines (Hallauer et al., 1971), has been commonly used to
prepare most insect cell cultures (Schneider, 1967; Vago, 1967; Hirumi
and Maramorosch, 1971). Fetal bovine serum, which is used as a sup-

plement of most insect cell culture media, is often contaminated with various types of viruses (Molander et al., 1968, 1971; Merril et al., 1972; Chu et al., 1973; Kniazeff, 1973; Ludovici and Holmgren, 1973).

Many insects act as vectors of animal and plant pathogenic viruses which persist in vector cells. A large number of viruses are also known to be pathogenic to the host insects. Furthermore, various types of virus-like particles whose biological nature is still uncertain have been detected in aphids (Moericke, 1963; Parrish and Briggs, 1966), Drosophila (Kernaghan et al., 1964; Akai et al., 1967; Kernaghan and Ehrman, 1968; Perotti and Bairati, 1968; Rae and Green, 1968; Philpott et al., 1969; Felluga et al., 1971; Gartner, 1971, 1972; Herman et al., 1971; Miquel et al., 1972; Tandler, 1972), leafhoppers (Lee, 1965; Herold and Munz, 1967), planthoppers (Ammar et al., 1970), mosquitoes (Scherer and Hurlbut, 1967; Lebedeva et al., 1973; Pudney et al., 1973; Richardson et al., 1974; Wagner et al., 1974), and chironomids (Federici et al., 1974). Since most insect cell cultures have been derived from batches of minced embryos or larvae, vertically transmitted viruses or virus-like particles could have been introduced to the cell cultures by being present in the source material. Also, accidental contamination with a virus should not be overlooked. This is particularly true in a laboratory using mosquito cells and various types of arthropod-borne vertebrate pathogenic viruses (arboviruses).

B. INCIDENCE IN FRUIT FLY (Drosophila) CELLS

Akai et al. (1967) observed virus-like particles in both the nuclei and cytoplasm of Drosophila melanogaster "cell lines" derived from imaginal discs which were cultured in vivo in the abdomen of Drosophila virilis for nearly 1 year. The particles were slightly elliptical in shape (approx. 35 × 45 nm) as determined by means of thin section electron microscopy.

Wehman and Brager (1971) cultured the wing imaginal discs, derived from late instar larvae of D. melanogaster, in Schneider's Drosophila medium, using hanging-drop culture techniques. Slightly elliptical virus-like particles (having a 38.6-nm mean major axis and 32.2-nm mean minor axis) were detected in the nuclei of discs after 7 days in culture. No virus-like particles were seen in 3-day-old cultures. Highly electron-opaque particles appeared in the nucleus in 5-day-old cultures, but these lacked clear definition. The virus-like particles did not increase in number in the discs after 8–10 days in culture.

Virus-like particles also were observed in sublines of Schneider's (1972) D. melanogaster cell lines 1, 2, and 3 (Williamson and Kernaghan, 1972). The spherical virus-like particles (approx. 43 nm diam.) were seen in both nuclei and cytoplasm of the cultured cells. Recently, similar types of virus-like particles were detected by means of electron microscopy in the nuclei (Fig. 1) and cytoplasm (Fig. 2) of the same *Drosophila* cell line received directly from Schneider's laboratory (H. Hirumi, unpublished data, 1975). In addition, two other types of intracytoplasmic particles were seen (Figs. 3 and 4). Although the nature of these virus-like particles is still uncertain, the presence of the intranuclear particles may confound studies on the nucleic acids in which *Drosophila* cells are employed (Williamson and Kernaghan, 1972).

The origin of these particles has not been established. However, the intranuclear virus-like particles came most likely from the initial tissue explants used to prepare the primary cultures since a widespread occurrence of similar particles has been demonstrated in the following organs of *Drosophila:* the Malphigian tubule and muscle cells (Kernaghan *et al.*, 1964), larval imaginal discs, brain tumor and adult gut cells (Akai *et al.*, 1967), the midgut of larvae (Filshie *et al.*, 1967), melanotic tumors (Perotti and Bairati, 1968), adult trachea, midgut, connective tissue, paragonia and nurse cells of the ovary (Rae and Green, 1968), oenocytes and glial cells of young and old adults (Philpott *et al.*, 1969), fat body and tracheolar cells of larvae (Felluga *et al.*, 1971), brain nerve cells (Herman *et al.*, 1971), adult midgut cells (Gartner, 1971, 1972), salivary glands, accessory glands, and muscles and nerves (Miquel *et al.*, 1972) of *D. melanogaster.* Similar particles have also been found in the testis of *D. virilis* (Tandler, 1972).

C. INCIDENCE IN MOSQUITO CELLS

During the past decade, a number of continuous cell lines have been established from many species of mosquitoes (Grace, 1966; Singh, 1967; Peleg, 1968, 1969; Varma and Pudney, 1969; Schneider, 1969, 1971, 1973; Hsu *et al.*, 1970; Kitamura, 1970; Bhat and Singh, 1970; Sweet and McHale, 1970; Pudney and Varma, 1971; Bhat, 1973; Bhat and Guru, 1973). Their primary application is in the study of arthropod-borne animal-pathogenic viruses (Buckley, 1971; Yunker, 1971; Peleg, 1971, 1972; Singh, 1971, 1972; Singh *et al.*, 1973; Grace, 1973; Raghow *et al.*, 1973a,b; Stollar *et al.*, 1973; Yunker and Cory, 1975). The presence of virus contaminants in mosquito cell lines presents a hazard to the study of arboviruses. Recently, viral contamination of Singh's *Aedes albopictus* TCA-15 cell line received attention in several laboratories, and this suggests that contamination is a common occurrence among its sublines (Hirumi, 1974; Hirumi *et al.*, 1975; Cunningham *et al.*, 1975). The existence of virus-like particles in other cell lines also has been reported.

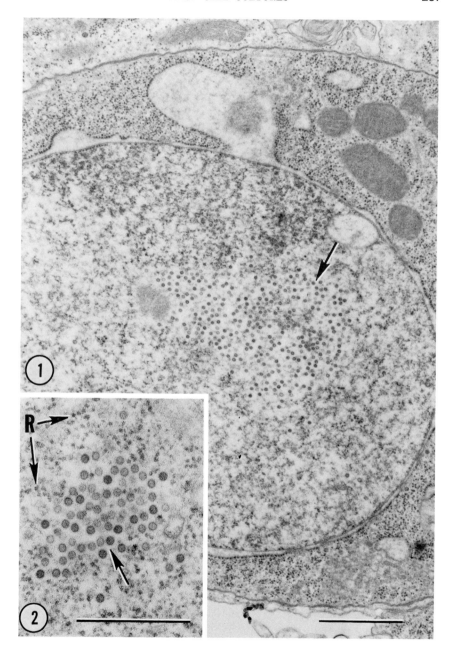

Figs. 1 and 2. Virus-like particles (approx. 42–45 nm diam.) in Schneider's *Drosophila melanogaster* cells. Fig. 1. Intranuclear particles (arrow). ×22,500. Fig. 2. Intra-cytoplasmic particles (arrow). R, ribosomes. ×62,000.

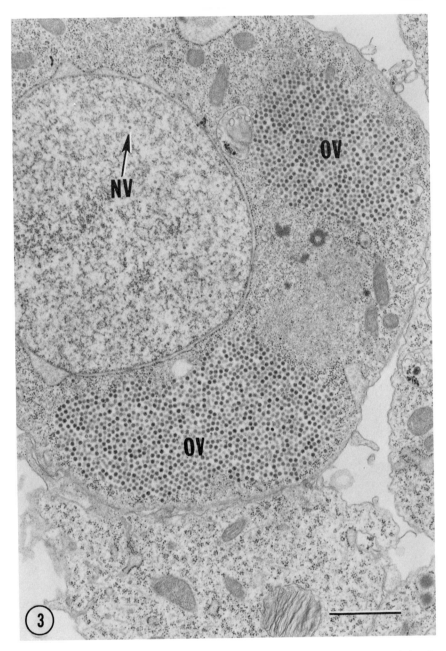

Fig. 3. Intracytoplasmic virus-like particles (OV) (approx. 60–65 nm diam.) in the *D. melanogaster* cell. The particles are morphologically similar to the orbi-type virus-like particles seen in Singh's *Aedes albopictus* cell line (see Fig 11). Note scattered intranuclear virus-like particles (NV). ×19,000.

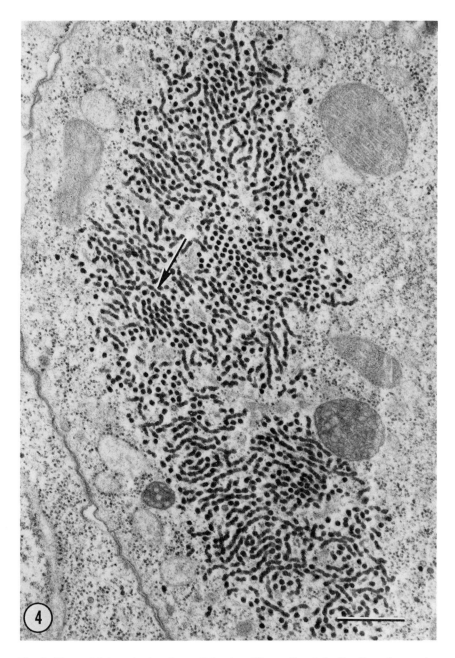

Fig. 4. Unusual intracytoplasmic particles (av. 38 nm diam.) in the *D. melanogaster* cell. The particles appear in a beaded form (arrow). ×35,500.

1. Singh's Aedes albopictus Cell Line

The *A. albopictus* cell line was established by Singh (1967) from a large number of minced larvae. A culture of the cell line was brought to the Yale Arbovirus Research Unit (YARU), Yale University, and subsequently distributed to many laboratories. This cell line has become valuable because of its wide range of susceptibility to arboviruses. Furthermore, its ability to induce cytopathic effects (CPE), including plaque induction, with certain arboviruses has made this cell line a useful tool for quantitative assay of the viruses (Paul *et al.*, 1969; Suitor, 1969; Suitor and Paul, 1969; Buckley, 1971; Yunker, 1971; Singh, 1971, 1972; Cory and Yunker, 1972; Davey *et al.*, 1973; Djinawi and Olson, 1973; Raghow *et al.*, 1973a,b).

Suitor and Paul (1969) first suggested a possible virus contamination of a subline of the *A. albopictus* culture. Subsequently the presence of a Chikungunya-like virus in this subline was detected by means of serology (Sweet, 1970). Spontaneous syncytium formation was observed independently in another subline of *A. albopictus* culture, and its viral nature was suspected (Hirumi, 1974).

Recently, Hirumi *et al.* (1975) examined the syncytium-associated subline, Suitor's subline, and a presumably virus-free subline. These will be referred to as sublines (SL) 1, 2, and 3, respectively. The *A. albopictus* syncytium-inducing agents, ALSA1 and ALSA2, were extracted from SL1 and SL2 cells, respectively. They induced syncytia in SL3 cells (Fig. 5) at 22°, 27°, 28°, 29°, and 32°C, but not at 37°C. The optimum temperature for syncytium induction was 27°C. No syncytia were induced in other cells tested, such as cells of African green monkey kidney (Vero); the cabbage looper (Lepidoptera), *Trichoplusia ni*, TN-368 (Hink, 1970); leafhopper (Homoptera), *Agallia constricta*, AC-20 (Chiu and Black, 1967); and mosquito (Diptera), *Aedes aegypti* (Singh, 1967), *Aedes w-albus* (Singh and Bhat, 1971), *Anopheles gambiae* (M. G. R. Varma, unpublished data, 1974), and *Culex quinquefasciatus* (Hsu *et al.*, 1970).

ALSA1 also produced plaques in SL3 and Vero cells. The plaque-forming ability coincided with the syncytium-inducing ability up to the ninth passage in SL3 cells. However, no plaques were formed thereafter, although ALSA1 still induced syncytia in SL3 cells. Incubation periods of the syncytium induction in SL3 cells varied from 18 hours to 15 days after inoculation, depending on the concentration of ALSA1 inocula. When the cells were moderately infected, the cultures often recovered from the polykaryogenesis, increasing the number of morphologically normal cells. These cultures remained healthy through a number of subculti-

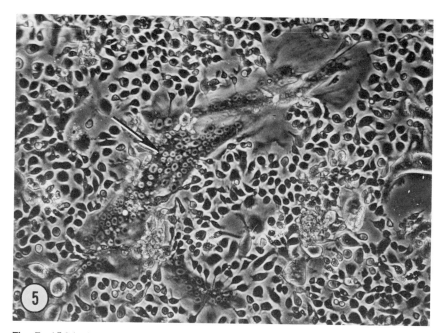

Fig. 5. ALSA1-inoculated *A. albopictus* SL3 cell cultures, forming syncytia (arrow), 4 days postinoculation. Phase contrast. ×200.

vations until an unknown stimulus reactivated the polykaryogenesis. This finding indicates that the contamination may be easily overlooked if cultures are infected with a low dose of the agent and are frequently subcultured.

A complement-fixation (CF) test of ALSA1 antigens showed a clear-cut reaction of the antigens with a Chikungunya (an alphavirus of the Togaviridae) hyperimmune mouse serum (Ross strain) that had been prepared at YARU in 1964. A subsequent CF cross-box titration test with newly prepared Chikungunya and O'nyong-nyong (an alphavirus) antigens and their hyperimmune mouse serum and ascitic fluid demonstrated that ALSA1 was related to the two alphaviruses. However, there was only a one-way reaction between ALSA1 and the other two viruses. Thus, the syncytial agent is definitely separable from Chikungunya and O'nyong-nyong viruses although it is serologically related to the alphaviruses.

Electron microscopy examinations of the syncytia induced by ALSA1 revealed the presence of five different types of virus-like particles during early passages (up to passage 4) in SL3 cells (Table I). These particles

TABLE I

Virus-Like Particles Detected in Singh's *Aedes albopictus* and *A. aegypti* Cells Which Are Associated with Syncytium Formation[a]

Capsid sym- metry	Size (nm)	Envelope	Location	*A. albopictus* SL3 cells[b] inoculated with		*A. aegypti* cells	Morphologic type
				ALSA1[c]	ALSA2[d]		
Isometric	18–20	Naked	Nucleus	+	+	–	Parvovirus
Isometric	18–20	Naked	Cytoplasm	+	–	–	Picornavirus
Isometric	60–65	Naked	Cytoplasm	+	++	–	Orbivirus
Isometric	50–55	Enveloped	Cytoplasm	+	++	+	Togavirus
Isometric	55–60	Naked (tail 140)	Cytoplasmic vacuoles	+	–	–	Bacteriophage

[a] Results obtained by means of thin section electron microscopy.
[b] *Aedes albopictus* subline, presumably free of viral contamination.
[c] Extracts of the *A. albopictus* subline SL1 cells, associated with syncytium formation.
[d] Extracts of the *A. albopictus* subline SL2 cells, associated with syncytium formation.

morphologically resembled the toga-, parvo-, orbi-, picorna-, and bacterial viruses. Among those, the toga-, parvo- and orbi-type agents were consistently seen in all materials examined up to the eleventh passage. The same three types of the particles were also consistently detected in syncytia induced by ALSA2 in SL3 cells (Table I). The findings clearly indicated that ALSA1 and ALSA2 were not a single viral agent.

Virions of alphaviruses are assembled on the host cell surface, budding from the plasma membranes, whereas flavivirus virions bud from intracytoplasmic membranes (Horzinek, 1973). In ALSA1-induced syncytia, the togavirus-like particles (having an envelope and approximately 50–55 nm diam.) appeared within the cisternae of the endoplasmic reticulum (Fig. 6). No budding forms were observed in the syncytia. However, prior to syncytium formation, numerous togavirus-like particles, budding from the cell surface membranes, were seen in ALSA1-inoculated cultures (Fig. 7) (Hirumi et al., 1975). A 25% chloroform treatment of ALSA1 completely inactivated the syncytium-inducing ability as well as propagation of the togavirus-type agent, while the treatment did not inhibit replication of the parvo- and orbi-type particles (H. Hirumi, unpublished data, 1975).

Based on the serological and electron microscopic observation, Hirumi et al. (1975) suggested that the togavirus-like agent may be responsible for the syncytium formation in the A. albopictus cells and that the agent most likely belongs to the alphavirus group. It was also pointed out that the agent could not be identified as "Chikungunya virus" because (a) all alphaviruses tested, including Chikungunya virus, failed to induce syncytia in Singh's A. albopictus cells (Buckley, 1971; Raghow et al., 1973a,b) in contrast to certain flaviviruses and Bunyamwera virus (Suitor and Paul, 1969; Paul et al., 1969; Yunker, 1971; Singh, 1971, 1972; Djinawi and Olson, 1973), and (b) only a one-way reaction occurred with Chikungunya virus as tested with the CF cross-box titration.

The origin and identification of the togavirus-like agent is unresolved. High incidence of bovine diarrhea virus infection in cattle has been reported (Kniazeff, 1973). The virus was often isolated from fetal bovine sera (Fogh et al., 1971; Molander et al., 1971). Although the bovine diarrhea virus has not been taxonomically classified, it may belong to the flavivirus group (Fraenkel-Conrat, 1974). The A. albopictus cells have been carried in culture media supplemented with fetal bovine sera since the cell line was established in 1967. Thus, it may be worthwhile to examine the serological relationship between the bovine diarrhea virus and the togavirus-like agent detected in the mosquito cells.

The parvovirus-like particles (approximately 18–20 nm diam.) were seen in nuclei of the ALSA1- and ALSA2-infected A. albopictus cells (Fig.

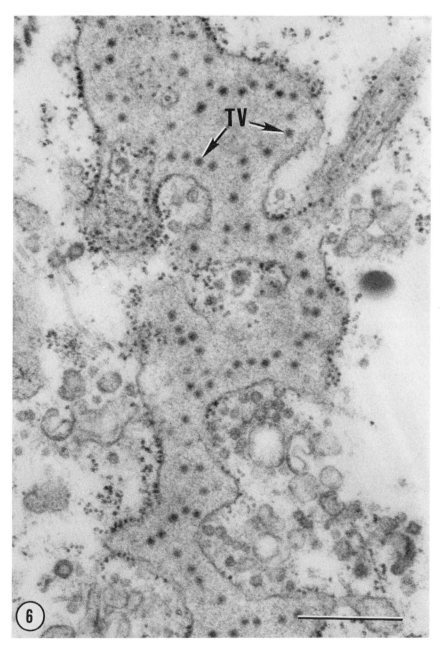

Fig. 6. Portion of an ALSA1-induced syncytium in Singh's *A. albopictus* cell culture, 8 days postinoculation. "Fuzzy spheres" (TV) are seen in irregularly elongated cisterna of the endoplasmic reticulum. The morphology of enveloped particles (approx. 50–55 nm diam.) is similar to that of the togavirus. ×56,800.

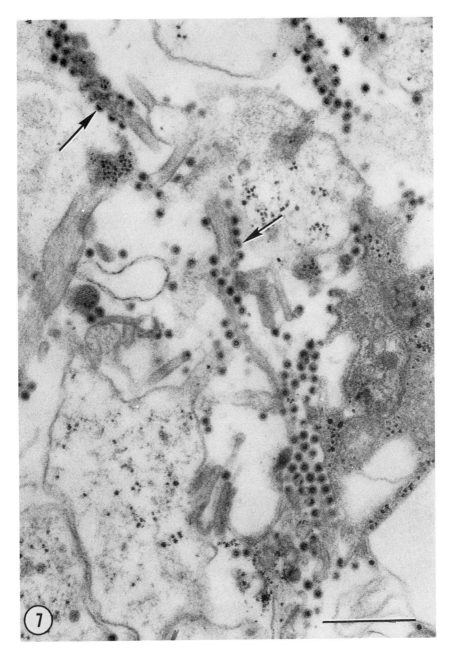

Fig. 7. Portion of ALSA1-infected *A. albopictus* SL3 cells, 42 hours postinoculation. Note that togavirus-type particles, budding from cell surface membranes (arrows), are seen prior to appearance of detectable syncytia in the culture. ×49,000.

8). The morphology of the particles strikingly resembled that of intranuclear particles observed in the midgut epithelium of apparently healthy *A. aegypti* larvae (Figs. 9 and 10) and ALSA1-inoculated *A. albopictus* larvae (H. Hirumi, unpublished data, 1975). Lebedeva *et al.*, (1973) reported the presence of similar particles in diseased larvae of *A. aegypti*. The nature of the particles and their distribution among mosquito species are unknown. If the intranuclear particles are widely present in mosquito larvae, the parvovirus-type agent found in Singh's *A. albopictus* cells may have come from the original tissues, since the cell line was initiated from a large amount of larval tissue fragments. Trypsin, a source of porcine parvovirus contamination (Hallauer *et al.*, 1971), may be another source of insect cell contamination, although no animal pathogenic parvoviruses are known to be infectious to mosquito cells.

   In other insects, parvoviruses (densonucleosis viruses) were previously isolated from *Galleria mellonella* and *Jumonia coenis* moths (Kurstak, 1972; Tinsley and Longworth, 1973). Since the *Galleria* parvovirus was used in this laboratory in the past, pathogenicity of ALSA1 in *G. mellonella* larvae was tested (H. Hirumi, unpublished data, 1975). Highly concentrated ALSA1 did not kill the larvae when inoculated intraperitoneally, whereas the *Galleria* parvovirus did. Serological comparisons between ALSA1 and *Galleria* parvovirus have not been made.

   The orbivirus-like particles (approx. 60–65 nm diam.) often formed crystalline or paracrystalline arrays in the cytoplasm of the ALSA1- and ALSA2-infected SL3 cells (Fig. 11). Scattered particles were also observed in membrane-bound cytoplasmic vacuoles as well as in homogeneous aggregates of electron-opaque matrices. The morphology of the particles was similar to that found in Schneider's *Drosophila* cells (Fig. 3). Richardson *et al.* (1974) recently detected similar virus-like particles (approximately 56 nm diam.) along with smaller particles in *Culex tarsalis* mosquitoes infected with the zoophytic sowthistle yellow vein virus. The two isometric virus-like agents were seen in the salivary gland, fat body, and nerve tissues, although none of the original virus was observed. Based on serial injection tests, the viral nature of the agents was suggested. The same type of isometric virus-like particles (average 62 nm diam.) were observed in hemocytes, midgut epithelium, nerve, muscles, fat body, and epidermis of entomopoxvirus-infected chironomid *Goeldichironomus holoprasinus* larvae (Federici *et al.*, 1974).

   These findings suggest that this type of virus-like particle may exist widely in various organs of certain insects in nature and may be introduced into cell culture systems of these insects.

   Dual and triple infections with the toga-, parvo-, and orbi-type agents were often detected in large *A. albopictus* SL3 syncytia induced by

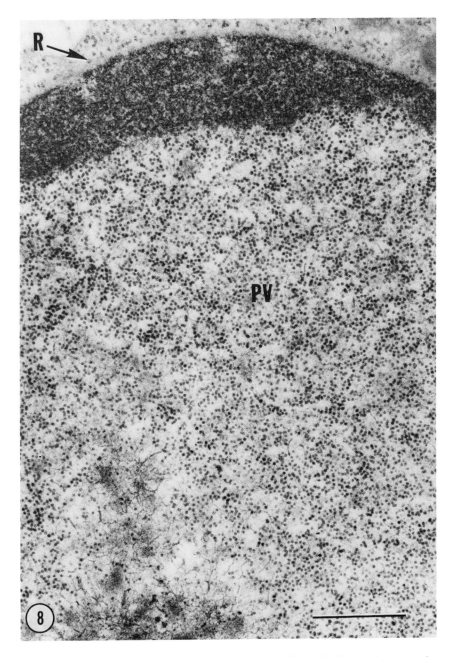

Fig. 8. Portion of an ALSA1-infected *A. albopictus* SL3 cell. The nucleus contains numerous parvovirus-type particles (PV) (18–20 nm diam.). R, ribosome. ×50,100.

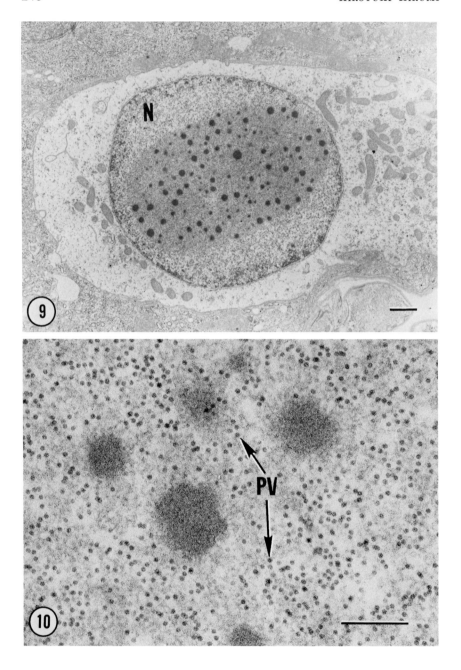

Figs. 9 and 10

ALSA1 and ALSA2 (Fig. 12). These findings suggest a possible interaction between these agents. It is also possible that the broad spectrum of susceptibility of the A. albopictus cells and their ability to produce CPE with a number of togaviruses may be partially due to the interactions between these existing viral contaminants and the experimentally introduced viruses. The consistent appearance of the toga-, parvo-, and orbi-type agents in both ALSA1 and ALSA2 indicated that negative results of serological tests against known togaviruses may not give a guarantee that the sublines tested are free from viral contaminants.

The picornavirus-like particles (approximately 18–20 nm diam.) observed in the A. albopictus cells infected with ALSA1 (Fig. 13) were smaller than the Nodamura virus (28–29 nm diam.) isolated from Culex tritaeniorhynchus (Scherer and Hurlbut, 1967) and the picornavirus of mosquitoes (30.2 ± 0.4 nm diam.) isolated from Aedes taeniorhynchus larvae infected with mosquito iridescent virus (Wagner et al., 1974). At present, the question still remains as to whether aggregates of the picornavirus-like agent detected in the A. albopictus cells are intracytoplasmic accumulations of the parvovirus-like particles or a different virus entirely. Further studies of their nucleic acid type are needed.

Bacteriophage contaminations have often been encountered in fetal bovine sera (Merril et al., 1972; Chu et al., 1973). Thus, it was suggested that the bacterial virus-like particles seen in a sample of the ALSA1-infected SL3 cells at an early passage (Fig. 14) might have come from the fetal bovine serum used as a supplement in the culture media and were not propagated in the A. albopictus cells (Hirumi et al., 1975).

Cunningham et al. (1975) independently observed a similar incidence of viral contamination of A. albopictus cells associated with syncytium formation. The starter culture was obtained from the same laboratory as the original subline (SL1) described above. Virus-like particles which were icosahedral in shape (approximately 60 nm in thin section and 65 nm in negatively stained preparations) were detected. The particles appeared in paracrystalline arrays in the cytoplasm or in budding forms on the cell surface membrane. Based on serological examinations, the authors identified the agent as Chikungunya virus. The morphology of the particles illustrated resembled that of the orbi-type particles described above.

Fig. 9. Midgut epithelial cells of apparently normal A. aegypti larvae. A nucleus (N) contains parvovirus-like particles and many aggregates of electron-opaque matrices in various sizes. ×72,000.

Fig. 10. Portion of the nucleus shown in Fig. 11 at a higher magnification. Note clearly defined parvovirus-like particles (PV). ×86,800.

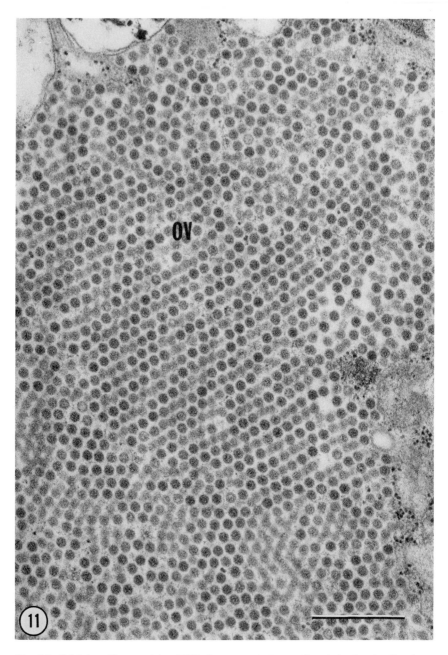

Fig. 11. Orbivirus-like particles (OV) (approx. 60–65 nm diam.) in the *A. albopictus* SL3 cell inoculated with ALSA1. The particles are morphologically similar to those seen in the *D. melanogaster* cell (see Fig. 3). ×49,700.

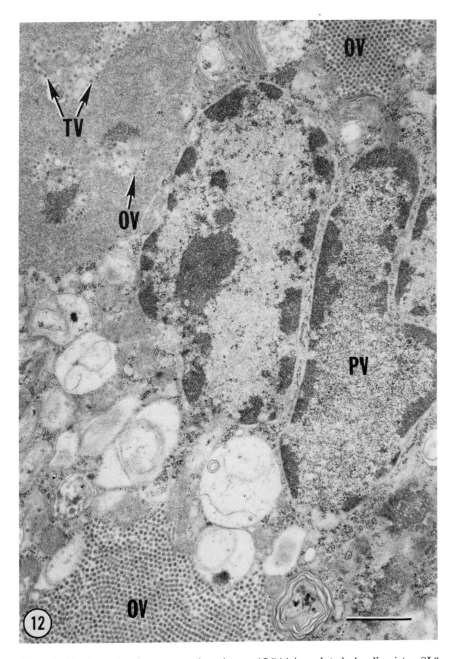

Fig. 12. Portion of a large syncytium in an ALSA1-inoculated *A. albopictus* SL3 cell culture, 7 days postinoculation. Note triple infection with the toga (TV)-, parvo (PV)-, and orbi (OV)-type virus-like particles. ×17,500.

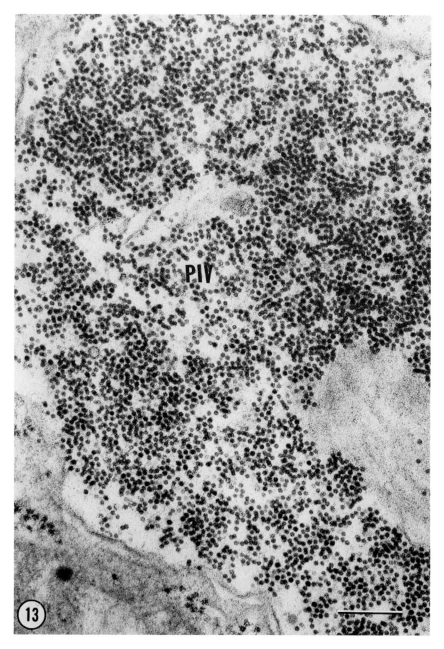

Fig. 13. Intracytoplasmic accumulation of picornavirus-like particles (PIV) in an ALSA1-inoculated *A. albopictus* SL3 cell, 7 days postinoculation. ×86,800.

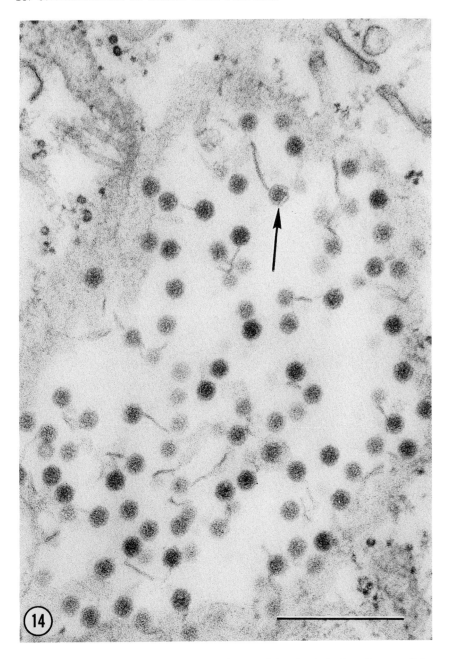

Fig. 14. Portion of a cytoplasmic vacuole containing bacteriophage-type particles (arrow) (approx. 55–60 nm diam.) with a tail-like structure (140 nm length) in an *A. albopictus* SL3 cell inoculated with ALSA1 in the third passage, 5 days post-inoculation. ×68,500.

2. Singh's Aedes aegypti Cell Line

The *A. aegypti* cell line was established at the time the *A. albopictus* cell line became available (Singh, 1967). Different spectrums of susceptibility to certain togaviruses (Buckley, 1971; Yunker, 1971; Singh, 1971, 1972) made both cell lines useful in comparative studies of the arthropod-borne viruses.

Syncytium formation was observed in recently received cultures of the *A. aegypti* cells using a phase microscope prior to exposing the cells to our own laboratory environment (H. Hirumi, unpublished data, 1975). To date, no extensive study of the *A. aegypti* syncytium-inducing agent (AASA1) has been carried out. However, a preliminary study demonstrated that the fluid phase of the cultures induced syncytia in the *A. albopictus* SL3 cells at 27°C. Electron microscopic examination of the original *A. aegypti* cells, as well as AASA1-inoculated SL3, demonstrated the presence of togavirus-like particles which were morphologically similar to those observed in the ALSA1- and ALSA2-infected SL3 cells. (Table I).

3. Grace's Aedes aegypti Cell Line

Filshie *et al.* (1967) observed spherical virus-like particles (42–46 nm diam.) in the *A. aegypti* cells (Grace, 1966). The particles lacked a clearly defined nucleoid and appeared in the cells inoculated with either Murray Valley encephalitis or Japanese B encephalitis virus. The nature and origin of the particles are unknown. Filshie *et al.* (1967) also observed the same type of virus-like particles in Grace's moth *Antheraea eucalypti* cells (Grace, 1962).

4. Peleg's Aedes aegypti Cell Line

Kitamura *et al.* (1973) detected virus-like particles in the cytoplasm of Peleg's *A. aegypti* cells (Peleg, 1966). The particles were observed during attempts to adapt the cell line to a culture medium which was free of calf serum. No details concerning their morphology and site are presently available; however, it was stated that the particles did not induce CPE in the host cells. Recently, Stollar *et al.* (1975) isolated a viral agent, which induced syncytia in Singh's *A. albopictus* cells, from Peleg's *A. aegypti* cell line. The viral agent has been suggested to be a flavivirus of the Togaviridae.

5. Singh and Bhat's Aedes w-albus Cell Line

Cytoplasmic inclusions containing virus-like particles (approximately 45 nm diam.) were detected in "uninoculated" *A. w-albus* cells (Singh and Bhat, 1971) by means of thin section electron microscopy (S. M. Buckley, G. Kemp, J. Casals, and M. Lipman, personal communication 1974). At present, the nature and origin of the particles also are unknown.

6. Pudney and Varma's Anopheles stephensi Cell Line

In the *Anopheles stephensi* cell line (Pudney and Varma, 1971), long rod-shaped virus-like particles (75–650 nm length by 17–18 nm diam.) resembling the tobacco mosaic virus were observed by Varma *et al.* (1970). They also reported the presence of icosahedral virus-like particles (65–70 nm diam.) in primary cultures of the tick *Hyalomma dromedarii* cells.

III. Mycoplasmal Contamination

A. POSSIBLE SOURCES

Mycoplasma, the common name of microorganisms which belong to the order Mycoplasmatales in the class Mollicutes, are small prokaryotic cells. The organisms are pleomorphic and vary in size from 10 to 100 μm and have no cell wall. They are widely distributed in nature among humans, animals, and plants. On appropriate agar media they form small "fried-egg" type colonies. Because of the unique biological properties, the organisms have been extensively studied during the past two decades. Basic information concerning their biological nature, serological types, ultrastructural aspects, classification, and prevalence in biological materials has been well elucidated (Hayflick, 1969; Maramorosch, 1973).

A mycoplasmal contaminant was first isolated from mammalian cell cultures by Robinson *et al.* (1956). Since then, numerous mycoplasmal contaminations of vertebrate cell cultures have been reported (Brown and Officer, 1968; Fogh *et al.*, 1971; Barile, 1973a; Kenny, 1973). Many mycoplasmal species (Serological Group 1: *Mycoplasma arginini, M. arthritidis, M. gallinarum, M. hominis,* and *M. orale* type 1; Serological Group 2: *Mycoplasma canis, M. fermentans, M. hyorhinis,* and *M. pulmonis;* Serological Group 3: *Acholeplasma laidlawii* and *A. granularum;* and Serological Group 4: *Mycoplasma gallisepticum*) were isolated from uninoculated vertebrate cell cultures (Barile, 1973a; Kenny, 1973). There are four major groups of mycoplasmal contaminants based on their origin. The first group, which contains the human oral and genital species of mycoplasmas, such as *M. orale* type 1, *M. hominis*, and *M. fermentans*, accounts for the largest group of contaminants. The second group comprises the bovine mycoplasmas, such as *M. arginini*, *Acholeplasma laidlawii*, and unclassified *Mycoplasma* and *Acholeplasma* species, which most likely originated from commercial bovine sera. The third group is represented by the swine species, especially *M. hyorhinis*, that might come from commerical trypsin. The fourth group consists of mycoplasmal species that originated from tissues used to initiate the primary cell cultures (Barile, 1973a).

Mycoplasmal contaminants often cause changes in vertebrate cell cultures, including alternations in cell physiology and morphology, chromosomal aberrations, and an increase or decrease in virus yields. Some contaminants also produce CPE, including plaques (Barile, 1973a; Kenny, 1973). The detection, isolation, identification, prevention, and elimination of these mycoplasmal contaminants in cell cultures confront workers in most laboratories where cell culture systems are used. Nevertheless, standardized techniques in solving the problems are now available (Fogh *et al.*, 1971; Barile, 1973a,b; Kenny, 1973; Hayflick, 1973).

The major groups of mycoplasmal contaminations originating from human sources, commercial sera, and trypsin can be expected to occur in invertebrate cell culture systems to a similar extent. Although not many insect mycoplasmas have been reported, certain insect species are known to be carriers of mycoplasma-like organisms (Maramorosch *et al.*, 1970; Davis and Whitcomb, 1971). Thus, tissue-associated mycoplasmal contaminations of these insect cells in cultures should not be overlooked. This is especially important when leafhopper cell cultures are used in the study of plant-pathogenic mycoplasma-like organisms carried by leafhopper vectors.

B. INCIDENCE IN VARIOUS INSECT CELLS

Limited information concerning mycoplasmal contamination in insect cell cultures is presently available.

1. Cockroach Cells

Unidentified microorganisms thought to be mycoplasmas were isolated from primary as well as established cell line cultures of the cockroach

Leucophaea madera (Quiot *et al.*, 1973). The fluid phase of cell cultures showing a "degeneration phenomena" was filtrated through Millipore filters of 0.45, 0.30, and 0.22 μm and placed in normal cell cultures as well as mycoplasmal agar and liquid media. Newly infected cells degenerated 6–10 days postinoculation. The filtrate formed "fried-egg" type colonies on the agar medium. Thin section electron microscopy of the colonies and pellets prepared from the liquid cultures revealed the presence of microorganisms whose morphology was similar to that of the mycoplasma. The organisms had no serological relationships with *Mycoplasma laidlawii* (presently classified as *Acholeplasma laidlawii*), *M. orale, M. hominis* (types I and II), *M. pneumonia,* S. H., *M. fermentans,* or *M. salivarium.* It was suggested that the organism was a mycoplasma originally present in the cockroach tissues.

2. Various Cell Lines

Suspicious mycoplasmal contamination was noticed in the codling moth *Carpocapsa pomonella* CP-1268 cell line (Hink and Ellis, 1971), by means of phase microscopy (H. Hirumi, unpublished data, 1975). Numerous intercellular, pleomorphic microorganisms bound by a unit membrane and lacking a cell wall were observed in all suspected cultures (Fig. 15). The organisms contained ribosome-like granules which were located at the cell periphery and DNA-like strands which transversed electron-lucent central areas, lacking nuclear envelopes. The morphology of the organisms resembled that of microorganisms belonging to the Mycoplasmatales.

The contamination rapidly spread to other insect cell cultures. Grace's *Aedes aegypti,* Singh's *A. aegypti, Anopheles stephensi* (Schneider, 1969), *D. melanogaster, Agalia constricta, Macrosteles fascifrons* (H. Hirumi, unpublished data, 1975), *Antheraea eucalypti, Carpocapsa pomonella* CP-169 and CP-1268, *Spodoptera frugiperda* (J. L. Vaughn, unpublished data, 1968), and *Trichoplusia ni* cell cultures were tested for mycoplasmal contamination by inoculating them on mycoplasmal agar plates (Baltimore Biological Laboratory, Maryland). All insect cells listed were carried in the Hirumi–Maramorosch Leafhopper Medium (HML) (Hirumi and Maramorosch, 1964, 1971) in which the cells easily adapted and continued to grow actively. The agar plates inoculated with the culture media and fresh medium were incubated at 30°C for 20 days. "Fried-egg" type colonies (Fig. 16) were isolated from 33 out of 73 cultures tested, but none were detected in the fresh culture medium (Table II). The isolated organisms were positive to methylene blue azure staining. Electron micros-

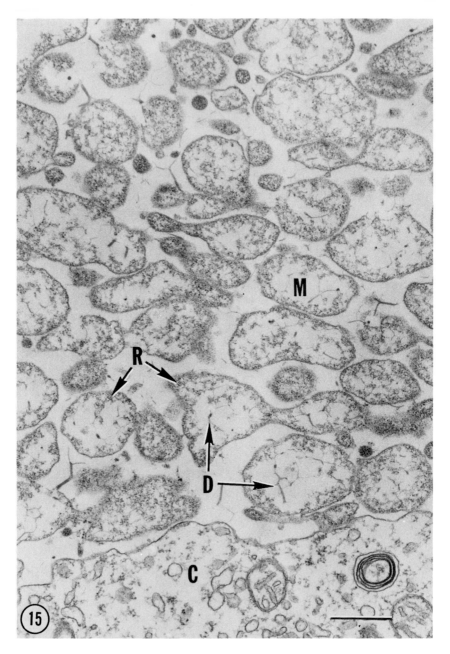

Fig. 15. Extracellular mycoplasma contaminants (M) in a codling moth *Carpocapsa pomonella*, CP-1268 cell culture. The mycoplasmas are distinguishable from the moth cells (C) on the basis of their internal structures, such as peripherally located ribosomes (R) and DNA-like strands (D). ×33,600.

TABLE II
Isolation of Mycoplasmas from Various Insect Cell Cultures[a]

Cell line	Number isolated	Number tested
Diptera		
Aedes aegypti (Grace)	1	10
Aedes aegypti (Singh)	0	2
Aedes albopictus	1	7
Anopheles stephensi	3	4
Drosophila melanogaster	0	1
Homoptera		
Agallia constricta	13	14
Macrosteles fascifrons	0	3
Lepidoptera		
Antheraea eucalypti	2	10
Carpocapsa pomonella CP-169	2	5
Carpocapsa pomonella CP-1268	3	6
Spodoptera frugiperda	5	6
Trichoplusia ni	3	6
Fresh cell culture media[b]	0	22

[a] Isolated on BBL mycoplasma agar medium at 30°C.
[b] Hirumi–Maramorosch leafhopper (HML) cell culture medium.

copy of thin sections of the colonies after the second passage demonstrated that the colonies were composed of pleomorphic organisms (Fig. 17). The basic morphology of the organisms strikingly resembled that of mycoplasmal species, such as *M. pneumoniae* and *M. salivarium*, grown in agar media (Knudson and MacLeod, 1970).

A total of six isolates were identified serologically by two different laboratories* and were found to be *Acholeplasma laidlawii* (three isolates) and *Acholeplasma* species (three isolates). On the basis of their morphology and growth patterns, the other isolates probably belong to the *Acholeplasma* group. Since commercial bovine sera are frequently contaminated with *Acholeplasma laidlawii* (Barile, 1973a), the detected mycoplasmal contaminants most likely came from the fetal bovine serum. These findings suggest that the mycoplasmal contaminants can be introduced to insect cell cultures regardless of the type of cells and that the same precautions are necessary with insect cell cultures as with vertebrate cell cultures in order to prevent such contamination.

* The identification was made through the courtesy of Dr. M. F. Barile, NIH, Bethesda, Maryland, and Dr. G. E. Kenny, University of Washington, Seattle, Washington.

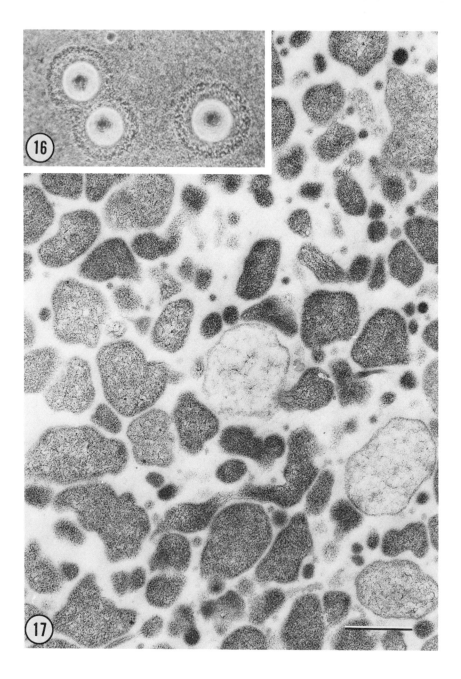

Figs. 16 and 17

IV. Symbiotic Bacteria Contamination

Many insects contain, and transovarially transmit, microorganisms whose morphology differs from that of the mycoplasmas. Some of the organisms which inhabit insects intracellularly are known to be symbiotic in the hosts. They are pleomorphic and have a cell wall. The presence of these so-called "rickettsia-like" organisms in various insect organs has been demonstrated by means of electron microscopy (Koch and King, 1966; Maramorosch et al., 1968, 1970; Brooks, 1970; Brinton and Burgdorfer, 1971; Davis and Whitcomb, 1971; Maillet, 1971; Schwemmler et al., 1971). The organisms observed in situ in the CNS of a leafhopper Agallia constricta and the ovarian tissues of the cat flea Ctenocephalides felis are illustrated in Figs. 18 and 19, respectively (H. Hirumi, unpublished data, 1975). In primary cultures of a leafhopper Macrosteles fascifrons derived from minced embryonic tissues, following the procedures described earlier (Hirumi and Maramorosch, 1964), intercellular pleomorphic microorganisms were detected by means of electron microscopy after 2–3 weeks cultivation (Fig. 20). The ultrastructural aspects of the organisms, particularly the cell walls, resembled "rickettsia-like" organisms rather than mycoplasmas. No mycoplasma was isolated from the primary cultures (H. Hirumi, unpublished data, 1975).

Since the symbiotic bacteria are transmitted transovarially, it is feasible that they can be introduced to insect cell cultures derived from ovarian or whole embryonic insect tissues inhabited by symbiotic bacteria. To date, however, little attention has been given to this type of contamination of insect cell culture systems.

V. Extrinsic Cell Contamination

During the past two decades, many laboratories have often encountered serious problems with inter- and intraspecies cell contamination of vertebrate cell cultures. Various effective methods of identifying cell contaminants in vertebrate systems have been developed by a number of investigators. It is beyond the scope of this chapter to review the details

---

Fig. 16. "Fried-egg" type colonies of mycoplasmas isolated from the contaminated C. pomonella, CP-1268 culture, shown in Fig. 15, 8-day-old culture on agar medium at the second passage. The organism was identified as Acholeplasma laidlawii by means of serology. Phase contrast. ×200.

Fig. 17. Thin section of a colony shown in Fig. 16. The colony is composed of pleomorphic organisms whose basic morphology is similar to that of known mycoplasmas grown on agar media. ×18,250.

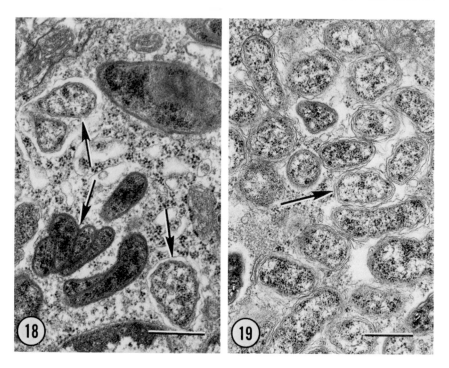

Figs. 18 and 19. Intracellular "rickettsia-like" microorganisms having thin cell walls (arrows) in the CNS of a leafhopper *Agallia constricta* (Fig. 18, ×30,000) and the ovarian tissue of a cat flea *Ctenocephalides felis* (Fig. 19, ×26,400) *in situ.*

concerning contamination of vertebrate cells. The reader is referred to an excellent review by Stulberg (1973) and pertinent technical articles (Arrighi, 1973; Brautbar, 1973; Gartler and Farber, 1973; House, 1973; Lin and Uchida, 1973; Melnick, 1973; Shannon and Macy, 1973; Stulberg and Simpson, 1973; Wang and Federoff, 1973).

Problems similar to those experienced with vertebrate cell cultures may be encountered with invertebrate cell cultures. A large number of cell lines derived from various insect species have been maintained at different laboratories, some of which are presently carrying more than two cell lines. Exchanges of cell lines are now common practice among individual laboratories. More than two different types of insect cells can be maintained in the same culture medium. Furthermore, some insect cells, such as those of lepidopteran insects, contain a large number of small chromosomes whose individual identification is extremely difficult. In contrast, most dipteran cells, such as those of mosquitoes ($2n = 6$) and fruit flies ($2n = 8$), have rather small, readily countable chromosome numbers.

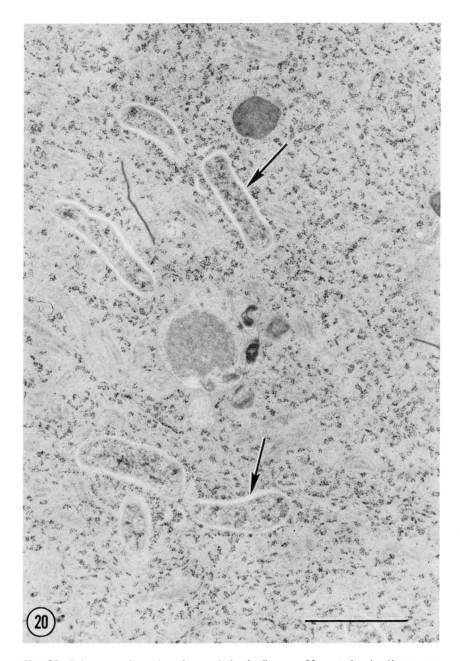

Fig. 20. Primary embryonic culture of the leafhopper *Macrosteles fascifrons*, containing intracytoplasmic "rickettsia-like" organisms having thin cell walls (arrows), 21-day-old culture. ×27,500.

However, their karyotypes are barely distinguishable from those of other species, using classical methods of cytogenetics, even if maintained in a diploid number *in vitro*. In addition, several different types of intraspecies cell lines have been established from certain insects, such as *A. aegypti, A. albopictus*, and *D. melanogaster*. It is expected that more such intraspecies cell lines will become available in the near future. These facts may acount for some extrinsic cell contamination of insect cells *in vitro*.

In the past, it was the experience of this laboratory that a small number of Singh's *A. albopictus* cells accidentally introduced into primary cultures of human amnion cells became dominant in the cultures. Recently, *A. albopictus* cells (approx. 2×10^4 cells/ml) were experimentally introduced into *T. ni* cell cultures (containing approx. 5×10^6 cells/ml). The *A. albopictus* cells, which were readily distinguishable from *T. ni* cells on the basis of their size and morphology, rapidly propagated in the culture and completely dominated the mixed cell population within 2 weeks (H. Hirumi, unpublished data, 1975).

Direct evidence of extrinsic cell contamination of an established insect cell subline has been presented by Greene and Charney (1971) and by Nichols *et al.* (1971). The characterization and identification of Grace's *Antheraea eucalypti* (Freeze 268 and 308), Grace's *Aedes aegypti* (Freeze 309 and 335), Singh's *A. albopictus* (Freeze 402), and Singh's *A. aegypti* (Freeze 415) were carried out utilizing agar gel immunodiffusion, isoenzyme analysis, and chromosome analysis. The findings clearly indicated that Grace's *A. aegypti* cells were not mosquito cells but moth cells. Greene and Charney (1971) suggested that the cell contamination may have occurred in the past during passage through a number of laboratories. It was also pointed out that, using the agar microimmunodiffusion test method and isoenzyme patterns of glucose-6-phosphate dehydrogenase, lactate dehydrogenase, malate dehydrogenase, 6-phosphogluconate dehydrogenase, and acid phosphatase, the *Antheraea eucalypti* cells can be distinguished from the *Aedes albopictus* cells as well as cells from mammalian, fish, and amphibian species (Greene and Charney, 1971, 1973b).

If the recently developed chromosome banding techniques (Arrighi, 1973; Lin and Uchida, 1973; Wang and Federoff, 1973) could be used in identifying insect chromosomes, they would provide an additional effective means for the detection and prevention of the extrinsic cell contamination in insect cell culture systems in the future.

References

Akai, H., Gateff, E., Davis, L. E., and Schneiderman, H. A. (1967). *Science* **157**, 810–813.

Ammar, E. D., Milne, R. G., and Watson, M. A. (1970). *J. Gen. Virol.* **6**, 315–318.

Arrighi, F. E. (1973). *In* "Tissue Culture: Methods and Applications" (P. F. Kruse, Jr. and M. K. Patterson, Jr., eds.), pp. 773–777. Academic Press, New York.

Barile, M. F. (1973a). *In* "Contamination in Tissue Culture" (J. Fogh, ed.), pp. 131–172. Academic Press, New York.

Barile, M. F. (1973b). *In* "Tissue Culture: Methods and Applications" (P. F. Kruse, Jr. and M. K. Patterson, Jr., eds.), pp. 729–735. Academic Press, New York.

Bhat, U. K. M. (1973). *Curr. Sci.* **42**, 66.

Bhat, U. K. M., and Guru, P. Y. (1973). *Exp. Parasitol.* **33**, 105–113.

Bhat, U. K. M., and Singh, K. R. P. (1970). *Curr. Sci.* **39**, 388–390.

Brautbar, C. (1973). *In* "Tissue Culture: Methods and Applications" (P. F. Kruse, Jr., and M. K. Patterson, Jr., eds.), pp. 758–763. Academic Press, New York.

Brinton, L. P., and Brugdorfer, W. (1971). *J. Bacteriol.* **105**, 1149–1159.

Brooks, M. A. (1970). *J. Invertebr. Pathol.* **16**, 249–258.

Brooks, M. A., and Kurtti, T. J. (1971). *Annu. Rev. Entomol.* **16**, 27–52.

Brown, A., and Officer, J. E. (1968). *In* "Methods in Virology" (K. Maramorosch and H. Koprowski, eds.), Vol. 4, pp. 531–564. Academic Press, New York.

Buckley, S. M. (1971). *Curr. Top. Microbiol. Immunol.* **55**, 133–137.

Chiu, R. J., and Black, L. M. (1967). *Nature (London)* **215**, 1076–1078.

Chu, F. C., Johnson, J. B., Orr, H. C., Probst, P. G., and Petricciani, J. C. (1973). *In Vitro* **9**, 31–34.

Coriell, L. L. (1973a). *In* "Contamination in Tissue Culture" (J. Fogh, ed.), p. 29–49. Academic Press, New York.

Coriell, L. L. (1973b). *In* "Tissue Culture: Methods and Applications" (P. F. Kruse, Jr. and M. K. Patterson, Jr., eds.), pp. 718–721. Academic Press, New York.

Cory, J., and Yunker, C. E. (1972). *Acta Virol. (Prague)* **16**, 90.

Cunningham, A., Buckley, S. M., Casals, J., and Webb, S. R. (1975). *J. Gen. Virol.* **27**, 97–100.

Davey, M. W., Dennett, D. P., and Dalgarno, L. (1973). *J. Gen. Virol.* **20**, 225–232.

Davis, R. E., and Whitcomb, R. F. (1971). *Annu. Rev. Phytopathol.* **9**, 119–154.

De Harven, E. (1973). *In* "Contamination in Tissue Culture" (J. Fogh, ed.), pp. 205–231. Academic Press, New York.

Djinawi, N. K., and Olson, L. C. (1973). *Arch. Gesamte Virusforsch.* **43**, 144–151.

Federici, B. A., Granados, R. R., Anthony, D. W., and Hazard, E. I. (1974). *J. Invertebr. Pathol.* **23**, 117–120.

Felluga, B., Johnson, V., and Liljeros, M. R. (1971). *J. Invertebr. Pathol.* **17**, 339–346.

Filshie, B. K., Grace, T. D. C., Poulson, D. F., and Reháček, J. (1967). *J. Invertebr. Pathol.* **9**, 271–273.

Fogh, J. (1973). *In* "Contamination in Tissue Culture" (J. Fogh, ed.), pp. 173–203. Academic Press, New York.

Fogh, J., Holmgren, N. B., and Ludovici, P. P. (1971). *In Vitro* **7**, 26–41.

Fraenkel-Conrat, H. (1974). *In* "Comprehensive Virology" (H. Fraenkel-Conrat and R. R. Wagner, eds.), pp. 3–62. Plenum, New York.

Gartler, S. M., and Farber, R. A. (1973). *In* "Tissue Culture: Methods and Applications" (P. F. Kruse, Jr. and M. K. Patterson, Jr., eds.), pp. 797–804. Academic Press, New York.

Gartner, L. P. (1971). *Experientia* **27**, 562–564.

Gartner, L. P. (1972). *J. Invertebr. Pathol.* **20**, 364–366.

Grace, T. D. C. (1962). *Nature (London)* **195**, 788–789.

Grace, T. D. C. (1966). *Nature (London)* **211**, 366–367.

Grace, T. D. C. (1973). *In* "Viruses and Invertebrates" (A. J. Gibbs, ed.), pp. 321–346. Amer. Elsevier, New York.

Greene, A. E., and Charney, J. (1971). *Curr. Top. Microbiol. Immunol.* 55, 51–61.

Greene, A. E., and Charney, J. (1973a). *In* "Tissue Culture: Methods and Applications" (P. F. Kruse, Jr., and M. K. Patterson, Jr., eds.), pp. 750–753. Academic Press, New York.

Greene, A. E., and Charney, J. (1973b). *In* "Tissue Culture: Methods and Applications" (P. F. Kruse, Jr. and M. K. Patterson, Jr., eds.), pp. 753–758. Academic Press, New York.

Hallauer, C., Kronauer, G., and Siegl, G. (1971). *Arch. Gesamte Virusforsch.* 35, 80–90.

Hayflick, L., ed. (1969). "The Mycoplasmatales and the L-Phase of Bacteria." Appleton, New York.

Hayflick, L. (1973). *In* "Tissue Culture: Methods and Applications" (P. F. Kruse, Jr. and M. K. Patterson, Jr., eds.), pp. 722–728. Academic Press, New York.

Herman, M. M., Johnson, M., and Miquel, J. (1971). *J. Invertebr. Pathol.* 17, 442–445.

Herold, F., and Munz, K. (1967). *J. Virol.* 1, 1028–1036.

Hink, W. F. (1970). *Nature (London)* 226, 466–467.

Hink, W. F., and Ellis, B. J. (1971). *Curr. Top. Microbiol. Immunol.* 55, 19–28.

Hirumi, H. (1974). *U.S.-Jap. Coop. Conf. Invertebr. Tissue Cult.: Appl. Fundam. Res., 1974* p. 39

Hirumi, H., and Maramorosch, K. (1964). *Science* 144, 1465–1467.

Hirmuri, H., and Maramorosch, K. (1971). *In* "Invertebrate Tissue Culture" (C. Vago, ed.), Vol. 1, pp. 307–339. Academic Press, New York.

Hirumi, H., Hirumi, K., Speyer, G., Yunker, C. E., Thomas, L. A., Cory, J., and Sweet, B. H. (1975). *In Vitro* (in press).

Horzinek, M. C. (1973). *J. Gen. Virol.* 20, 87–103.

House, W. (1973). *In* "Tissue Culture: Methods and Applications" (P. F. Kruse, Jr. and M. K. Patterson, Jr., eds.), pp. 739–744. Academic Press, New York.

Hsu, S. H., Mao, W. H., and Cross, J. H. (1970). *J. Med. Entomol.* 7, 703–707.

Kenny, G. E. (1973). *In* "Contamination in Tissue Culture" (J. Fogh, ed.), pp. 107–129. Academic Press, New York.

Kernaghan, R. P., and Ehrman, L. (1968). *J. Invertebr. Pathol.* 10, 432–434.

Kernaghan, R. P., Bonneville, M. A., and Pappas, G. D. (1964). *Genetics* 50, 262.

Kitamura, S. (1970). *Kobe J. Med. Sci.* 16, 41–50.

Kitamura, S., Imai, T., and Grace, T. D. C. (1973). *J. Med. Entomol.* 10, 488–489.

Kniazeff, A. J. (1973). *In* "Contamination in Tissue Culture" (J. Fogh, ed.), pp. 233–242. Academic Press, New York.

Knudson, D. L., and MacLeod, R. (1970). *J. Bacteriol.* 101, 609–617.

Koch, E. A., and King, R. C. (1966). *J. Morphol.* 119, 283–304.

Kurstak, E. (1972). *Advan. Virus Res.* 17, 207–241.

Lebedeva, O. P., Kuznetsova, M. A., Zelenko, A. P., and Gudz-Gorban, A. P. (1973). *Acta Virol. (Prague)* 17, 253–256.

Lee, P. E. (1965). *Virology* 25, 471–472.

Lin, C. C., and Uchida, I. A. (1973). *In* "Tissue Culture: Methods and Applications" (P. F. Kruse, Jr. and M. K. Patterson, Jr., eds.), pp. 778–781. Academic Press, New York.

Ludovici, P. P., and Holmgren, N. B. (1973). *Methods Cell Biol.* 6, 143–208.

Maillet, P. L. (1971). *Bull. Biol. Fr. Belg.* 105, 95–111.

Maramorosch, K., ed. (1973). "Mycoplasma and Mycoplasma-Like Agents of Human, Animal, and Plant Diseases," Ann. N.Y. Acad. Sci. No. 225. N.Y. Acad. Sci, New York.

Maramorosch, K., Shikata, E., and Granados, R. R. (1968). Trans. N.Y. Acad. Sci. [2] 30, 841–855.

Maramorosch, K., Granados, R. R., and Hirumi, H. (1970). Advan. Virus Res. 16, 135–193.

Melnick, P. J. (1973). In "Tissue Culture: Methods and Applications" (P. F. Kruse, Jr. and M. K. Patterson, Jr., eds.), pp. 808–821. Academic Press, New York.

Merril, C. R., Friedman, T. B., Attallah, A. F. M., Geier, M. R , Krell, K., and Yarkin, R. (1972). In Vitro 8, 91–93.

Miquel, J., Bensch, K. G., and Delbert, E. P. (1972). J. Invertebr. Pathol. 19, 156–159.

Moericke, V. V. (1963). Z. Pflanzenkr. Pflanzenschutz 70, 464–470.

Molander, C. W., Paley, A., Boone, C. W., Kniazeff, A. J., and Imagawa, D. T. (1968). In Vitro 4, 148.

Molander, C. W., Kniazeff, A. J., Boone, C. W., Paley, A., and Imagawa, D. T. (1971). In Vitro 7, 168–173.

Nichols, W. W., Bradt, C., and Bowne, W. (1971). Curr. Top. Microbiol. Immunol. 55, 61–69.

Parrish, W. B., and Briggs, J. D. (1966). J. Invertebr. Pathol. 8, 122–123.

Paul, S. D., Singh, K. R. P., and Bhat U, K. M. (1969). Indian J. Med. Res. 57. 339–348.

Peleg, J. (1966). Experientia 22, 555–556.

Peleg, J. (1968). Virology 35, 617–619.

Peleg, J. (1969). J. Gen. Virol. 5, 463–471.

Peleg, J. (1971). Curr. Top. Microbiol. Immunol. 55, 155–161.

Peleg, J. (1972). Arch. Gesamte Virusforsch. 37, 54–61.

Perotti, M. E., and Bairati, A., Jr. (1968). J. Invertebr. Pathol. 10, 122–138.

Philpott, D. E., Weibel, J., Atlan, H., and Miquel, J. (1969). J. Invertebr. Pathol. 14, 31–38.

Pudney, M., and Varma, M. G. R. (1971). Exp. Parasitol. 29, 7–12.

Pudney, M., McCarthy, D., and Shortridge, K. F. (1973). Proc. Int. Colloq. Invertebr. Tissue Cult., 3rd, 1971 pp. 337–345.

Quiot, J.-M., Giannotti, J., and Vago, C. (1973). Proc. Int. Colloq. Invertebr. Tissue Cult., 3rd, 1971 pp. 467–475.

Rae, P. M. M., and Green, M. M. (1968). Virology 34, 187–189.

Raghow, R. S., Davey, M. W., and Dalgarno, L. (1973a). Arch. Gesamte Virusforsch. 43, 165–168.

Raghow, R. S., Grace, T. D. C., Filshie, B. K., Bartley, W., and Dalgarno, L. (1973b). J. Gen. Virol. 21, 109–122.

Richardson, J., Sylvester, E. S., Reeves, W. C., and Hardy, J. L. (1974). J. Invertebr. Pathol. 23, 213–224.

Robinson, L. B., Wichelhausen, R. H., and Roizman, B. (1956). Science 124, 1147.

Sanders, F. K. (1973). In "Contamination in Tissue Culture" (J. Fogh, ed.), pp. 243–256. Academic Press, New York.

Scherer, W. F., and Hurlbut, H. S. (1967). Amer. J. Epidemiol. 86, 271–285.

Schneider, I. (1967). In "Methods in Developmental Biology" (F. H. Wilt and N. K. Wessells, eds.), pp. 543–554. Crowell-Collier, New York.

Schneider, I. (1969). J. Cell Biol. 42, 603–606.

Schneider, I. (1971). Curr. Top. Microbiol. Immunol. 55, 1–12.

268

HIROYUKI HIRUMI

highSchnieder, I. (1972). *J. Embryol. Exp. Morphol.* **27**, 353–365.
Schneider, I. (1973). *Proc. Int. Colloq. Invertebr. Tissue Cult., 3rd, 1971* pp. 121–134.
Schwemmler, W., Quiot, J.-M., and Amagier, A. (1971). *Ann. Soc. Entomol. Fr.* [N. S.] **7**, 423–438.
Shannon, J. E., and Macy, M. L. (1973). *In* "Tissue Culture: Methods and Applications; (P. F. Kruse, Jr. and M. K. Patterson, Jr., eds.), pp. 804–807. Academic Press, New York.
Singh, K. R. P. (1967). *Curr. Sci.* **36**, 506–508.
Singh, K. R. P. (1971). *Curr. Top. Microbiol. Immunol.* **55**, 127–133.
Singh, K. R. P. (1972). *Advan. Virus Res.* **17**, 187–206.
Singh, K. R. P., and Bhat, U. K. M. (1971). *Experientia* **27**, 142–143.
Singh, K. R. P., Goverdhan, M. K., and Bhat, U. K. M. (1973). *Indian J. Med. Res.* **61**, 1134–1137.
Stollar, V., Shenk, T. E., Stollar, B. D., Stevens, T. M., and Schlesinger, R. W. (1973). *Proc. Int. Colloq. Invertebr Tissue Cult., 3rd, 1971* pp. 367–379.
Stollar, V., and Thomas, V. L. (1975). *Virology* **64**, 367–377.
Stulberg, C. S. (1973). *In* "Contamination in Tissue Culture" (J. Fogh, ed.), pp. 1–27. Academic Press, New York.
Stulberg, C. S., and Simpson, W. F. (1973). *In* "Tissue Culture: Methods and Applications" (P. F. Kruse, Jr. and M. K. Patterson, Jr., eds.), pp. 744–749. Academic Press, New York.
Suitor, E. C., Jr. (1969). *J. Gen. Virol.* **5**, 545–546.
Suitor, E. C., Jr., and Paul, F. J. (1969). *Virology* **38**, 482–485.
Sweet, B. H. (1970). *Arthropod-Borne Virus Inform. Exch.* No. 20, 40–41.
Sweet, B. H., and McHale, J. S. (1970). *Exp. Cell Res.* **61**, 51–63.
Tandler, B. (1972). *J. Invertebr. Pathol.* **20**, 214–215.
Tinsley, T. W., and Longworth, J. F. (1973). *J. Gen. Virol.* **20**, 7–15.
Vago, C. (1967). *In* "Methods in Virology" (K. Maramorosch and H. Koprowski, eds.), Vol. 1, pp. 567–602. Academic Press, New York.
Varma, M. G. R., and Pudney, M. (1969). *J. Med. Entomol.* **6**, 432–439.
Varma, M. G. R., Pudney, M., and Bird, R. G. (1970). *J. Parasitol.* **56**, Sect. II, Part 1, 351 (Resume No. 646).
Wagner, G. W., Webb, S. R., Paschke, J. D., and Campbell, W. R. (1974). *J. Invertebr. Pathol.* **24**, 380–382.
Wang, H. C., and Federoff, S. (1973). *In* "Tissue Culture: Methods and Applications" (P. F. Kruse, Jr. and M. K. Patterson, Jr., eds.), pp. 782–787. Academic Press, New York.
Wehman, H. J., and Brager, M. (1971). *J. Invertebr. Pathol.* **18**, 127–130.
Williamson, D. L., and Kernaghan, R. P. (1972). *Drosophila Inform. Serv.* **48**, 58–59.
Yunker, C. E. (1971). *Curr. Top. Microbiol. Immunol.* **55**, 113–126.
Yunker, C. E., and Cory, J. (1975). *Appl. Microbiol.* **29**, 81–89.

14

Fundamental Studies on Insect Icosahedral Cytoplasmic Deoxyribovirus in Continually Propagated *Aedes aegypti* Cells

J. D. PASCHKE and S. R. WEBB

I. Introduction

Mosquito iridescent virus (MIV) belongs to a group of large icosahedral DNA-containing viruses which replicate in the cytoplasm of infected host cells. The descriptive term that encompasses the morphological and biological features of these viruses, i.e., icosahedral cytoplasmic deoxyribovirus (ICDV), was proposed by Stoltz (1971). This grouping includes virus of reptiles, fish, mammals, and fungi, as well as the similar viruses of insects.

Williams and Smith (1957) first described the "iridescent viruses" of insects, so-called because the patently infected hosts exhibit an iridescent color resulting from light defraction by the paracrystalline array of virus in infected cells. This group of insect viruses has been reviewed by Bellett (1968) and the ICDV's by Kelly and Robertson (1973).

The MIV was first described by Clark *et al.* (1965) from the black salt-marsh mosquito *Aedes taeniorhynchus*. Other iridescent viruses were isolated from the mosquitoes *Aedes cantans* and *Aedes annulipes* in Czechoslovakia at about the same time (Weiser, 1965), and which appeared similar to the virus described from *A. taeniorhynchus*. In addition to these original records, other MIV isolates have been made from *Aedes stimulans* (Anderson, 1970), *A. cantans* (Tinsley *et al.*, 1971), and *A. detritus* (Hasan *et al.*, 1970; Vago *et al.*, 1969).

The original MIV isolated by Clark *et al.* (1965) imparted an orange iridescence to the infected host. Subsequent to the discovery of this strain, Woodard and Chapman (1968) discovered a single larva with an atypical blue iridescence in a laboratory colony of *A. taeniorhynchus* which had been inoculated with the orange strain of MIV. This blue MIV isolate, and the original MIV isolate, could be transmitted per os through serial passages in *A. taeniorhynchus* larvae with an average infection rate of 21 and 16%, respectively. Both viruses were also transmitted to progeny by transovarial transmission.

It became apparent that the two isolates required some designation other than the hue of the iridescence of patently infected larvae or of a pellet of purified virus. They were, respectively, designated RMIV (regular) and TMIV (turquoise) (Matta and Lowe, 1970).

Because of its potential use for the biological control of mosquitoes, we undertook investigation of MIV and the infection caused by the virus in *A. taeniorhynchus*. One of the main problems encountered was the lack of a reproducible bioassay since laboratory inoculations of insectary-reared mosquitoes resulted in approximately 10–15% infection.

Also, we believed that to adequately elucidate the pathology caused by MIV we would need to not only conduct studies on infected larvae, but also investigate the pathology in the tissue culture system.

Webb *et al.* (1974) first described the infection of continuously propagated *Aedes aegypti* cells with MIV. Two cell lines derived from *A. aegypti* (Singh, 1967; Peleg, 1968) were shown to be permissive to infection as based on the expression of a cytopathic effect (CPE). The results of these findings served as a basis for establishing a titration of infectivity and the elucidation of the pathology resulting from MIV infection.

Kelly and Tinsley (1974) studied iridescent virus types 2 and 6 (Tinsley and Kelly, 1970), *Sericesthis* iridescent virus (SIV) (Steinhaus and Leutenegger, 1963) and *Chilo* iridescent virus (CIV) (Fukaya and Nasu, 1966), respectively, in *A. aegypti* cells in addition to Grace's *Antheraea eucalypti* cells (Grace, 1962). Their report of infection of mosquito cell lines with iridescent viruses coincided with the report of Webb *et al.* (1974).

Webb *et al.* (1975a,b) have described the titration of MIV in *A. aegypti* cells and they elucidated the pathology involved in the infection. Their results reviewed below will be contrasted with available information from other studies similar in nature.

II. Assay of MIV in *Aedes aegypti* Cells

Singh's (1967) and Peleg's (1968) *A. aegypti* cells exhibit a CPE resulting from infection by either TMIV or RMIV (Webb *et al.*, 1974). In general, the CPE is first observed by phase microscopy approximately 2.5 days postinoculation (PI) as a granular appearance and increased vacuolization of the cytoplasm of infected cells. There is also a tendency for adjacent cells to form clusters. By 3 days PI, infected cells formed tenuous cytoplasmic filaments which contained refractile bodies, and by 5 days PI the infected cells began to detach from the substrate on which they were growing. These detached cells were subsequently shown to be infected (Webb *et al.*, 1975b).

Both RMIV and TMIV caused similar cell response upon infection and as a dose-dependent relationship with CPE evidenced at 25°, 31°, and 35°C. The CPE occurred only rarely at 21°C and only in Peleg's aegypti cells. Other cell lines, *A. aegypti*, Mos-20A (Varma and Pudney, 1969), *Aedes albopictus* (Singh, 1967), and *An. eucalypti* (Grace, 1962), were refractile to infection by MIV strains under the experimental conditions.

First, Webb *et al.* (1975a) found that the maximum titer was exhibited by 5 days PI and did not show further increase up to 15 days. Second, incubation of MIV with the cells for 1 or 2 hours, or over the entire observational period, did not significantly alter the tissue culture infective dose$_{50}$ (TCID$_{50}$).

Table I summarizes the results (Webb *et al.*, 1975a) of titrations of MIV in Linbro tissue culture trays. It will be noted that no differences in infectivity between RMIV and TMIV were detected. There was, however, a difference in the responses between Peleg's aegypti cells and those established by Singh, the latter being somewhat less responsive based on the TCID$_{50}$'s. Temperatures higher than 31°C caused a nonspecific CPE and therefore 31°C was considered optimum for titration.

The average —log TCID$_{50}$ at 31°C in both cell lines using purified RMIV and TMIV was 3.4 and 2.6, respectively. This corresponds to a virion-to-cell ratio (v:c) of 10 and 55, respectively, at the 50% end point. The standard errors of the estimated end points were 0.15 and 0.16 with 95% confidence limits for the TCID$_{50}$ of ±0.30 for the Peleg's cells and ±0.32 in Singh's cell line.

Virus neutralization by either homologous or heterologous antisera from RMIV and TMIV specifically reduced CPE (Table II).

Plaque assays were negative using techniques described by Cory and

TABLE I

Infectivity Titration of RMIV and TMIV on Singh's and Peleg's *A. aegypti* Cells

| | $-\log_{10} \text{TCID}_{50}$ | | | |
| | Peleg's *A. aegypti* cells inoculated with | | Singh's *A. aegypti* cells inoculated with | |
Incubation temperature	RMIV	TMIV	RMIV	TMIV
21°C	0	0	0	0
25°C	2.5	2.7	2.4	2.4
31°C	3.4[a]	3.4	2.6[b]	2.6
35°C	3.4	3.2	2.6	2.5

[a] $-\log_{10} \text{TCID}_{50} = 3.4$ is equivalent to a virion:cell ratio of 10 at the time of inoculation.

[b] $-\log_{10} \text{TCID}_{50} = 2.6$ is equivalent to a virion:cell ratio of 55.

Yunker (1972) for arbovirus titration in *A. albopictus* cells. No plaques were formed using either Peleg's or Singh's cell lines.

A preliminary qualitative bioassay of infectivity of progeny cell-

TABLE II

Neutralization Titration of RMIV and TMIV by Homologous and Heterologous Antiserum on Peleg's *A. aegypti* Cells

| | $-\log_{10} \text{TCID}_{50}$ (neutralization index[a]) | |
Serum	RMIV	TMIV
None	3.2	3.0
Normal rabbit[b]	3.2 (0)	3.0 (0)
Anti-RMIV[c]	1.3 (1.8)	1.2 (1.8)
Anti-TMIV[c]	1.3 (1.8)	1.2 (1.8)

[a] Neutralization indexes calculated by the method of Reed and Muench (1938).

[b] Normal rabbit serum was not necessarily from the same rabbit in which the antiserum was produced.

[c] Antisera to RMIV and TMIV were not absorbed to the respective cell lines before incubation with virus.

produced virus was conducted by inoculation of Singh's and Peleg's cell lines with the medium for 7-day-old infected cultures, by cocultivating infected cells with each of the respective lines and by rearing first stage *A. taeniorhynchus* larvae from each of the infected cell lines with either cells and media, or media only. The results of these assays indicated cell-

TABLE III

Infectivity Tests of 7-Day RMIV-Infected *A. aegypti* **Cultures,**[a] **Cell-Derived RMIV and Larval-Derived RMIV to Cultured Cells**[b] **with Larvae**[c]

Inoculum sample	Inoculated specimen (% infected cultures or larvae)		
	Singh's *A. aegypti* cell line	Peleg's *A. aegypti* cell line	*A. taeniorhynchus* larvae
RMIV			
From larvae	100	100	10
From Singh's	0	ND	0
From Peleg's	ND[i]	0	0
Singh's			
Culture[d]	0	ND	0
Medium[e]	0	ND	0
Suspended cells[f]	0	ND	ND
Attached cells[g]	0	ND	ND
Culture with inoculum[h]	ND	ND	0.2
Peleg's			
Culture	ND	0	0
Medium	ND	0	0
Suspended cells	ND	0	ND
Attached cells	ND	0	ND
Culture with inoculum	ND	ND	0.2

[a] Complete infectivity tests have not been run with TMIV.
[b] Each test was replicated a minimum of four times.
[c] A minimum of 1000 larvae were used with each inoculum.
[d] Culture medium, attached and detached cells from T-flask culture exhibiting CPE 7 days PI with larval-derived RMIV from which the inoculum had been removed after a 2-hour adsorption period.
[e] Medium decanted from infected culture (*d*).
[f] Suspended cells from infected culture (*d*).
[g] Attached cells from infected culture (*d*).
[h] RMIV-infected culture as (*d*) from which the original larval-derived inoculum had not been removed.
[i] ND = not determined.

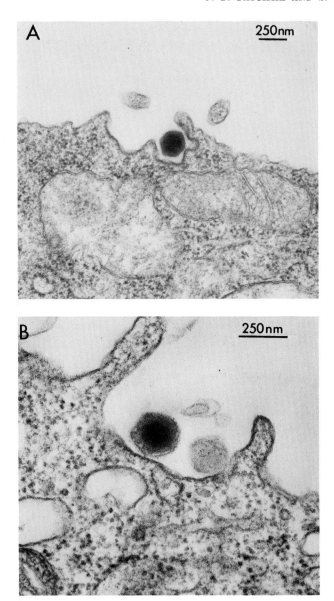

Fig. 1. Attachment and internalization of MIV virions by *A. aegypti* cells 5 minutes PI. Virions attach to the plasma membrane, many times at one of the vertices of the particle (A and B). They are progressively surrounded by pseudopodlike projections of the cell surface (B) until they are completely surrounded (C). From Webb *et al.* (1975b).

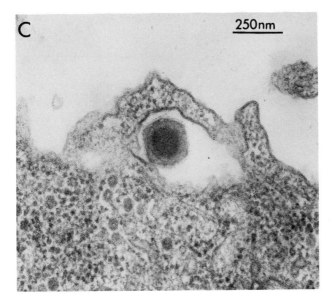

Fig. 1C

produced MIV to be noninfective. Infection of the mosquito larvae occurred only in those test larvae allowed to feed in 7-day infected cultures from which the original inoculum had not been removed (Webb *et al.*, 1975a). These results are summarized in Table III.

III. Infection and Replication of MIV in *Aedes aegypti* Cells

Webb *et al.* (1975b) investigated the infectious process involved in the infection of both Singh's and Peleg's *A. aegypti* cell lines. These studies involved electron microscopic observations supplemented with observations made by phase microscopy with limited study using immunofluorescent staining techniques.

In their study, viropexis was the only process observed by which MIV was internalized by *A. aegypti* cells. Cells fixed 5 minutes PI and incubation at 31°C showed various stages in the formation of phagocytic vesicles around attached MIV virions (Fig. 1A,B,C).

Attachment of MIV at 4°C was much less efficient than at 31°C, contrary to normal expectations. Titration of three different RMIV preparations adsorbed to Peleg's *A. aegypti* cells resulted in a $-\log_{10}$ TCID$_{50}$

of −0.5 at 4°C while at 31°C it was −3.2. Thus attempts to synchronize the attachment and subsequent adsorption process were unsuccessful.

Fifteen minutes PI, virions were located in the cell but lacking any visible vesicular membrane (Fig. 2) while others were confined to vesicles within the cell cytoplasm and others were still attached at the cell surface (Fig. 3A). By 30 minutes PI, the virion-containing vesicles were well within the cytoplasm of the cells. In some cases, the vesicles structurally resembled lysosomes (Fig. 3B) and contained some virions that appeared to be degraded while some remained intact. Sixty minutes PI, most internalized virions were associated with smooth endoplasmic reticulum (Fig. 3C,D) where it appeared that the vesicular membrane was degraded and some of the associated virions in a state of uncoating.

Within 24 hours PI, ultrastructural changes in the infected cell cytoplasm were observed. Definitive areas devoid of cell organelles, and with electron density distinctly different from that of the balance of the cell, could be observed. These areas, the viroplasma, contained no virions and corresponded to those areas of infected cells containing viral antigens as detected by fluorescent antibody staining. Another change in these infected cells was the aggregation of chromatin at the periphery of the nucleus at the nuclear envelope. Like the viroplasm, margination became more evident 2 or more days PI (Fig. 4).

As infection progressed, cells exhibited increased vacuolation, enlarged viroplasma with mature virions in surrounding cytoplasm, and budding at the cell membrane (Fig. 5). Progeny virions budded from the cell surface and became enveloped by the plasma membrane (Fig. 6).

As early as 2.5 days PI, and continuing into the late stages of infection, virions were observed budding into cell vacuoles (Fig. 7A,B,C) and occasionally some were apparently located in intracisternal spaces.

Large areas of the cell surface appeared to be involved with the budding process, and most virions appeared to be "mature" but some apparently aberrant particles were observed with electron-translucent cores.

In one instance possible nuclear involvement in MIV replication was suggested by the presence of a mature virion located within the nucleus of an infected cell. Confirmation of the particle's location was supported by serial sections through the nucleus and the occurrence of the virion in each of the serial sections.

The MIV-infected cells agglutinated in the presence of antiserum to the originally inoculated virus. Control cells remained dispersed as did normal and infected cells in the presence of normal rabbit serum.

Because the process of maturation of MIV in tissue culture was different than observed *in vivo*, Webb *et al.* (1975b) conducted centrifugation studies on virus derived from cell culture in contrast to that derived from

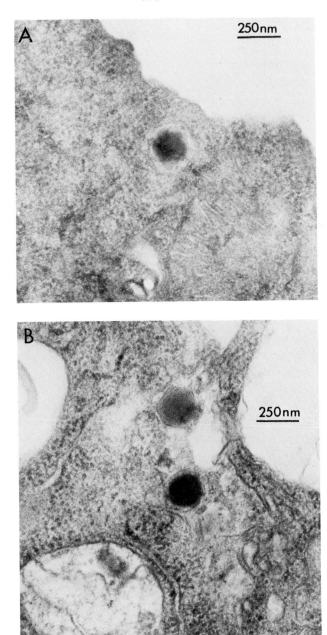

Fig. 2. Examples of intact MIV virions in the cytoplasm of inoculated *A. aegypti* cells 15 minutes PI. Note the absence of definable phagocytic vesicle.

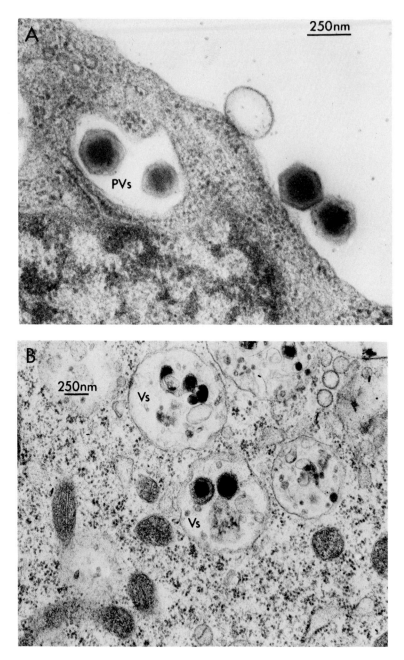

Fig. 3A and B. The MIV virions within *A. aegypti* cells 15–60 minutes PI. (A) 15 minutes PI virions observed in phagocytic vesicles. (B) Lysosomelike vesicles contain partially degraded virions 30 minutes PI. PVs, phagocytic vesicle; Vs, lysosomelike vesicle.

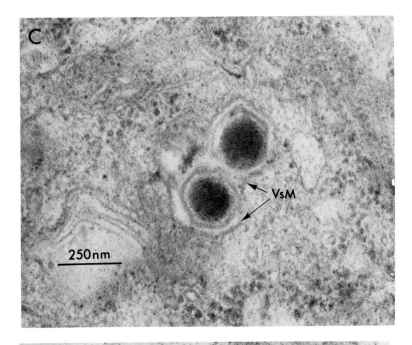

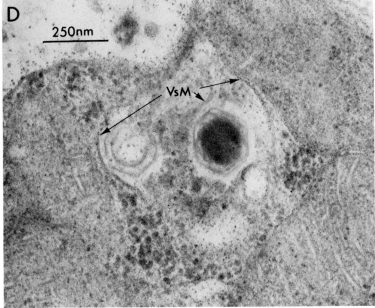

Fig. 3C and D. 60 minutes PI intact virions observed associated with membrane remanents of the phagocytic vesicle (C) and partially degraded virions with translucent cores still associated with remanents of the membrane of phagocytic vesicles. VsM, vesicular membrane. (A)–(D) from Webb *et al.* (1975b).

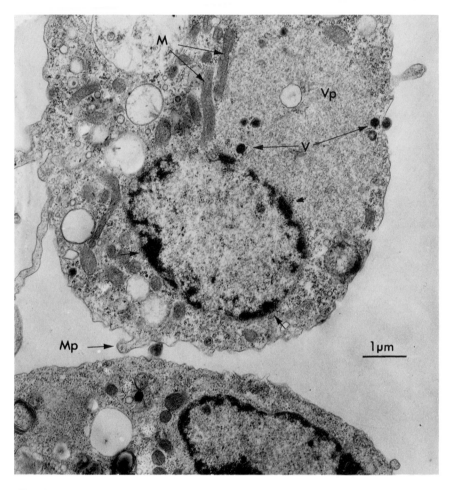

Fig. 4. *Aedes aegypti* cell 2.5 days PI with MIV showing characteristic margination of the chromatin (arrow), viroplasm, and progeny virions, some of which are in the process of budding from the cell surface. Note also the degenerate appearing mitochondria at the periphery of the viroplasm and the plasma membrane protrusions. Vp, viroplasm; V, virions; M, mitochondria; Mp, membrane protrusions. From Webb *et al.* (1974).

the insect host. Obvious physical differences between the two virus preparations existed.

Sedimentation profiles (Fig. 8) on sucrose density gradients of virus from 7-day cultures revealed only one major component not present in profiles of respective controls. This fraction was comprised of enveloped progeny virions and sedimented considerably above the position at which

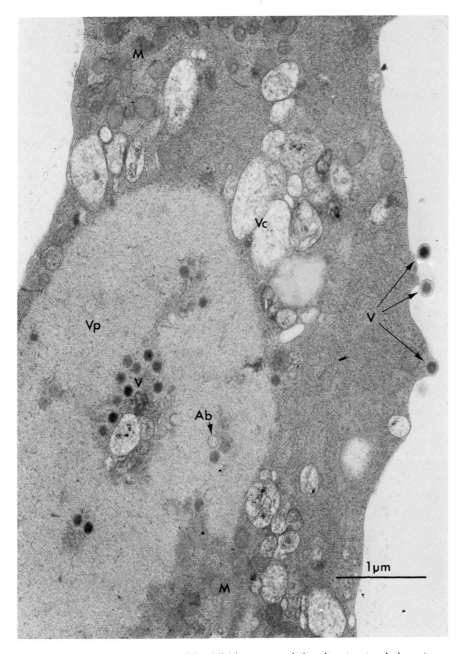

Fig. 5. *Aedes aegypti* cell 6 days PI exhibiting many of the ultrastructural characteristics of MIV infection. They include the enlarged viroplasm, increased vacuolization, mitochondria some of which appear to be degenerate, and most of which are restricted to definitive areas of the cytoplasm, and numerous virions in the viroplasm. In addition there are abberant forms visible in the viroplasm. Budding virions are evident at the cell surface. Vp, viroplasm; Vc, vacuolization; M, mitochondria; V, virions; Ab, abberant forms. From Webb *et al.* (1975b).

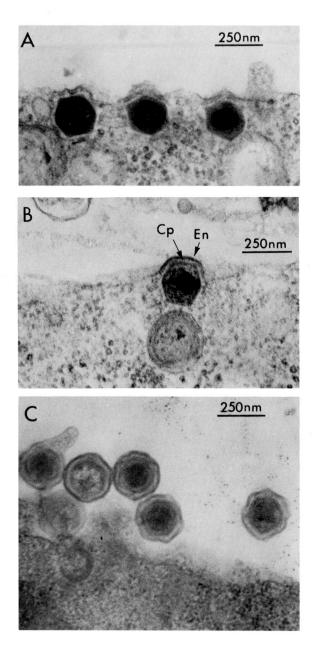

Fig. 6. Infected *A. aegypti* cells showing the budding of MIV from the plasma membrane. Virions appear oppressed to the plasma membrane and some apparent structural changes of the membrane are suggested (A). The virions progressively protrude from the cell and are characterized by the more dense enveloping membrane (B). Virions released (C) from the cell denoting the still dense envelope formed from the plasma membrane and which conforms to the shape of the virus capsid. Cp, capsid; En, envelope. (A) and (B) from Webb *et al.* (1975b).

the nonenveloped RMIV from mosquito larvae appeared. Top component (Wagner *et al.*, 1973) was not detected due to lack of sufficient quantities of this component. Electron microscopic observations support the presence of nonenveloped MIV within the cells, and the sedimentation profiles indicate most of the virus produced by cultured cells becomes enveloped. The same profiles were obtained using TMIV.

Webb *et al.* (1975b) proposed the probable sequence of events during infection as indicated in Fig. 9.

IV. Discussion

A. COMPARISON OF MIV WITH OTHER INSECT ICDV'S *IN VITRO*

The titration method described by Webb *et al.* (1975a) is the only reproducible quantitative bioassay for MIV. Bioassay of the virus in the mosquito host results only in qualitative differences, because no typical dose response of the test insects has been observed (Linley and Nielsen, 1968a,b; Paschke *et al.*, 1972). Our investigations (J. D. Paschke and S. R. Webb, unpublished) of alternate hosts for use in bioassay procedures were negative, as were the results of Chapman *et al.* (1966). This is in contrast to SIV, CIV, and TIV which can be assayed in the larvae of insects other than the normal host.

Bellett (1965) developed an assay for SIV using the proportion of Graces' (1962) *A. eucalypti* cells which stained with fluorescent antibody as a criterion of infection. Unfortunately he could not depend on CPE as an indication of virus multiplication, nor was plaque formation achieved. However, *in vitro* titration remained the most sensitive method of assaying SIV (Bellett, 1965).

Webb *et al.* (1975a) hypothesized that lack of plaque formation, as well as infectivity to cells and larvae, resulted from the process by which MIV matured and was released from infected Singh's or Peleg's *A. aegypti* cells. Only one cycle replication and CPE could be demonstrated using insect-derived virus as inoculum.

It has been shown that the cell-derived virus acquires a portion of the plasma membrane as the virion buds from an infected cell. It has also been established that the outer capsid protein of MIV is required for infection (Wagner *et al.*, 1975). Removal of the capsid protein renders the virus core noninfective. Webb *et al.* (1975a) therefore assumed that the outer capsid protein is essential for virion attachment to the cells.

It was also observed that, if allowed to grow, infected cell cultures would

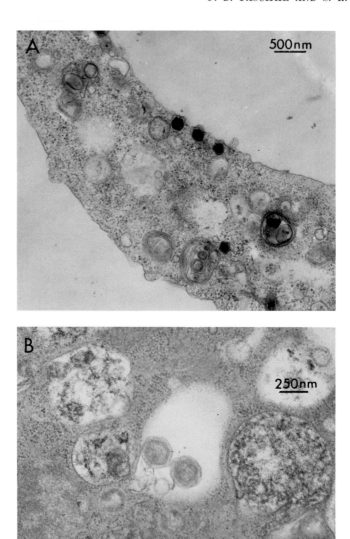

Fig. 7. The MIV virions budding into vacuoles of the cytoplasm of infected *A. aegypti* cells. (A) Extremely large vacuole, 2.5 days PI; (B) and (C) 6 days PI. (C) from Webb *et al.* (1975b).

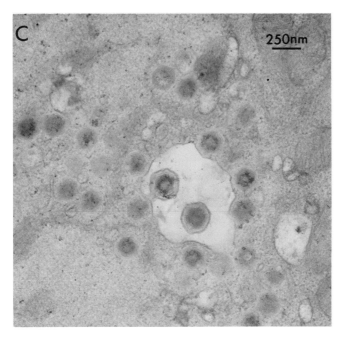

Fig. 7C

soon be overgrown with apparently normal cells. This "recovery" of the cultures obscured the originally observed CPE.

The virions which are enveloped upon release from the infected cells may be rendered noninfective to neighbor noninfected cells because the enveloped virion is recognized as "self" and thus attachment and/or viropexis is effectively inhibited. Infection is therefore restricted to those cells which are initially infected by nonenveloped insect-derived virus. Without lateral transmission (from cell to cell), no plaque formation occurs. Furthermore, no indication of infectivity was observed when cell-produced virus was assayed against *A. taeniorhynchus* larvae. This further suggests a loss of infectivity of cell-derived virus.

Kelly and Tinsley (1974) did not assay cell-derived virus (SIV or CIV) in the cell cultures used in the study or in the alternate host *Galleria mellonella*, so no comparison can be made to their system. Bellett (1965) assayed SIV released from *A. eucalypti* cells or as cell-associated virus and found little difference in infectivity. The SIV released from the infected cells is enveloped by a portion of the cell plasma membrane in much the same fashion observed in the case of MIV.

Mitsuhashi (1966) propagated CIV in a hemocyte cell line of *Chilo*

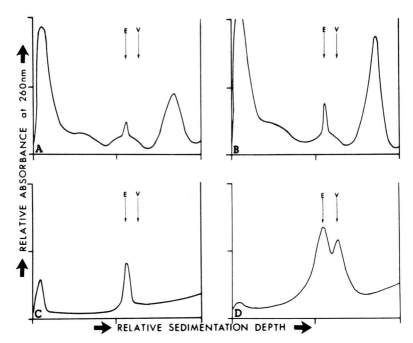

Fig. 8. Sedimentation profiles from sucrose density gradient analysis of MIV derived from either infected *A. aegypti* cells or from patently infected *A. taeniorhynchus* larvae. Singh's (A) and Peleg's (B) cultures 7 days PI with RMIV, and a mixture of the enveloped form from each of the above cultures sedimented alone (C) or with nonenveloped form (larval derived) as a marker (D). E, position of enveloped virions of RMIV; V, position of nonenveloped form. Sedimentation is from left to right. From Webb *et al.* (1975).

suppressalis and determined that cell-associated virus was infective in diapausing larvae of *C. suppressalis*. Used culture medium inoculated into similar test insects produced infection within a week.

The apparent lack of infectivity of cell-associated and/or cell-released MIV virions, no plaque formation, and the apparent recovery of cell cultures after one cycle of virus growth is unique to the MIV–*A. aegypti* cell system employed by Webb *et al.* (1975b). Because of the unusual results regarding the biological activity of cell-released or cell-associated virus, continued studies are in progress to unequivocally substantiate the lack of infectivity and determine whether or not aberrant forms are produced in *A. aegypti* cells.

The five insect cell lines investigated by Webb *et al.* (1975a) were evaluated to determine their suitability as substrates on which to titrate MIV and not specifically for their susceptibility to virus infection. The

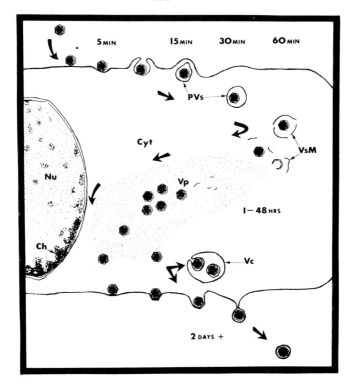

Fig. 9. Schematic representation of the infection and replication of MIV in cultured *A. taeniorhynchus* cells. PVs, phagocytic vesicles; Cyt, cytoplasm; VsM, vesicular membranes; Vp, viroplasm; Nu, nucleus; Ch, chromatin; Vc, vacuoles. From Webb *et al.* (1975b).

refractile nature of *A. albopictus, A. aegypti* (Mos-20A), and *A. eucalypti* cells to infection by MIV does not conclusively rule out their susceptibility to MIV, but it does rule them out from the described titrimetric procedure.

Singh's *A. aegypti* cells are suitable for titration of MIV, but their lower susceptibility when compared with Peleg's *A. aegypti* line precludes their use since the latter provide a more sensitive and reliable assay system.

Internalization of MIV by *A. aegypti* cells occurred by a process of viropexis, with no other process observed (Webb *et al.*, 1975b). Kelly and Tinsley (1974) described similar events in the infection of *A. aegypti* cells with SIV and CIV.

In our study (Webb *et al.*, 1975b), attachment was observed as early as 5 minutes PI and, by 15 minutes PI, virions were observed in the cytoplasm either in phagocytic vesicles or without vesicular membranes, while some still remained attached to the plasma membrane. Unfortunately, Kelly and Tinsley (1974) presented no observations on the internalization processes earlier than 1 hour PI. The actual mechanism by which MIV virions are taken up by the cells is suggested by the observations we made wherein pseudopodlike projections of the plasma membrane appeared to progressively surround the particles.

Kelly and Tinsley (1974) observed complete virus particles in the cytoplasm, either with no restricting membrane around the particles, or enclosed in "vacuoles" which they suggested "may be, or subsequently become lysosomes." In our study (Webb et al., 1975b), phagocytic vesicles were observed in the cytoplasm of infected cells, some of which frequently contained structurally degraded virions. Fusion of virus-containing phagocytic vesicles with lysosomes was not observed, but the accumulation of intact virus and its degraded forms in lysosomelike vesicles indicated that the process did occur and is in general agreement with the observations of Kelly and Tinsley (1974).

The uncoating process of MIV in the infected cells could not be unequivocally demonstrated, but observations indicated that the phagocytic vesicles degrade to release the intact virion into the cell cytoplasm after which there is a gradual degradation of the virions and release of the infectious component (Webb et al., 1975b).

Kelly and Tinsley (1974) did not observe the gradual disruption of the phagocytic vesicles and release of the virions, but they did observe "naked" virus particles in the cytoplasm which were sometimes partially stained, "possibly in the process of uncoating."

In the study of MIV infection of A. aegypti cells, it was detected that progeny virions were present as early as 48 hours PI. Kelly and Tinsley (1974) did not detect appreciable progeny virions until 144 hours postinfection. Few particles were detected at 96 and 120 hours postinfection. Using fluorescent antibody techniques, they detected virus antigens present in infected cells as early as 96 hours postinfection and suggested that insufficient antigen was present or absent, at earlier times, to allow observable staining. In their paper, electron microscope observations were not presented for sampling times between 1 and 96 hours PI.

Microtubule formation or paracrystalline inclusion formation as noted by Kelly and Tinsley (1974) were not observed in our study of MIV-infected cells (Webb et al., 1975b). The release of mature virions in our system was similar to that described for SIV by Bellett and Mercer (1964) and by Hukuhara and Hashimoto (1967) for TIV and CIV. Adequate observations are lacking concerning the egress of virus by processes other than the predominant budding, but some observations suggested that mature virions may bud into vacuoles. Whether or not MIV virions are transported out of the cell in this structure in a similar fashion as described by Kelly and Tinsley (1974) for SIV and CIV was not determined. The overall pattern of replication of SIV and CIV in *A. eucalypti* cells was similar to that observed in Peleg's *A. aegypti* cells (Kelly and Tinsley, 1974).

In addition to the differences between MIV and the SIV–CIV infection of *A. aegypti* cells as mentioned above, it is significant that the optimal temperatures for replication of MIV was determined (Webb *et al.*, 1975a) to be 31°C compared to a 21°C optimum for SIV and CIV in the same cell lines. Kelly and Tinsley (1974) determined 33°C to be nonpermissive for the latter iridescent virus infection of Peleg's *A. aegypti* cells. Webb *et al.* (1975a) achieved infection with MIV at 35°C but with nonspecific cytopathic effects after short incubation at this temperature.

Bellett (1965) determined the optimal temperature for replication of SIV in *A. eucalypti* cells to be 20°C. The proportion of infected cells was significantly reduced at temperatures much above 24°C. This is in contrast to the results of Kelly and Tinsley (1974) which resulted in only slightly reduced infection at 24°C.

The maturation and release of MIV from infected cell culture is similar in many respects to some of the other ICDV's as well as other animal viruses propagated in cell lines. However, the process is quite different from that observed *in vivo*. In the latter, infected cells normally contain large crystalline arrays of virions and usually we have not observed particles in the process of release from these cells by budding. Virus release occurred only when infected cells lyse or were physically disrupted. One exception to this has been noted in cases of infected hemocytes which did not exhibit large crystalline arrays of MIV. In these instances, the hemocytes contained numerous virions and an occasional virion was detected in the process of budding from the plasma membrane (J. D. Paschke, unpublished).

B. COMPARISON OF MIV AND FROG ICDV'S *IN VITRO*

Recently Kelly (1975) reported on the infection and replication of frog virus 3 (FV-3) in chick embryo fibroblasts and drew some analogies with other ICDV's, notably the insect-pathogenic iridescent viruses. In this study, Kelly (1975) found morphogenesis of FV-3 in primary chick embryo fibroblasts to be similar to that reported for the same virus in baby hamster kidney cells (BHK 21/13) by Bingen-Brendel *et al.* (1971) and FV-1 in fat head minnow cells (FHM) by Darlington *et al.* (1966).

Uptake of cell-derived FV-3 was reported (Bingen-Brendel *et al.*, 1971) to be by a process of pinocytosis in the BHK 21/13 cell system and the same process was the predominant means by which FHM cells took up FV-3 (Houts et al., 1974). Kelly (1975) suggested that in addition FV-3 may enter chick embryo fibroblast cells by a process of envelope fusion of the virion (~15% enveloped particles, Kelly, 1975) because of intact "naked" particles observed in the cytoplasm of cells as early as 1 hour postinfection. This is contrary to the observations of Houts et al. (1974) who unequivocally stated that their virus inoculum contained no enveloped particles, yet they observed intact particles free in the cytoplasm of FHM cells as early as 5 minutes PI. Our studies of MIV infection of A. aegypti cells, as well as those of Kelly and Tinsley (1974) using SIV and CIV, support the latter observation and do not support a hypothesis of envelope fusion since the iridescent virus preparations employed were insect derived and therefore nonenveloped.

Houts et al. (1974) suggest that virions free in the cytoplasm result from direct penetration into the cell or alternatively by the rapid release of virions from the phagocytic vacuoles directly into the cytoplasmic matrix. They did not observe such a release. However, we observed what appeared to be a degradation of the membrane surrounding internalized MIV particles which suggests that the process may occur. The overall pattern of MIV virion internalization resembles that reported for FV-3 (Bingen-Brendel et al., 1971; Darlington et al., 1966; Houts et al., 1974; Kelly, 1975).

Margination of nuclear chromatin was observed in our studies of MIV in A. aegypti cells (Webb et al., 1974, 1975b) and appeared similar to that reported by Darlington et al. (1966), Campadelli-Fiume et al. (1973), and in the recent contribution of Kelly (1975) for FV-3 infection of FHM, BHK, and chick fibroblast cells.

The viroplasm resulting from MIV infection of A. aegypti cells appeared similar to that reported by Bingen-Brendel et al. (1971) and Darlington et al. (1966) in FV-3 infection of BHK and FV-1 infection of FHM cells, respectively. In the case of MIV infection, mitochondria were observed at the periphery of the viroplasm and appeared to be somewhat degenerate; in general the cytopathology resembled that of the above-mentioned frog virus infections and not that reported by Kelly and Tinsley (1974) for SIV and CIV infection of A. aegypti cells in which electron-opaque areas devoid of mitochondria were seen.

Lunger (1969) noted budding of frog ICDV's into cytoplasmic vacuoles, and Webb et al. (1974) observed virions in close proximity to the membrane surrounding vacuoles of MIV-infected A. aegypti cells. At the time of their report, it had not been determined that these virions were budding into the cytoplasmic vacuoles, and it is only recently (Webb *et al.*, 1975b) that the process was observed. Kelly (1975) noted budding of FV-3 progeny virions into cytoplasmic vacuoles in infected chick embryo fibroblast cells and also noted the rare occurrence of multienveloped particles in the cytoplasm, apparently a result of intracellular budding. African swine fever virus, another ICDV, has been observed (Breeze and DeBoer, 1966) to produce similarly multienveloped virions, but these have never been observed with the insect-pathogenic ICDV's.

Aedes aegypti cells infected with MIV do not accumulate virus to form crystalline arrays (Webb *et al.*, 1975b) as has been noted in FV infection of the tissue culture systems employed (Darlington *et al.*, 1966; Bingen-Brendel, 1971; Kelly, 1975). The MIV buds continually from the cells as it is produced. Less extensive budding was reported (Darlington *et al.*, 1966) in the release of FV-1 virions from infected FHM or BHK cells. Bingen-Brendel *et al.* (1971) reported budding of FV-3 from BHK cells, but to what extent is not evident. It appears that the extent of budding may be associated with the formation of paracrystalline arrays, less budding of progeny virions taking place in those cases where crystals are formed.

The process of budding by MIV from infected *A. aegypti* cells (Webb *et al.*, 1975b) is apparently similar to that reported by Kelly (1975) for FV-3. The lipid bilayer appears thickened and definitely more electron-opaque than the adjacent cell membrane. Thoroughly rinsed MIV-infected *A. aegypti* cells are agglutinated in the presence of the respective virus antisera and do not react with normal rabbit sera (S. R. Webb, unpublished observations). This indicates "insertion or dislocation of virus structural polypeptides" as suggested by Kelly (1975). Unfortunately, enveloped particles were not reacted with antisera to determine any serological cross reaction between viral and/or cell proteins.

Our results are in agreement with those of Kelly (1975) regarding the opacity of the FV-3 envelope being a transient phenomenon since sucrose density gradient purified preparations of the enveloped progeny MIV from cell cultures lacked the opacity noted during the budding process (S. R. Webb and J. D. Paschke, unpublished observations).

In the above discussion, we have attempted to draw analogies between the infection of cell cultures by insect ICDV's and the similar frog viruses. Similarities and certain discrepancies do exist, but, in general, MIV infection of *A. aegypti* cells resembles that observed in the infection of other cell lines by members of the insect-pathogenic ICDV's and also the well-studied frog virus FV-3. One of the major biological discrepancies is the lack of infectivity by cell-produced MIV.

The study of MIV, or for that matter other ICDV's, in tissue culture systems will not, for obvious reasons, provide useful information in the interpretation of further studies of the *gross pathology* of iridescent viruses in the insect host as suggested by Kelly (1975). Such studies will however eventually lead to a better understanding of the physiopathology involved in viral infection of insect cells.

Acknowledgments

This investigation was supported in part by USPHS, NIH, NIAID Research Grant AI-09972. This paper is Purdue Agricultural Experiment Station Publication No. 5862.

References

Anderson, J. F. (1970). *J. Invertebr. Pathol.* **15**, 219–224.
Bellett, A. J. D. (1965). *Virology* **26**, 127–131.
Bellett, A. J. D. (1968). *Advan. Virus Res.* **13**, 225–247.
Bellett, A. J. D., and Mercer, E. H. (1964). *Virology* **24**, 645–653.
Bingen-Brendel, A., Tripier, F., and Kirn, A. (1971). *J. Microsc. Paris* **11**, 249–258.
Breeze, S. S., and DeBoer, C. J. (1966). *Virology* **28**, 420–428.
Campadelli-Fiume, G., Bragaglia, M. M., Costanzo, F., Foa, L., and LaPlaca, M. (1973). *J. Gen. Virol.* **21**, 197–200.
Chapman, H. C., Clark, T. B., Woodard, D. B., and Kellen, W. R. (1966). *J. Invertebr. Pathol.* **8**, 545–546.
Clark, T. B., Kellen, W. R., and Lum, P. T. M. (1965). *J. Invertebr. Pathol.* **7**, 519–520.
Cory, J., and Yunker, C. E. (1972). *Acta Virol. (Prague)* **16**, 90.
Darlington, R. W., Granoff, A., and Breeze, D. C. (1966). *Virology* **29**, 149–156.
Fukaya, M., and Nasu, S. (1966). *Appl. Entomol. Zool.* **1**, 69–72.
Grace, T. D. C. (1962). *Nature (London)* **195**, 788–789.
Hasan, S., Croizier, G., Vago. C., and Duthoit, J. L. (1970). *Ann. Zool. Ecol. Anim.* **2**, 295–299.
Houts, G. E., Gravell, M., and Granoff, A. (1974). *Virology* **58**, 589–594.
Hukuhara, T., and Hashimoto, Y. (1967). *J. Invertebr. Pathol.* **9**, 278–281.
Kelly, D. C. (1975). *J. Gen. Virol.* **26**, 71–86.
Kelly, D. C., and Robertson, J. S. (1973). *J. Gen. Virol.* **20**, 17–41.
Kelly, D. C., and Tinsley, T. W. (1974). *Microbios* **9**, 75–93.
Linley, J. R., and Nielson, H. T. (1968a). *J. Invertebr. Pathol.* **12**, 7–16.
Linley, J. R., and Nielson, H. T. (1968b). *J. Invertebr. Pathol.* **12**, 17–24.
Lunger, P. D. (1969). In "Biology of Amphibian Tumours" (M. Mizell, ed.), Recent Results Cancer Res., Suppl., pp. 296–309. Springer-Verlag, Berlin and New York.
Matta, J. F., and Lowe, R. E. (1970). *J. Invertebr. Pathol.* **16**, 38–41.
Mitsuhashi, J. (1966). *Appl. Entomol. Zool.* **1**, 199–201.
Paschke, J. D., Campbell, W. R., and Webb, S. R. (1972). *Monogr. Virol.* **6**, 20–24.
Peleg, J. (1968). *Virology* **35**, 617–619.
Reed, L. J., and Muench, H. (1938). *Am. J. Hyg.* **27**, 493–497.
Singh, K. R. P. (1967). *Curr. Sci.* **36**, 506–508.
Steinhaus, E. A., and Leutenegger, R. (1963). *J. Insect. Pathol.* **5**, 266–270.
Stoltz, D. B. (1971). *J. Ultrastruct. Res.* **37**, 219–239.

Tinsley, T. W., and Kelly, D. C. (1970). *J. Invertebr. Pathol.* **16**, 470–472.

Tinsley, T. W., Robertson, J. S., Rivers, C. F., and Service, M. W. (1971). *J. Invertebr. Pathol.* **18**, 427–429.

Vago, C., Rioux, J. A., Duthoit, J. L., and Dedet, J. P. (1969). *Ann. Parasitol. Hum. Comp.* **44**, 667–676.

Varma, M. G. R., and Pudney, M. (1969). *J. Med. Entomol.* **6**, 432–439.

Wagner, G. W., Paschke, J. D., Campbell, W. R., and Webb, S. R. (1973). *Virology* **52**, 72–80.

Wagner, G. W., Webb, S. R., Paschke, J. D., and Campbell, W. R. (1975). *Virology* **64**, 430–437.

Webb, S. R., Paschke, J. D., Wagner, G. W., and Campbell, W. R. (1974). *J. Invertebr. Pathol.* **23**, 255–258.

Webb, S. R., Paschke, J. D., Wagner, G. W., and Campbell, W. R. (1975a). *J. Invertebr. Pathol.* **26**, 205–212.

Webb, S. R., Paschke, J. D., Wagner, G. W., and Campbell, W. R. (1975b). *J. Invertebr. Pathol.* (in press).

Weiser, J. (1965). *Bull. W.H.O.* **33**, 586–588.

Williams, R. C., and Smith, K. M. (1957). *Nature (London)* **179**, 119–120.

Woodard, D. B., and Chapman, H. C. (1968). *J. Invertebr. Pathol.* **11**, 296–301.

15

The Production of Viruses for Insect Control in Large Scale Cultures of Insect Cells

JAMES L. VAUGHN

I. Introduction

The study of viruses associated with insects was one motivating force in insect cell culture research from its beginning. In fact, the first person (Trager, 1935) to successfully establish primary cultures of insect cells, other than germ cells, infected these historic cultures with a nuclear polyhedrosis virus (NPV). Since the motivating idea behind most early studies of insect viruses was to use these pathogens to control harmful insect pests, it was only logical that the use of insect cell cultures to produce the viruses should be appealing.

The advantages of virus production in cell cultures over the production in insects are many. First, cleanliness is perhaps the most important advantage. Cell cultures can be tested to assure that they are free of viruses, mycoplasma, or other microorganisms. Once available, these uncontaminated cell lines can be maintained much easier and cheaper than living insects. During production, the conditions for cell growth and virus replication can be closely controlled to assure a uniform final product that is free of any microorganisms except the desired virus. In addition to contaminating microorganisms, insect-produced viruses contain large amounts of insect protein and cuticle; this material from some insects is highly allergenic to humans and must be removed from the virus before incorporation into the final product. Since viruses produced in tissue culture would contain no contaminating microorganisms, no insect cuticle, and little insect protein, the procedures needed to harvest the polyhedra and remove the contaminants are simple and inexpensive compared to those required for insect-produced virus.

Second, stability is an advantage of the tissue culture system. Available cell lines can be cloned and a cell strain selected which produces a virus of high virulence and uniformity. Such cell strains can be stored frozen for long periods with little or no change.

Third, many of the probable producers of insect viruses, i.e., companies engaged in the production of agricultural chemicals and vaccines, already have plant facilities and personnel experienced in some type of tissue culture. These facilities can easily be used to culture insect cells when needed. Therefore, large investments in new facilities and new personnel needed to rear the large numbers of insects required to produce the virus by this method would not be needed.

Despite these obvious advantages favoring the tissue culture system, all insect viruses produced for pest control are produced in living insects. In reviewing this subject, Ignoffo and Hink (1971) wrote: "Application of laboratory technology to commercial systems has yet to be demonstrated and judging from our present rate of development will not be realized for some time." They said "the need must be coupled with major developments in insect tissue culture [sic] technology," and they listed four areas in which such developments were needed: "(1) development of prolific, high yield per volume cell lines; (2) simplification of culture media; (3) propagation of high pathogen titers in selected cell lines; and (4) design and development of plant-scale equipment and routine production procedures" (Ignoffo and Hink, 1971). It is my purpose in this paper to redefine the status of insect cell cultures for the production of viruses by reviewing the recent research on these points. Most of the research on this problem has been to develop methods for the propagation of the NPV's of the Lepidoptera, and I will confine my remarks to those research methods. However, it should be pointed out that such groups as the World Health Organization are interested in the use of viruses to control mosquitoes. Some research on the replication of viruses to control these insects is being done and the results of such research are reported by Paschke and Webb in Chapter 14.

II. The Development of Cell Lines

The first and most critical of the four points listed above is the development of the required cell lines. Although the production of large scale primary cultures is possible with some vertebrate cells, similar methods for the production of large scale cultures of insect cells are not possible because of the size and anatomy of insects. Therefore, continuously propagated cell lines are a necessity. Ignoffo and Hink listed valid cell lines from 16 insect species in their review. Five of these are important agricultural pests: *Heliothis zea, Spodoptera frugiperda, Trichoplusia ni,* and *Chilo suppressalis.* Control of the first three with NPV was being considered at the time of the review. However, all virus propagation was done in the insect host. Also, among the 16 insect species listed were four mosquito species which are vectors of human diseases: *Aedes aegypti, Aedes albopictus, Aedes vittatus,* and *Anopheles stephensi,* but in 1971 there was little evidence to indicate future control of these pests with viruses. The cell lines were used principally to study the replication of arboviruses.

Since the publication of the review, considerable progress has been made in the replication of NPV's in cell lines. Goodwin *et al.* (1970) demonstrated that the NPV from *S. frugiperda* would replicate in a cell line from that insect (IPLB-21). Subsequently another cell line IPLB-1254, which would replicate the Spodoptera NPV, was developed from the same insect (Goodwin *et al.,* 1973). In our laboratory, the IPLB-21 line has been maintained on a hemolymph-containing medium which is obviously unsuited for large scale culture. However, the cell line has been adapted to a hemolymph-free medium by Gardiner and Stockdale (1975).

The cell lines from *H. zea* listed by Ignoffo and Hink are not capable of complete replication of the NPV from *H. zea.* Infection with purified DNA resulted in the production of unidentified infectious material, but no polyhedra were produced in the cells (Ignoffo *et al.,* 1971). However, cell lines have been developed from that insect which will replicate the *H. zea* NPV with the production of infectious polyhedra (Goodwin *et al.,* 1973). The cell lines from *T. ni,* (TN-368), were later shown to replicate the *T. ni* NPV (Faulkner and Henderson, 1972) and a promising NPV isolated from the Alfalfa looper *Autographa californica* (Vail *et al.,* 1973).

Thus, there are cell lines now available which are capable of replicating the *H. zea, T. ni, S. frugiperda,* and *A. californica* NPV's. These viruses are the ones that have shown the most promise for use in controlling their respective host insects. They are representative of the cell lines obtained from Lepidoptera. The IPLB-21, IPLB-1254, and the TN-368 cell lines are the only ones for which data on production capability have been studied and thus are the basis for the remainder of this report.

III. Development of Simplified, Less Expensive Media

The media used in the early studies on insect cell culture were all supplemented with insect hemolymph which, as mentioned earlier, made them unsuitable for large scale cultures. However, that was only one of the problems. The medium used for the culture of these cell lines and, in fact, for most cell lines from Lepidoptera was developed by Grace (1962) or some modifications of it. This medium contains 44 different ingredients and is costly to prepare. In addition the first supplemented media that did not contain insect hemolymph contained 10% fetal bovine serum, 10% whole chicken egg ultrafiltrate and 1% bovine plasma albumin (Yunker et al., 1967). All these added to the cost considerably. Thus, there was definite need for developing a simpler, less expensive media for large scale work.

The first successful alternate medium was a modification of the Yunker, Vaughn, and Cory medium, which reduced the level of the supplements. In this medium, the egg ultrafiltrate and the bovine plasma albumin were omitted, and 0.3% yeastolate and 0.3% lactalbumin hydrolysate were added. The level of fetal bovine serum was reduced to 4%. These changes reduced the cost of medium for the TN-368 line by about 50% (Hink et al., 1974).

Another approach to simplifying the media and reducing the cost has been to replace many of the purified chemical constituents in the minimal Grace's medium with less-pure natural materials, to test various other components, and to omit those not required for cell growth. The most successful medium developed with this approach thus far is the BML-TC/7A medium of Gardner and Stockdale (1975). The composition of their medium is shown in Table I. This medium contains only 10 components and I estimate could be prepared in 4–6 hours, as compared to Grace's medium which takes 12–18 hours. Since the labor costs for preparing the medium are much more than the cost of the raw materials, considerable progress has been made in simplifying and reducing the cost of media. It would appear that, subject to minor modifications, we have achieved the maximum results possible until some further major discoveries are made in insect cell metabolism and that the most beneficial advances will now be made in other areas.

TABLE I

Formula for BML-TC/7A Medium of
Gardiner and Stockdale[a]

Component	Amount/liter
$NaH_2PO_4 \cdot 2H_2O$	1.145 g
KCl	4.61 g
$CaCl_2 \cdot 2H_2O$	1.32 g
$MgCl_2 \cdot 6H_2O$	2.28 g
$MgSO_4 \cdot 7H_2O$	2.78 g
$NaHCO_3$	0.35 g
Lactalbumin hydrolysate	10.00 g
Tryptose broth	5.00 g
Glucose	2.00 g
Grace's vitamins (1000X)	1.0 ml

[a] From Gardiner and Stockdale (1975).

IV. Propagation of High Titers in Selected Cell Lines

One possible area of advancement is the third point raised by Ignoffo and Hink, namely, the propagation of high titers of virus in selected cell lines. Table II shows a summary of the results of several attempts to produce viruses in various cell lines. The average yield shown here for the NPV from *A. californica* is 1.46×10^7 polyhedra/ml of medium. From these data and the data on cell yield, it can be calculated that an average of 10–11 polyhedra/cell is produced in TN-368 cells and 2 polyhedra/cell in the 1254 line. The TN-368 line produces an average of 22–23 polyhedra of the *T. ni* NPV/cell.

To evaluate this data in terms of commercial production, in the TN-

TABLE II

NPV Virus Yields from Insect Cell Lines

Virus	Cell line	Cells/ml	Polyhedra/ml	Reference
T. ni	TN-368	1.9×10^6	44.4×10^6	Faulkner and Henderson (1972)
A. californica	TN-368	1.9×10^6	19.2×10^6	Vail *et al.* (1973)
A. californica	IPLB-1254	4.5×10^6	10.0×10^6	
S. frugiperda	IPLB-1254	4.5×10^6	12.4×10^6	

368 cell line, it would require 15.8 liters of culture to produce the *A. californica* virus necessary to treat an acre of leafy vegetables at the rate of 50 larval equivalents (3×10^{11} polyhedra) per acre. Corresponding amounts of culture are 30 liters for *A. californica* virus in IPLB-1254 cells and 6.9 liters for *T. ni* virus in the TN-368 cells.

It has already been shown that many more than 10 polyhedra are produced in some cells. Faulkner and Henderson (1972) reported that microscopic observation of TN-368 cells infected with the *T. ni* virus contained as few as 5 polyhedra and as many as 200 polyhedra/nucleus. Vail *et al.* (1973) reported that TN-368 cells infected with *A. californica* virus produced an average yield of 64 polyhedra/cell, but that many cells contained well over 100 polyhedra/nucleus. These results were obtained from static cultures in standard tissue culture flasks. It is obvious that the efficiency of the volume culture system needs to be improved and at least two means of accomplishing this seems to have reasonable chances of success.

First, strains of highly susceptible and productive cells need to be selected by cloning the existing cell lines. Grace (1967) demonstrated that improved susceptibility to virus could be obtained by the judicious selection of cell clones. He was able to increase the susceptibility of his *Antheraea* cell line from 0.1% to 20%. However, no effort has yet been made by others to use this potent method to increase the virus yield of other cell culture systems.

Second, the condition of the cells when the culture is inoculated with virus appears to have a great influence on the yield of polyhedra. Cultures inoculated when the majority of cells are not dividing, i.e., in the stationary phase, produce few if any polyhedra. This seems to be true even if fresh medium is added after the virus has been allowed to adsorb. At the other extreme, if the virus is added too early, the yield of polyhedra is reduced because the number of cells per milliliter of medium is low. Although no one has obtained sufficient data to establish the proper time for inoculation, it would seem to be when the cells have nearly reached the maximum number but are still dividing. Thus, methods for synchronizing cell growth to obtain cultures with a high number of rapidly dividing cells would be necessary for maximum yields.

V. Design and Development of Plant-Scale Equipment and Routine Production Procedures

The last of the four areas to be considered is the design and devlopment of plant-scale equipment. Hopefully, this will require the least amount of research effort. By the late 1950's, when insect tissue culture was just developing into a useful tool for virus work, it had already been established that cultures of vertebrate cells were the method by which the poliovirus vaccine was to be produced. Viral vaccines against measles virus and adenovirus can also be produced in tissue cultures. Cell culturists now speak of producing vertebrate cells in terms of 50–100 gm/week.

Thus, the instrumentation and methodology for the large scale production of animal cells is being developed to meet the needs of medicine and public health. There seems to be every indication that the technology already developed or currently being developed for the large scale culture of vertebrate cells can be used, with only minor modification, for the culture of insect cells. The current technology can be divided into two types of culture, those in which the cells are grown attached to a surface of some type and those in which the cells grow suspended in the culture medium by some method. Kruse and Patterson (1973, Section VI) describe several of these procedures. They range from simple batch cultures in roller bottles, to multisurface chambers, to continuous suspension cultures by means of a Nephelostat.

Some of these systems have already been tested and found suitable for growing insect cells. It was demonstrated in our laboratory (Vaughn, 1968) that the Grace *Antheraea* cell line, which normally adheres loosely to the surface, could be grown in suspension culture in standard spinner flasks. In these flasks, the cells were maintained in suspension by a slowly rotating magnetic bar suspended from the top of the vessel and driven by an external magnet. The cells were protected from possible damage from shearing forces by the addition of 0.15% methyl cellulose (400 centipoise). The population doubling time was equal to that of these cells in static cultures, but the final concentration of cells achieved was only about one-half that obtained in static cultures.

Recently another system has been developed which is referred to as a spin-filter vessel. The construction and geometry of the rotating filter is such that cells or other particles will not accumulate on the surface or within the filter. Thus medium and/or other additives can be added or removed without cell loss. The filter and stirrer are rotated by an external magnetic stirrer. Ports are available through which electrodes can be installed to monitor pH or dissolved oxygen. These variables can also be automatically controlled through additional ports. Hink *et al.* (1974) have demonstrated that his TN-368 cell line grows well in this system. Methyl cellulose or Darvan No. 2 was added to reduce clumping. Bovine serum albumin also reduced clumping and increased the growth rate and final cell density. Aeration was required to achieve the maximum cell growth. When these variables were controlled, an inoculum of 1.3×10^5 cells/ml produced 1.9×10^6 cells/ml in 56 hours. Thus, the growth rate and final cell density were comparable to static cultures of this cell line.

VI. Conclusions

If in summary we look again at the four areas of research listed by Ignoffo and Hink, it is obvious that our knowledge about virus production in large scale tissue culture has increased considerably in the 3 years since they prepared their review. Cell lines have been developed which will replicate the NPV's infectious for many serious insect pests. In addition to those discussed in this paper from important agricultural pests, a cell line has been developed from the forest tent caterpillar *Malacosma disstria*, which will support the replication of two NPV's of two important forest insect pests, the spruce budworm *Choristoneura fumiferana* and the western oak looper *Lambdina fescellaria sommiaria* (Sohi and Bird, 1971; Sohi and Cunningham, 1972). As mentioned earlier many cell lines are available from mosquitoes and, should a suitable virus for controlling this public health pest be found, production methods could be developed.

Remarkable progress has been made in developing suitable media for use in these systems. While it is not known whether other cell lines would grow in the simplified media described here, these media will at least provide the basis for development of more suitable media if needed. There seems little doubt that the technology being developed for use with vertebrate cells can be used for insect cells. Some modifications will no doubt be necessary, but basically the technology is compatible with our requirements.

It is in the production of high virus titer yields that the most research is needed and where the benefits will be the greatest. Highly selected clones need to be obtained and protocols for their efficient use need to be developed. Even here the methods for cell cloning are already available; they need only be applied to insect cell lines. Thus, I think that the production of virus in large scale insect cultures is commercially feasible and that the benefits of such systems, already demonstrated for vaccine production in vertebrate cells, can be obtained with insect cells.

References

Faulkner, P., and Henderson, J. F. (1972). *Virology* **50**, 920–924.
Gardiner, G. R., and Stockdale, H. (1975). *J. Invertebr. Pathol.* **25**, 363–370.
Goodwin, R. H., Vaughn, J. L., Adams, J. R., and Louloudes, S. J. (1970). *J. Invertebr. Pathol.* **16**, 284–288.

Goodwin, R. H., Vaughn, J. L., Adams, J. R., and Louloudes, S. J. (1973). *Misc. Publ. Entomol. Soc. Amer.* **9**, 66–72.

Grace, T. D. C. (1962). *Nature (London)* **195**, 788–789.

Grace, T. D. C. (1967). *In Vitro* **3**, 104–117.

Hink, W. F., Strauss, E., and Mears, J. L. (1974). *In Vitro* **9**, 371 (abstr.).

Ignoffo, C. M., and Hink, W. F. (1971). *In* "Microbiol Control of Insects" (H. D. Burgess and N. W. Hussey, eds.), pp. 541–580. Academic Press, New York.

Ignoffo, C. M., Shapiro, M., and Hink, W. F. (1971). *J. Invertebr. Pathol.* **18**, 131–134

Kruse, P. F., Jr., and Patterson, M. K., Jr., eds. (1973). "Tissue Culture: Methods and Applications." Academic Press, New York.

Sohi, S. S., and Bird, F. T. (1971). *Proc. 4th Annu. Meet. Soc. Invertebr. Pathol* S. I. P. Newsletter No. 3, p. 9 (abstr.).

Sohi, S. S., and Cunningham, J. C. (1972). *J. Invertebr. Pathol.* **19**, 51–61.

Trager, W. (1935). *J. Exp. Med.* **61**, 501–514.

Vail, P. V., Jay, D. L., and Hink, W. F. (1973). *J. Invertebr. Pathol.* **22**, 231–237.

Vaughn, J. L. (1968). *Proc. Int. Colloq. Intertebr. Tissue Cult., 2nd,* 1967 pp. 119–125

Yunker, C. E., Vaughn, J. L., and Cory, J. (1967). *Science* **155**, 1565–1566.

16

Plant Pathology Applications

KARL MARAMOROSCH

I. History

Applications of invertebrate tissue culture for the study of filterable plant pathogens, specifically viruses, spiroplasmas, mycoplasmalike agents, and rickettsialike agents, have trailed behind work carried out with mammalian cells and disease agents affecting higher animals. The reasons for this slow progress were numerous. Until 1955, insect tissue culture was in its infancy. In that year an informal conference on insect tissue culture was held at MacDonald College of McGill University in Canada. The participants concluded that insect cells have a limited capacity to grow *in vitro*. Only certain types of tissues, such as insect gonads, and only cells from immature stages were believed to be suitable for *in vitro* cultivation. Besides, media for insect tissue culture were believed to require insect hormones, as well as trehalose and insect hemolymph, as essential ingredients in order to maintain cell growth.

Two decades later, nearly 160 insect cell lines have been developed and several have been grown in commercial media, such as TC199-MK, found suitable for their continuous cultivation (McIntosh *et al.*, 1973), although devoid of trehalose and hemolymph.

The realization that certain filterable plant disease agents are not viruses, but microorganisms resembling wall-less mycoplasmas, or very small, walled rickettsiae, came with the recognition of such agents in Japan in 1967 (Doi *et al.*, 1967; Ishiie *et al.*, 1967; Nasu *et al.*, 1967).

Until then, such disease agents were grouped together with viruses. Nevertheless, successful maintenance of the aster yellows agent was reported in nymphal tissues of a leafhopper vector *in vitro* (Maramorosch, 1956). Subsequently, during the period 1962–1965, methods were developed in France (Vago and Flandre, 1963), the United States (Hirumi and Maramorosch, 1964a,b,c; Mitsuhashi and Maramorosch, 1964a), and Japan (Mitsuhashi, 1965) to culture leafhopper embryonic, and even nymphal, and imaginal tissues. These studies were of notable importance in helping to achieve the study of filterable plant disease agents in cultured cells of vectors. The main purpose of these attempts was to provide rapid means for identification of viruses and mycoplasmalike agents that routinely required prolonged greenhouse and insectary tests. Another applied type of work concerned the accurate quantitative determination of the concentration of plant viruses in plants, as well as in insect vectors, not possible by available techniques. Even in the rare instances in which local lesion tests have been available, as in the case of potato yellow dwarf virus (Black, 1969), the lesions on plant hosts were not suitable for precise estimation of virus concentration, requiring an excess of virions to cause lesions in plants. Basic aspects that could profit from invertebrate tissue culture systems concerned quantitative assay of viruses and mycoplasmalike organism (MLO), and studies of interactions between the disease agents and insect host on a cellular level.

Some of the goals have already been achieved while others have not. For reasons that are obscure, the inoculation of insect cells with MLO has not yet been successful, and systems other than those employing leafhopper vector cells have lagged behind. The plant viruses studied so far have failed to produce plaques that could be observed directly under a light microscope. Nevertheless, by applying immunofluorescence techniques, accurate quantitative studies on a cellular level of several plant pathogenic viruses have been possible. With the breakthrough in cell cloning (McIntosh and Rechtoris, 1974) and with the application of simplified media, it can be expected that invertebrate tissue culture will become widely used in plant pathology.

II. Tissue Explants *in Vitro*

The first demonstration that a plant-pathogenic agent, now recognized as a mycoplasmalike organism, can be maintained *in vitro* in tissues of its vector was reported in 1956 (Maramorosch, 1956). The test consisted of an attempt to maintain living tissues from nymphs of the sixspotted leafhopper *Macrosteles fascifrons*, a vector of the aster yellows MLO, for a period of time adequate for its multiplication. Nymphs of *M. fascifrons* were confined to aster yellows-diseased plants to acquire the disease agent. After 2 days the nymphs were surface-sterilized by dipping them into a potassium permanganate solution, 10 mg in 100 ml of distilled water, with a few drops of a wetting agent. Each nymph was then cut into 10–12 pieces, and these in turn were washed twice in the medium and placed in hanging drops on Maximov slides. The medium contained mineral salts, bovine albumin, casein hydrolysate, and the antibiotics streptomycin and penicillin, as well as mycostatin. Using needle inoculation of stock leafhopper vectors as bioassay, it was found that the MLO of aster yellows completed their incubation in the leafhopper tissues *in vitro*. The injected insects became active vectors, following needle inoculation. The tissues did not deteriorate and most cells remained alive. No MLO was recovered from the culture medium. The results thus demonstrated that the aster yellows agent could complete its incubation not only in leafhopper vectors *in vivo*, but also in insect tissues *in vitro*. Multiplication of the MLO obviously did occur in vector tissues under the conditions of the experiment. Later attempts of the author to inoculate nymphal tissues under similar culture conditions with aster yellows MLO *in vitro* failed (unpublished). Until now, although monolayers and even cloned cell lines of leafhopper vectors became available, it has not been possible to successfully inoculate them with MLO *in vitro*. On the other hand, inoculation with plant-pathogenic viruses, developed by Black and co-workers (Black, 1969), provided a useful tool for the study of viruses that infect plants as well as insect vectors.

Although the aster yellows MLO were recovered from tissues maintained *in vitro*, the site of multiplication and storage in the leafhopper vector was not known. An attempt was made to recover MLO from organs maintained in culture media for 14 days (Hirumi and Maramorosch, 1963a). Insects were dissected, their organs washed in sterile salt solution, and then transferred to culture media. After 14 days, MLO was recovered from salivary glands, and gut, but not from Malpighian tubules, ovaries, and testes maintained *in vitro*.

One of the difficulties encountered in this work was the frequent contamination of cultured material with fungi. These were more difficult to eliminate than the bacterial contaminants, usually susceptible to penicillin and streptomycin. To prevent fungal contamination, a new method was devised to grow leafhopper vectors under aseptic conditions. Four species were obtained in such manner by Mitsuhashi and Maramorosch in 1963, and their tissues found suitable for the preparation of tissue cultures. Besides, the aseptic insects were also suitable for feeding on plant tissues maintained *in vitro* (Mitsuhashi and Maramorosch, 1963).

Inoculation of plant tissue cultures with aster yellows MLO was eventually achieved using a combination of these techniques (Mitsuhashi and Maramorosch, 1964b).

III. Primary Cultures of Cells

Two major breakthroughs were made in 1964: the establishment that the most suitable cells for *in vitro* culture were those obtained during the blastokinetic stage of leafhopper embryo (Hirumi and Maramorosch, 1964a) and the finding that cells from stages other than the embryo also can be grown *in vitro* (Mitsuhashi and Maramorosch, 1964a). When embryonic tissue fragments in the blastokinetic movement stage were placed in flasks with culture medium, a number of cells became attached to the glass during the initial 2 hours of cultivation. The number of cells increased about tenfold during the next 2 days. Fibroblast and epithelial-type cells continued to grow for more than 40 days. At first, subculturing of these cells was hampered by the lack of proper media. Nevertheless, certain experiments were possible with the primary cultures, for instance, the inoculation with a plant-pathogenic virus. This was achieved by Mitsuhashi with *Nephotettix cincticeps* tissues inoculated *in vitro* with the rice dwarf virus (Mitsuhashi, 1965). Granulation of cells in the cell sheets and other changes, indicating degeneration, were described. Electron micrographs revealed the presence of virions as well as of numerous particles of the size of the virus cores.

These pioneering experiments led Black and his co-workers to inoculate primary cultures of the vector *Agallia constricta* with wound tumor virus. Initial inocula were obtained from virulifeous leafhoppers that were homogenized and their extracts filtered through Millipore membranes. Although no plaques were obtained, the use of fluorescent antibody permitted the detection of the viral antigen on the third day of cultivation. Using the insect inoculation technique as bioassay, Reddy and Black (1966) calculated a 10^2 increase in titer during 8 days of incubation.

Shortly thereafter, Mitsuhashi and Nasu (1967) were able to obtain cells that no longer showed signs of degeneration *in vitro* and that supported the multiplication of the rice dwarf virus. Mature virions, observed by electron microscopy, were dispersed in viral matrices. Near the periphery of such matrices, virions formed linear arrangements within sheathlike substructures.

Another breakthrough was reported the same year, when it was found by electron microscopy examination that wound tumor virus can infect and multiply in nonvector cells *in vitro* (Hirumi and Maramorosch,

1968). This finding contradicted previous speculations (Black, 1969) that only vector cells can be used for maintaining plant-pathogenic viruses *in vitro*. The finding was not only of practical, but also of basic importance. There had been a tendency to interpret results of *in vitro* cultivation as evidence for multiplication *in vivo*. One such example was the statement that observations of potato yellow dwarf virus (PYDV) multiplication in tissue culture "provided the first critical evidence that PYDV multiplies in its vector" (Black, 1969, p. 84). In fact, the mentioned tests merely provided evidence for multiplication in vector cells *in vitro*, but not evidence that PYDV multiplied in its vector.

Several other species of leafhopper vectors have provided cells for *in vitro* cultures. Shikata *et al.* (1970) succeeded in maintaining embryonic tissues of the following: *Laodelphax striatellus*, a vector of rice stripe, rice black streaked dwarf, and northern cereal mosaic viruses; *Inazuma dorsalis* and *Nephotettix apicalis* vectors of rice dwarf and rice transitory yellowing viruses. Surprisingly, the cells of these three vector species required different compositions of tissue culture media for optimal cell proliferation.

IV. Cell Lines

The year 1967 marks the start of a new phase in applications of insect tissue culture to plant pathological research. Chiu and Black (1967) announced the successful establishment of continuous cultures of *A. constricta* cells in monolayers. The cells were obtained from embryonic explants and the authors used the blastokinetic stage, found earlier as the most suitable for *in vitro* cultivation (Hirumi and Maramorosch, 1964a). In their earlier attempts, Chiu and Black had used the M & M medium developed by Mitsuhashi and Maramorosch (1964a), but then they substituted it with the *Drosophila* medium of Schneider (1964). The cultures that resulted were heterogenous, but epithelial-type cells were predominant. In addition to *A. constricta*, monolayers of cells from three other leafhopper vector species were successfully grown in continuous cultures: *Agallia quadripunctata*, *Aceratagallia sanquinolenta*, and *Agalliopsis novella*. When cells were subcultured for 55 passages, it became apparent that established cell lines of leafhoppers can finally be obtained.

Chiu and Black (1967) inoculated *A. constricta* monolayers with wound tumor virus, obtained from virus-carrying insects, infected primary explants, or plant tumors. It was shown earlier that all could provide infective inoculum (Chiu *et al.*, 1966). Since no cytopathic effects could be observed by phase contrast microscopy, specific staining with fluorescent antibody was used. The minimal concentration of virions required for the infection of cell monolayers was approximately 10^5/ml. The cell-infecting unit value was determined as 405 virions (Chiu and Black, 1969). Virus adsorption was maximal after 2 hours at 30°C, and the number of fluorescent cells increased linearly with the relative virus concentration of the inoculum. This indicated that a single virus particle produced each infection. Using for the bioassay the needle injection technique developed by Maramorosch et al. (1949), the minimum infective dose was calculated as 398 virions (Gamez and Black, 1968).

The report by Mitsuhashi (1967) was the first announcement of successful in vitro inoculation of cells from an invertebrate nonhost by a virus. Mitsuhashi infected primary cell cultures of N. cincticeps with the Chilo iridescent virus. Within 24 hours virus infection became apparent and electron micrographs of sectioned cells confirmed the presence of virions in the cultured cells.

Shortly thereafter, the interactions at the cellular level between the plant-pathogenic wound tumor virus and cultured cells of a nonvector leafhopper Macrosteles fascifrons were studied in vitro by Hirumi and Maramorosch (1968). Electron microscopic observation of cultured cells, fixed 7 days after virus inoculation, revealed typical sites of wound tumor virus viroplasms, consisting of fibrous material and resembling viroplasms observed in infected insects and plants. Virions were scattered throughout the viroplasmic matrix area. Small aggregations of virions appeared in various areas of the cell cytoplasm, but none was observed within nuclei. These findings demonstrated that the plant- and insect-pathogenic wound tumor virus was able to multiply to a certain extent in cells of a nonvector insect. Similarly, cells of many species of animals support multiplication of viruses that cannot infect the host animals in vivo. This is particularly common with embryonic cells, such as chick embryo. The susceptibility of M. fascifrons in vivo to viruses for which this species does not act as vector was not tested, but corn stunt spiroplasma was found to be retained, and probably to multiply, in the nonvector leafhoppers (Maramorosch, 1952). It was assumed at the time that multiplication at a low level might have occurred. Similarly, the MLO causing the aster yellows disease was maintained in nonvector corn leafhoppers Dalbulus maidis. Attempts in recent years to inoculate monolayers of cells derived from M. fascifrons and D. maidis with MLO have not been successful up to now.

When continuous cell lines of nonvector species became available, A. sanquinolenta cells were inoculated with wound tumor virus and the number of infected cells found to be up to 2 log units lower than with vector cells (Chiu and Black, 1969). The susceptibility of nonvector cells to potato yellow dwarf virus was found to be about 10% of that of vector cells (Liu and Black, 1969).

Susceptibility of leafhopper cell monolayers varies from nearly 100% to less than 1%, depending on the viruses as well as on the cells used. Since the leafhopper cells have not been cloned until very recently, their heterogeneity could account for some of the discrepancies reported. The virus source, as well as the mode of preparation of the inoculum, present another variable. When *A. sanquinolenta* cell monolayers were incubated with potato yellow dwarf virus, less than 1% of cells was infected, as determined by immunofluorescent staining (Chiu *et al.*, 1970). A linear relationship, as in the case of wound tumor virus, was recorded with potato yellow dwarf virus concentration in the inoculum and the incidence of infected cells in monolayers. The rhabdo-type potato yellow dwarf virus has been located by electron microscopy at the perinuclear membranes (MacLeod *et al.*, 1966). Fluorescent staining of monolayers is also at first confined to the nucleus and occurs in spots near the periphery of nuclei.

A detailed study was made of factors affecting infectivity assays of plant-pathogenic viruses on vector cell monolayers (Kimura and Black, 1971). Infectivity assays on vector cell monolayers provided not only more rapid, but also more sensitive means than infectivity assays by needle inoculation of vectors. The latter always took several weeks, and they required the testing of inoculated vectors on plants (Maramorosch *et al.*, 1949). The optimum seeding density on monolayers was 2.9×10^5 cells/0.1 ml. The effect of the addition of histidine–$MgCl_2$ was noted in preserving the infectivity of wound tumor virus, when stored at $-80°C$. Cells of leafhopper vectors have now also been stored at this temperature for prolonged periods and later revived. Such preservation provides a convenient method for the storing, as well as the shipping of cells, and it minimizes the danger of losing these cell lines. Besides, perhaps even more important, it preserves the cells so that actively cultured lines, which change and become polyploid, can be compared with the original, parent lines. Kimura and Black (1971) reported that the optimum pH value for leafhopper cell inoculum was 6.5 when a glycine–$MgCl_2$–phosphate solution was used, and 6.0 for a histidine–$MgCl_2$ solution. Assay by monolayer detected wound tumor virus at a 10^{-2} lower concentration than assay by insect injection. For obscure reasons, no infectivity was found in inocula prepared at a dilution of 10^{-1}.

Further tests by Kimura and Black (1972a) provided quantitative data on the number of virions needed to produce wound tumor virus infection *in vitro*. While tobacco mosaic virus, the most infectious of all plant viruses, requires 20,000–200,000 particles for each local lesion on a

tobacco leaf, only about four virions were required per cell in the *in vitro*
system employing wound tumor virus and vector cell monolayers. Further
improvements in the cell culture technique might eventually result in
the nearly 1:1 relationship, theoretically obtainable with bacterial viruses
on bacterial lawns.

Cell monolayers were also used for accurate measurement of wound
tumor virus concentration in extracts from insect vectors. Virus could
be assayed 8 days after it was acquired by leafhoppers. Between the
eighth and twenty-fifth day the concentration increased 50,000–100,000-
fold, then reached a plateau level (Reddy and Black, 1972). The growth
curve on monolayers was determined by focus assay.

Aoki and Takebe (1969), and Takebe and Otsuki (1969), using plant
protoplasts, were able to infect up to 30% of plant cells *in vitro*. The
doubling time for tobacco mosaic virus was estimated at 30–60 minutes
(Kimura and Black, 1972b). Later Otsuki *et al.* (1972) improved the
synchronous infection, obtaining 80% of protoplasts infected with tobacco mosaic virus. The rate of multiplication of viruses can be measured
more accurately in *in vitro* systems than *in vivo*, and this applies to *in
vivo* systems in leaves, as well as in living insect vectors. This was clearly
demonstrated by Kimura and Black (1972b) who calculated the minimal
doubling time of wound tumor virus to be less than 60 minutes at
30°C. Doubling time of wound tumor virus estimated in insect vectors,
on the other hand, was approximately 20 hours.

The infectivity of wound tumor virus strains and isolates was compared
using cell monolayers (Liu *et al.*, 1973). The strain that was maintained
for many years by passing alternately through plants and vectors, as
well as a strain that lost its ability to infect insects when propagated
for many years in plants by grafting, was used in the experiments. In
addition, a strain that had lost its affinity to vectors only partially was
also tested. By combining infective assays on monolayers and particle
counts in the electron microscope, the following results were obtained.
The exvectorial or vectorless strain that no longer was transmissible by
leafhoppers was also unable to infect cell monolayers. The authors found
that this loss of infecting ability for cell monolayers of vectors occurs
during 5–9 years of exclusive propagation in plants. Apparently this loss
occurs in steps, and not as a sudden, single step mutation.

The leafhopper tissue culture medium was further improved by Martinez-Lopez and Black in recent years (1974). Best results were obtained
at 28°C, pH 6.43, osmotic pressure of 360 mOsm and component concentrations per liter of $CaCl_2 \cdot 2H_2O$, 0.36 g; $MgSO_4 \cdot 7H_2O$, 1.2 g;
KH_2PO_4, 0.27 g; KCl, 1.6 g; NaCl, 1.0 g; $NaHCO_3$, 0.9 g; dextrose, 9.0
g; lactalbumin hydrolysate, 10.0 g; yeast hydrolysate, 8.0 g; histidine-HCl

(monohydrate), 4.8 g; histidine (free base), 3.45 g; and 100 ml of fetal bovine serum, heat treated at 56°C for 30 minutes. This medium reportedly resulted in shorter doubling times, more cells in mitosis, and higher susceptibility to virus inoculation. The studies discussed so far were performed with the double stranded RNA-containing wound tumor virus. This virus resembles the rice dwarf virus closely (Reddy et al., 1974) and, to a lesser extent, the human-pathogenic reoviruses.

A different plant-pathogenic virus, the potato yellow dwarf virus, has been studied in cell monolayers since 1970 (Chiu et al., 1970). This virus is transmitted biologically by leafhopper vectors and, most likely, also multiplies in its insect vectors. Morphologically the virus resembles rhabdoviruses. Hsu and Black (1973a) found that the effect of pH on potato yellow dwarf inoculum was mediated through the virus, and not through the plant or insect host. A pH of 7.0 was better than 6.5 or 7.5 for the mechanical inoculation of leaves, while for leafhopper cell monolayers the optimum pH was 5.3–5.9. The rate of multiplication of potato yellow dwarf virus in monolayers of cultured vector cells was measured by Hsu and Black (1974). An eclipse period was observed during the first 9 hours postinoculation and a sharp increase of infectivity recorded between 9 and 29 hours postinoculation. A plateau was reached at 29 hours. The virus increase spanned 4 log units and the doubling time was 80 minutes during this period.

V. Aphid Tissue Culture

While tissue and cell cultures of leafhopper vectors of plant-pathogenic viruses were successfully developed, culminating in cell lines and monolayer cultures used for basic and applied studies, progress with cells of other groups of vectors was much slower. Aphid tissue culture has been attempted with some success, but no other invertebrate vectors have been cultured up to now. The first aphid tissue cultures were obtained by Tokumitsu and Maramorosch (1966) from the pea aphid Acyrthosiphon pisum, a vector of pea enation mosaic virus. Aphids reared in a greenhouse insectary and parthenogenetic apertous females were surface sterilized by immersion in 70% ethanol for 5 seconds. The insects were then placed in a sterilized petri dish and dissected in a Ringer–Tyrode solution. Besides the surface-sterilized insects, aseptically reared nymphs were also used for cultures. Trypsin at 0.25% was used to dissociate the tissues and to obtain dispersed cells. In some instances, adult aphids were dissected and nymphs removed by cesarean section, and cut into small pieces. Several kinds of cells migrated from explants and some survived up to 10 days attached to the glass surface. Eventually they detached and degenerated. The longest survival was up to 20 days, but cell division was not observed. Even though no cell lines were obtained from aphids, primary cultures of another vector species, *Hyperomyzus lactucae*, were successfully employed in a study of saw thistle yellow virus a few years later (Peters and Black, 1970). Infection of cells was demonstrated by fluorescent antibody staining. The first infection was noticed 37 hours postinoculation and the number of infected cells reached a maximum after 48 hours.

VI. Conclusions

Many leafhopper-borne viruses require two hosts, a plant and an arthropod, for their maintenance in nature. Several viruses belonging to this group have been shown to multiply in alternate plant and invertebrate hosts, and they thus constitute a link between typical animal and plant viruses. A few aphid-transmitted viruses also are transmitted biologically and probably multiply in aphid vectors. For many years, virus–host interactions of such biologically transmitted viruses could be studied only in plant or insect hosts *in vivo*, and to a limited extent in plant tissues *in vitro*. The development of invertebrate tissue culture methods, as well as of plant protoplasts, now permits the study of these viruses on a cellular level.

The manipulation of cells of higher plants, in a manner heretofore possible only with microorganisms and animal cells, opened new possibilities for experimental virus work. First, tissues of higher plants were dispersed under submerged conditions, subcultured, and maintained in cell suspensions for many weeks, (Maramorosch *et al.*, 1958). More recently, protoplasts of plant cells stripped of the cell walls by means of enzymes were found highly susceptible to virus inoculation *in vitro* (Otsuki *et al.*, 1972). The rapid progress in the cultivation of leafhopper vector cell lines and the less spectacular progress in respect to other vectors already provided accurate quantitative methods for assaying several plant-pathogenic viruses. Hopefully, other plant virus vectors, such as white flies, eriophyid mites, and nematodes, will eventually be cultured in a similar manner *in vitro*, so as to provide accurate quantitative means for virus research. It can also be expected that the wall-less microorganisms, such as spiroplasmas and mycoplasmalike organisms, as well as the walled, filterable microorganisms that resemble rickettsiae, will be grown in tissues of their respective invertebrate vectors (Maramorosch, 1973). The success of such approaches will depend on many factors, including improved techniques, better culture media, and, last but not least, close collaboration between entomologists, plant pathologists, virologists, and tissue culture experts.

VII. Summary

In 1956 the first successful attempt was made to maintain the agent of aster yellows disease in leafhopper vector tissues *in vitro*. During the following years, methods were developed for maintaining leafhopper tissues and embryonic cell cultures, and in 1965 the first leafhopper cell line was established from *Agallia constricta*, a vector of wound tumor virus. Other leafhopper cell lines have been established in recent years. Aphid tissue explants have been maintained *in vitro*, but no cell line has been obtained. Leafhopper cell lines have served mainly to study plant-pathogenic viruses. Attempts to infect such cells *in vitro* with mycoplasmalike or rickettsialike agents of plant diseases have not yet been successful. Invertebrate cell lines have provided rapid diagnostic tests as well as served for basic studies of plant viruses in which infectivity, immunofluorescence, autoradiography, and electron microscopy were employed. No plaques have been produced by plant viruses in invertebrate cells *in vitro*. Future research might provide cell lines of other types of vectors, such as aphids, eriophyid mites, and nematodes. Attempts will be made to obtain plaques in cell monolayers for quantitative assays. Cloning of cell lines might provide homogenous and highly susceptible lines. Improved and simplified media will permit the wider use of invertebrate tissue culture for the study of plant-pathogenic viruses.

REFERENCES

Aoki, S., and Takebe, I. (1969). *Virology* 39, 439–448.
Black, L. M. (1969). *Annu. Rev. Phytopathol.* 7, 73–100.
Chiu, R. J., and Black, L. M. (1967). *Nature (London)* 215, 1076–1078.
Chiu, R. J., and Black, L. M. (1969). *Virology* 37, 667–677.
Chiu, R. J., Reddy, D. V. R., and Black, L. M. (1966). *Virology* 30, 562–566.
Chiu, R. J., Liu, H. Y., MacLeod, R., and Black, L. M. (1970). *Virology* 40, 387–396.
Doi, Y., Terenaka, M., Yora, K., and Asuyama, H. (1967). *Nippon Shokubutsu Byori Gakkaiho* 33, 259–266.
Gamez, R., and Black, L. M. (1968). *Virology* 34, 444–451.
Hirumi, H., and Maramorosch, K. (1963a). *Contrib. Boyce Thompson Inst.* 22, 141–152.
Hirumi, H., and Maramorosch, K. (1963b). *Ann. Epiphyt.* 14, Num. Hors Ser. III, 77–79.
Hirumi, H., and Maramorosch, K. (1964a). *Science* 144, 1465–1467.

316 KARL MARAMOROSCH

Hirumi, H., and Maramorosch, K. (1964b). *Exp. Cell Res.* 36, 625–631.
Hirumi, H., and Maramorosch, K. (1964c). *Contrib. Boyce Thompson Inst.* 22, 343–352.
Hirumi, H., and Maramorosch, K. (1968). *Proc. Int. Colloq. Invertebr. Tissue Cult.,* *2nd, 1967* pp. 203–217.
Hsu, H. T., and Black, L. M. (1973a). *Virology* 52, 187–198.
Hsu, H. T., and Black, L. M. (1973b). *Virology* 52, 284–286.
Hsu, H. T., and Black, L. M. (1974). *Virology* 59, 331–334.
Ishiie, T., Doi, Y., Yora, K., and Asuyama, H. (1967). *Nippon Shokubutsu Byori Gakkaiho* 33, 267–275.
Kimura, I., and Black, L. M. (1971). *Virology* 46, 266–276.
Kimura, I., and Black L. M. (1972a). *Virology* 48, 852–854.
Kimura, I., and Black, L. M. (1972b). *Virology* 49, 549–561.
Liu, H. Y., and Black, L. M. (1969). *Phytopathology* 59, 1038.
Liu, H. Y., Kimura, I., and Black, L. M. (1973). *Virology* 51, 320–326.
McIntosh, A. H., and Rechtoris, C. (1974). *In Vitro* 10, 1–5.
McIntosh, A. H., Maramorosch, K., and Rechtoris, C. (1973). *In Vitro* 8, 375–378.
MacLeod, R., Black, L. M., and Moyer, F. H. (1966). *Virology* 29, 540–552.
Maramorosch, K. (1952). *Phytopathology* 42, 663–668.
Maramorosch, K. (1956). *Virology* 2, 369–376.
Maramorosch, K. (1973). *Proc. Int. Colloq. Invertebr. Tissue Cult., 3rd, 1971* pp. 501–509.
Maramorosch, K., Brakke, M. K., and Black, L. M. (1949). *Science* 110, 162–163.
Maramorosch, K., Nickell, L. G., Littau, V. C., and Grace, T. D. C. (1958). *Anat. Rec.* 131, 579.
Martinez-Lopez, G., and Black, L. M. (1974). *Phytopathology* 64, 1040–1041.
Mitsuhashi, J. (1965). *Jap. J. Appl. Entomol. Zool.* 9, 107–114.
Mitsuhashi, J. (1967). *J. Invertebr. Pathol.* 9, 432–434.
Mitsuhashi, J., and Maramorosch, K. (1963). *Contrib. Boyce Thompson Inst.* 22, 165–173.
Mitsuhashi, J., and Maramorosch, K. (1964a). *Contrib. Boyce Thompson Inst.* 22, 435–460.
Mitsuhashi, J., and Maramorosch, K. (1964b). *Virology* 23, 277–279.
Mitsuhashi, J., and Nasu, S. (1967). *J. Appl. Entomol. Zool.* 2, 113–114.
Nasu, S., Sugiura, M., Wakimoto, T., and Iida, T. T. (1967). *Nippon Shokubutsu Byori Gakkaiho* 33, 343.
Otsuki, Y., Shimomura, T., and Takebe, I. (1972). *Virology* 50, 45–50.
Peters, D., and Black, L. M. (1970). *Virology* 40, 847–853.
Reddy, D. V. R., and Black, L. M. (1966). *Virology* 30, 551–561.
Reddy, D. V. R., and Black, L. M. (1972). *Virology* 50, 412–421.
Reddy, D. V. R., Kimura, I., and Black, L. M. (1974). *Virology* 60, 293–296.
Schneider, I. J. (1964). *Exp. Zool.* 156, 91–104.
Shikata, E., Yamada, K., and Tokumitsu, T. (1970). *J. Fac. Agr., Hokkaido Univ.* 56, 292–302.
Takebe, I., and Otsuki, Y. (1969). *Proc. Nat. Acad. Sci. U.S.* 64, 843–848.
Tokumitsu, T., and Maramorosch, K. (1966). *Exp. Cell Res.* 44, 652–655.
Vago, C., and Flandre, O. (1963). *Ann. Epiphyt.* 14, Num. Hors Ser. III, 127–139.

PART D

Cell Lines and Culture Media

17

A Compilation of Invertebrate Cell Lines and Culture Media*

W. FRED HINK

I. Introduction

This compilation lists a total of 121 cell lines from 56 species of invertebrates. Sublines, strains, or clones of parent lines are not included in arriving at the total number of lines. This list is not complete. To enable us to update and include as many lines as possible in the next compilation, all new information on cell lines should be sent to the author. A synopsis of cell lines in this compilation is given in the tabulation below.

Order or class	Number of cell lines	Number of species
Lepidoptera	24	15
Diptera	63	21
Orthoptera	14	4
Homoptera	12	10
Hemiptera	1	1
Acarina	5	3
Gastropoda	2	2

* This is an updated and expanded version of the compilation which appeared in "Invertebrate Tissue Culture," Volume II (C. Vago, ed.), pp. 363–387, Academic Press, New York, 1972.

Some lines described in this tabulation may not be in existence. Other lines are not continuously subcultured but are stored frozen at specific passage levels. Questions regarding the current status of individual lines should be directed to the investigators who established each line.

Since there are now many invertebrate cell lines, it is important that each new line be given a designation. The use of designations will eliminate confusion that arises when only author, genus, and species are used to refer to a cell line. The Committee on Terminology of the Tissue Culture Association suggests that the designation consist of no more than four letters to indicate the laboratory of origin and the letters followed by numbers indicating the line, e.g., IPLB-21. A method of designating lines, which this author finds more meaningful, is as follows. The first two letters are the first letters of the generic and specific name of the invertebrate and these are followed by numbers representing the month and year the primary culture was initiated. For example, a primary culture from the honey bee *Apis mellifera* was set up in October 1974. A cell line from this primary culture would be AM-1074. If more than one line arose from primaries initiated in the same month, one could add a number after the letters, e.g., AM1-1074, AM2-1074, etc.

II. Abbreviations

BUN	Bunyamwera virus	JE	Japanese encephalitis virus
CE	California encephalitis virus	MVE	Murray Valley encephalitis virus
CHP	Chandipura virus		
CHIK	Chikungunya virus	NPV	Nuclear polyhedrosis virus
CIV	*Chilo* iridescent virus	PYPV	Potato yellow dwarf virus
CPV	Cytoplasmic polyhedrosis virus	RR	Ross River virus
CTF	Colorado tick fever virus	SF	Semliki Forest virus
CV	Cache Valley virus	SH	Snowshoe hare virus
DEN-1, -2, -3, -4	Dengue type 1, 2, 3, or 4 viruses	SIV	*Sericesthis* iridescent virus
		SLE	St. Louis encephalitis virus
EEE	Eastern equine encephalomyelitis virus	TAH	Tahyna virus
		TIV	*Tipula* iridescent virus
EHD	Epizootic hemorrhagic disease virus	VEE	Venezuelan equine encephalomyelitis virus
GAN	Ganjam virus	VS	Vesicular stomatitis virus
IVS	Indiana vesicular stomatitis virus	WN	West Nile virus
		WTV	Wound tumor virus
		YF	Yellow fever virus

III. Insecta

A. LEPIDOPTERA

Invertebrate species: *Antheraea eucalypti*
Common name: Australian emporer gum moth

Primary explant from which line was derived: Trypsinized pupal ovaries
Investigator who established the line: T. D. C. Grace
Date primary culture, from which line originated, was set up: August 1960
Morphology: Three distinct cell types—polygonal with finely granulated cytoplasm and 20–40 μm diameter; round to fibroblast-like in outline and 15–20 μm diameter; small and round with a clear cytoplasm and 10–15 μm diameter
Karyology: Chromosome numbers range from $2n$ to $128n$
Culture medium: Grace (1962) medium plus 5% *A. eucalypti* hemolymph; Yunker *et al.* (1967) medium
Growth characteristics of line: Population doubling time of 37 to 72 hours with maximum population of 2 to 3 × 10^6 cells/ml
Subculture interval: 7–10 days
Virus susceptibility: CIV, SIV, TIV, *Bombyx mori* NPV, BUN, CV, CE, IVS, JE, SH, SLE, TAH, VS, YF
Reference: Grace, T. D. C. (1962). Establishment of four strains of cells from insect tissues grown *in vitro*. *Nature (London)* **195**, 788–789.

Invertebrate species: *Antheraea eucalypti*
Common name: Australian emperor gum moth
Designation of line: RML-2 subline of Grace's *A. eucalypti* cells
Primary explant from which line was derived: Pupal ovarian tissue
Investigator who established the line: T. D. C. Grace
Date primary culture, from which line originated, was set up: 1960?
Number of subcultures to date: 450+
Morphology: Round to fusiform (type 2 of Grace, 1962)
Karyology: Polyploid
Culture medium: Grace (1962) medium plus 10% fetal calf serum (heat-inactivated), 10 mg/ml bovine plasma albumin (fraction V), and antibiotics, pH 6.4
Growth characteristics of line: Cells mostly attached to vessel after subcultivation, but becoming predominantly suspended with passage of time
Virus susceptibility: Growth, without CPE, of JBE, YF, SLE, VSI, CV, CE, BUN, SH, AND TAH (all insect-borne arboviruses)
History of line: Received from Grace at unknown passage level on August 31, 1965. Subsequently converted to grow in medium free of hemolymph
Reference: Grace, T. D. C. (1962). Establishment of four strains of cells from insect tissues grown *in vitro*. *Nature (London)* **195**, 788–789.
Submitted by: C. E. Yunker

Invertebrate species: *Antheraea pernyi*
Common name: Moth
Primary explant from which line was derived: Pupal ovary
Investigator who established the line: Helen Lee
Morphology: Similar to Grace's *A. eucalypti*
Culture medium: Yunker *et al.* (1967) medium
Growth characteristics of line: Suspension
Virus susceptibilty: Unknown
History of line: Received from Dr. Lee when she was at Unit of Invertebrate Pathology at Oxford
Reference: Unpublished
Submitted by: G. R. Gardiner

Invertebrate species: *Bombyx mori*
Common name: Silkworm
Primary explant from which line was derived: Larval ovaries
Investigators who established the line: C. Vago and S. Chastang
Number of subcultures to date: 12 as of original 1958 publication
Morphology: Fibroblast-like and epithelial-like cells
Culture medium: Vago and Chastang (1958) insect medium
History of line: No longer in existence
Reference: Vago, C., and Chastang, S. (1958). Obtention de lignées cellulaires en culture de tissus d'invertèbres. *Experientia* **14,** 110–111.

Invertebrate species: *Bombyx mori*
Primary explant from which line was derived: Minced larval ovaries
Number of subcultures to date: 22 as of original 1959 publication
Culture medium: Trager (1935) solution A plus 10% *B. mori* hemolymph
History of line: No longer in existence
Reference: Gaw, Z., Liu, N. T., and Zia, T. U. (1959). Tissue culture method for cultivation of virus grasserie. *Acta Virol. (Prague)* **3,** Suppl., 55–60.

Invertebrate species: *Bombyx mori*
Primary explant from which line was derived: Trypsinized larval ovaries
Investigator who established the line: T. D. C. Grace
Date primary culture, from which line originated, was set up: November 1967
Morphology: Most common cell is spindle-shaped and 12–25 μm wide by 50–70 μm long, another cell type is slightly spindle-shaped and 18–30 μm in diameter
Karyology: Many more than 100 chromosomes per cell

Culture medium: Grace (1962) medium plus 5% *A. eucalypti* hemolymph
Growth characteristics of line: Population doubling time of 48 hours with maximum population of 1.6 × 10⁶ cells/ml
Reference: Grace, T. D. C. (1967). Establishment of a line of cells from the silkworm *Bombyx mori. Nature (London)* **216**, 613.

Invertebrate species: *Bombyx mori*
Primary explant from which line was derived: Ovary
Date primary culture, from which line originated, was set up: March 1968
Culture medium: Grace (1962) medium plus 1% *A. pernyi* hemolymph
Growth characteristics of line: Population doubling time of 60 hours
Reference: T. D. C. Grace (unpublished)

Invertebrate species: *Chilo suppressalis*
Common name: Rice stem borer
Primary explant from which line was derived: Larval hemocytes
Investigator who established the line: J. Mitsuhashi
Date primary culture, from which line originated, was set up: February 1965
Number of subcultures to date: 42 as of June 1967
Morphology: Prohemocytes were predominant cell type
Culture medium: Mitsuhashi (1967) CSM-2F medium
Growth characteristics of line: Population doubling time of 108 hours with a maximum population of 4 × 10⁵ cells/ml
Virus susceptibility: CIV persistent infection
History of line: Line no longer in existence, lost in 1968
Reference: Mitsuhashi, J. (1967). Establishment of an insect cell strain persistently infected with an insect virus. *Nature (London)* **215**, 863–864.

Invertebrate species: *Choristoneura fumiferana*
Common name: Spruce budworm
Designation of line: IPRI-Cf 124
Primary explant from which line was derived: Minced trypsinized larvae
Investigator who established the line: S. S. Sohi
Date primary culture, from which line originated, was set up: May 1970
Morphology: Three distinct types: epithelial, elongated, and round
Culture medium: Grace (1962) medium plus 10% FBS and 5% *B. mori* hemolymph
Reference: Sohi, S. S. (1973). *In vitro* cultivation of larval tissues of

Choristoneura fumiferana (Clemens) (Lepidoptera:Tortricidae). *Proc. Int. Colloq. Invertebr. Tissue Cult., 3rd, 1971* pp. 75–92.

Invertebrate species: *Estigmene acrea*
Common name: Salt-marsh caterpillar
Designation of lines: EA1174A, EA1174H
Primary explant from which line was derived: Larval hemocytes
Virus susceptibility: *Amsacta moorei* entomopoxvirus and *Autographa californica* NPV
Reference: Granados, R. R., and Naughton, M. (1975). Replication of *Amsacta moorei* entomopoxvirus and *Autographa californica* nuclear polyhedrosis virus in hemocyte cell lines from *Estigmene acrea*. *Abstr., Int. Conf. Invertebr. Tissue Cult., 4th, 1975* p. 13.

Invertebrate species: *Heliothis zea*
Common name: Corn earworm or cotton bollworm
Designation of line: IMC-HZ-1
Primary explant from which line was derived: Minced adult ovaries
Investigator who established the line: W. F. Hink
Date primary culture, from which line originated, was set up: March 1967
Number of subcultures to date: 372 as of May 6, 1975
Morphology: Most cells (90%) are spherical or ellipsoidal with 9.9–26.2 μm diameter, 6% of cells have a single protoplasmic extension, and 4% are oblong binucleate cells
Karyology: Heteroploid
Culture medium: Hink and Ignoffo (1970) IMC-1 medium which is a slight modification of Yunker *et al.* (1967) medium
Growth characteristics of line: Population doubling time of 33 hours and a maximum population of $1-2 \times 10^5$ cells/ml
Virus susceptibility: *H. zea* NPV
Reference: Hink, W. F., and Ignoffo, C. M. (1970). Establishment of a new cell line (IMC-HZ-1) from ovaries of cotton bollworm moths, *Heliothis zea* (Boddie). *Exp. Cell Res.* **60**, 307–309.
Submitted by: W. F. Hink

Invertebrate species: *Heliothis zea*
Designation of line: IPLB-1075
Primary explant from which line was derived: Pupal ovary
Investigator who established the line: R. H. Goodwin
Date primary culture, from which line originated, was set up: November 1970

Number of subcultures to date: 130 as of April 1975
Morphology: Fibroblastic (separate cells)
Karyology: Polyploid
Culture medium: Goodwin (1975) IPL-45 medium, 100 ml basal medium
plus 3 ml turkey, chicken, and fetal calf serum; fiinal osmolality 345
mOsm
Growth characteristics of line: Split weekly at 1:4 ratio
Virus susceptibility: *H. zea* SEV NPV, *T. ni* MEV NPV, *A. californica*
MEV NPV
History of line: Inital medium was IPL-25 supplemented with 4 ml
turkey serum and 2 ml *B. mori* hemolymph, at early passage hemo-
lymph removed, and then to various combinations of sera (IPL-40
then IPL-45)
Reference: Goodwin, R. H. (1975). Insect cell culture: Improved media
and methods for initiating attached cell lines from Lepidoptera. *In
Vitro* (in press).
Submitted by: R. H. Goodwin

Invertebrate species: *Laspeyresia pomonella*
Common name: Codling moth
Designation of line: CP-1268
Primary explant from which line was derived: Minced embryo
Investigators who established the line: W. F. Hink and B. J. Ellis
Date primary culture, from which line originated, was set up: December
1968
Number of subcultures to date: 340 as of May 6, 1975
Morphology: Cell population consists of spindle-shaped, round, and cells
with single extensions
Karyology: Chromosome distribution is bimodal with 51% of cells
having 51–57 chromosomes and 35% having 102–110 chromosomes
Culture medium: Hink (1970) TNM-FH medium
Growth characteristics of line: Population doubling time of 24 hours with
a maximum population 1×10^7 cells/ml
Reference: Hink, W. F., and Ellis, B. J. (1971). Establishment and characterization of two new cell lines (CP-1268 and CP-169) from the
codling moth, *Carpocapsa pomonella. Curr. Top. Microbiol. Immunol.* **55,** 19–28.
Submitted by: W. F. Hink

Invertebrate species: *Laspeyresia pomonella*
Designation of line: CP-169
Primary explant from which line was derived: Minced embryos

Investigators who established the line: W. F. Hink and B. J. Ellis
Date primary culture, from which line originated, was set up: December
    1968
Number of subcultures to date: 236 as of September 1973, and then
    stored under N_2
Morphology: Most cells (70%) are spherical with 11.0–20.2 μm diame-
    ters, 22% have single protoplasmic extensions, and 7% are spindle-
    shaped
Karyology: Line is highly heteroploid with about 9% of population
    diploid and 72% contain more than 100 chromosomes
Culture medium: Hink (1970) TNM-FH medium
Growth characteristics of line: Population doubling time of 24 hours with
    a maximum population of 5×10^6 cells/ml
Reference: Hink, W. F., and Ellis, B. J. (1971). Establishment and char-
    acterization of two new cell lines (CP-1268 and CP-169) from the
    codling moth, *Carpocapsa pomonella*. *Curr. Top. Microbiol. Immu-
    nol.* **55,** 19–28.
Submitted by: W. F. Hink

Invertebrate species: *Lymantria dispar*
Common name: Gypsy moth
Designation of line: SCLd 135
Primary explant from which line was derived: Ovaries
Investigator who established the line: J. M. Quiot
Date primary culture, from which line originated, was set up: May 1974
Number of subcultures to date: 25 as of April 1975
Morphology: Cell population consists of spindle-shaped and round cells
    slightly attached to culture vessel
Culture medium: Quiot GM15 with 10% inactivated calf serum or Grace
    (1962) medium plus FBS
Growth characteristics of line: Population doubling time is 48 hours
Subculture interval: 7–10 days
Virus susceptibility: Entomopoxvirus of *Amsacta moorei;* CIV; *B. mori,
    Antheraea pernyi,* and *Galleria mellonella* NPV
Reference: Unpublished
Submitted by: J. M. Quiot

Invertebrate species: *Malacosoma disstria*
Common name: Forest tent caterpillar
Designation of line: IPRI 108
Primary explant from which line was derived: Larval hemocytes
Investigator who established the line: S. S. Sohi

Date primary culture, from which line originated, was set up: September
    1969
Morphology: Appear to be prohemocytes
Culture medium: Grace (1962) medium plus 20% FBS
Virus susceptibility: *Lambdina fiscellaria somniaria* and *C. fumiferana*
    NPV
Reference: Sohi, S. S. (1973). Establishment of cultures of *Malacosoma
    disstria* Hubner (Lepidoptera: Lasiocampidae) hemocytes in a
    hemolymph-free medium. *Proc. Int. Colloq. Invertebr. Tissue Cult.,
    3rd, 1971* pp. 27–39.

Invertebrate species: *Manduca sexta*
Common name: Tobacco hornworm
Investigator who established the line: P. Eide
Reference: Unpublished

Invertebrate species: *Papilio xuthus*
Common name: Swallowtail
Designation of lines: Px-58 and Px-64
Primary explant from which line was derived: Pupal ovaries
Investigator who established the line: J. Mitsuhashi
Date primary culture, from which line originated, was set up: July 1970
Number of subcultures to date: 290 as of April 1975
Morphology: Hemocyte-like floating cells of spherical, fusiform, tadpole-
    shaped, and amorphous shapes
Karyology: $2n$ (50) to $22n$. The mode has not yet been determined
Culture medium: Mitsuhashi (1972) MGM-431 medium
Growth characteristics of line: Population doubling time is about 48
    hours
Reference: Mitsuhashi, J. (1973). Establishment of cell lines from the
    pupal ovaries of the swallowtail, *Papilio xuthus* Linné (Lepidoptera,
    Papilionidae). *Appl. Entomol. Zool.* **8,** 64–72.
Submitted by: J. Mitsuhashi

Invertebrate species: *Samia cynthia*
Common name: Cynthia moth
Designation of lines: Several lines
Primary explant from which line was derived: Pupal hemocytes
Investigators who established the line: J. Chao and G. H. Ball
Date primary culture, from which line originated, was set up: September
    1968
Morphology: Probably prohemocytes
Karyology: Most cells are $4n$

Culture medium: Grace (1962) medium plus 10% FBS
Reference: Chao, J., and Ball, G. H. (1971). A cell line isolated from
hemocytes of *Samia cynthia* pupae. *Curr. Top. Microbiol. Immunol.*
55, 28–32

Invertebrate species: *Spodoptera frugiperda*
Common name: Fall armyworm
Designation of line: IPLB-21
Primary explant from which line was derived: Pupal ovary
Investigator who established the line: J. L. Vaughn
Date primary culture, from which line originated, was set up: January
1969
Number of subcultures to date: 115 as of April 1975
Morphology: Predominantly small spherical cells with a few fibroblast-
like cells
Karyology: Polyploid
Culture medium: Goodwin (1975) IPL-40 supplemented with 3% *B.*
mori hemolymph, 1% bovine serum albumin, and 5% fetal calf
serum
Growth characteristics of line: Population doubling time of 23 hours with
maximum density of 5–6 $\times$ 10^6 cells/ml. Grow attached to surface
Virus susceptibility: *S. frugiperda* and *A. californica* NPV
History of line: Initial cultures and first five passages were carried in
medium containing pen-strep. No antibiotics used in later passes.
Stock is maintained frozen with 14% FBS and 7% DMSO as
preservatives
Reference: Unpublished
Submitted by: J. L. Vaughn

Invertebrate species: *Spodoptera frugiperda*
Designation of line: IPLB-1254
Primary explant from which line was derived: Pupal ovary
Investigator who established the line: R. H. Goodwin
Date primary culture, from which line originated, was set up: November
1971
Number of subcultures to date: 112 as of April 1975
Morphology: Mixed cell types; majority are small spherical cells with
some fibroblast-like cells
Karyology: Polyploid
Culture medium: Goodwin (1975) IPL-40 supplemented with 3% fetal
calf serum, 4% turkey serum, and 3% chicken serum
Growth characteristics of line: Cells grow attached with a population

doubling time of 24 hours and maximum density of 4–5×10^6 cells/ml

Virus susceptibility: *A. californica, T. ni,* and *S. frugiperda* NPV

History of line: Line initiated with antibiotics and hemolymph in the medium. At the sixth pass the hemolymph was removed and now no antibiotics are used.

Reference: Unpublished

Submitted by: J. L. Vaughn

Invertebrate species: *Trichoplusia ni*

Common name: Cabbage looper

Designation of line: TN-368

Primary explant from which line was derived: Minced adult ovaries

Investigator who established the line: W. F. Hink

Date primary culture, from which line originated, was set up: March 1968

Number of subcultures to date: 970 as of May 6, 1975

Morphology: Cells are round or oval, or possess one, two, or three protoplasmic extensions. Protoplasmic extensions are up to 105 μm long

Karyology: A majority of the cells (90%) have 82–95 chromosomes but some have 160–180 chromosomes

Culture medium: Hink (1970) TNM-FH medium

Growth characteristics of line: Population doubling time is 16 hours with a maximum population of 2–3×10^6 cells/ml

Virus susceptibility: *A. californica, T. ni,* and *G. mellonella* NPV; *T. ni* CPV

Reference: Hink, W. F. (1970). Established insect cell line from the cabbage looper, *Trichoplusia ni. Nature (London)* **226,** 466–467.

Submitted by: W. F. Hink

Invertebrate species: *Trichoplusia ni*

Designation of lines: Several lines

Primary explant from which line was derived: Pupal ovary and fat body

Investigator who established the line: R. H. Goodwin

Number of subcultures to date: 6–12 as of publication below

Culture medium: Goodwin et al. (1973) IPL-25 medium supplemented with turkey and chicken serum

Virus susceptibility: *T. ni* NPV

Reference: Goodwin, R. H., Vaughn, J. L., Adams, J. R., and Louloudes, S. J. (1973). The influence of insect cell lines and tissue culture media on *Baculovirus* polyhedra production. *Misc. Publ. Entomol. Soc. Amer.* **9,** 66–72.

B. DIPTERA

Invertebrate species: *Aedes aegypti*
Common name: Yellow fever mosquito
Primary explant from which line was derived: Minced last instar larvae
Date primary culture, from which line originated, was set up: June 1963
Morphology: Most common cell type is spindle-shaped and 40–50 μm
   long by 8–10 μm wide and less common are round cells with 20 μm
   diameters. Small numbers of large round cells, 55–60 μm in diameter
   and cells with long extensions are also present
Karyology: Most cells are $32n$ while the next most frequent chromosome
   number is $16n$
Culture medium: Grace (1962) medium plus 5% *A. eucalypti*
   hemolymph
Growth characteristics of line: Population doubling time of about 50
   hours
Virus susceptibility: JE, Kunjin, *Malacosoma disstria* CPV, MVE,
   WN, YF
Reference: Grace, T. D. C. (1966). Establishment of a line of mosquito
   (*Aedes aegypti* L.) cells grown *in vitro. Nature* (*London*) **211,**
   366–367.

Invertebrate species: *Aedes aegypti*
Designation of line: ATC-10
Primary explant from which line was derived: Minced trypsinized larvae
Number of subcultures to date: 115 as of July 1969
Morphology: Mainly epithelial-like cells
Culture medium: Mitsuhashi and Maramorosch (1964) medium
Virus susceptibility: CHIK, CHP, DEN-2, EEE, IVS, NJVS, SF,
   Sindbis, SLE, VEE, WN, YF
Reference: Singh, K. R. P. (1967). Cell cultures derived from larvae of
   Aedes albopictus (Skuse) and *Aedes aegypti* (L.). *Curr. Sci.* **36,**
   506–508.

Invertebrate species: *Aedes aegypti*
Designation of line: 59
Primary explant from which line was derived: Homogenized embryos
Number of subcultures to date: 170+ as of publication below
Culture medium: Kitamura (1965) medium plus 10% FBS, 10% condi-
   tioned medium, and 1% chick embryo extract
Growth characteristics of line: Cells have three growth stages; a mono-
   layer is formed first, cells aggregate, and vesicles develop

Virus susceptibility: CIV, EEE, Kunjin, SE, Sindbis, TIV, WN
Reference: Peleg, J., and Shahar, A. (1972). Morphology and behavior
    of cultured *Aedes aegypti* mosquito cells. *Tissue & Cell* **4**, 55–62.

Invertebrate species: *Aedes aegypti*
Designation of line: 364
Primary explant from which line was derived: Homogenized embryos
Number of subcultures to date: 170+ as of publication below
Culture medium: Kitamura (1965) medium plus 10% FBS, 10% condi-
    tioned medium, and 1% chick embryo extract
Growth characteristics of line: Cells have three growth stages; a mono-
    layer is formed first, cells aggregate, and vesicles develop.
Virus susceptibility: EEE, JBE, SF, Sindbis
Reference: Peleg, J., and Shahar, A. (1972). Morphology and behavior
    of cultured *Aedes aegypti* mosquito cells. *Tissue & Cell* **4**, 55–62.

Invertebrate species: *Aedes aegypti*
Designation of line: Mos 20
Primary explant from which line was derived: Minced trypsinized larvae
Date primary culture, from which line originated, was set up: March
    1968
Number of subcultures to date: 207 as of June 1975
Culture medium: Mitsuhashi and Maramorosch (1964) medium
Reference: Varma, M. G. R., and Pudney, M. (1969). The growth and
    serial passage of cell lines from *Aedes aegypti* (L.) larvae in different
    media *J. Med. Entomol.* **6**, 432–439.

Invertebrate species: *Aedes aegypti*
Designation of line: Mos 20A
Primary explant from which line was derived: Minced trypsinized larvae
Number of subcultures to date: 332 as of June 1975
Culture medium: A 1:1 ratio of Mitsuhashi and Maramorosch (1964)
    medium and Varma and Pudney (1969) VP_{12} medium
Reference: Varma, M. G. R., and Pudney, M. (1969). The growth and
    serial passage of cell lines from *Aedes aegypti* (L.) larvae in different
    media. *J. Med. Entomol.* **6**, 432–439.

Invertebrate species: *Aedes aegypti*
Designation of line: Mos 29
Primary explant from which line was derived: Minced trypsinized larvae
Date primary culture, from which line originated, was set up: May 1968

Number of subcultures to date: 203 as of June 1975
Karyology: Most cells are diploid
Culture medium: Kitamura (1965) medium modified by Varma and
    Pudney (1969)
Growth characteristics of line: Population doubling time of 29 hours with
    a maximum population of 5×10^6 cells/ml
Virus susceptibility: TAH, WN
Reference: Varma, M. G. R., and Pudney, M. (1969). The growth and
    serial passage of cell lines from *Aedes aegypti* (L.) larvae in different
    media. *J. Med. Entomol.* **6**, 432–439.

Invertebrate species: *Aedes aegypti*
Designation of line: Mill Hill line
Primary explant from which line was derived: Larvae
Investigators who established the line: J. S. Porterfield, A. T. de Madrid,
    and Z. Marhol
Date primary culture, from which line originated, was set up: November
    1970
Number of subcultures to date: 50+
Morphology: Epithelial and a subline of vesicles
Karyology: Diploid
Culture medium: Leibovitz (1963) L15 medium, plus 10% tryptose phos-
    phate broth, plus 10% fetal calf serum initially, later reduced to
    3% FCS
Virus susceptibility: Nodamura, SF, YF
History of line: Good epithelial outgrowth and many vesicles present
    by day 33. Epithelial line established by trypsinization, vesicle line
    by shaking and subdivision of medium. Frozen in liquid N_2
Reference: Bailey, L., Newman, J. F. E., and Porterfield, J. S. (1975).
    The multiplication of nodamura virus in insect and mammalian cell
    cultures. *J. Gen. Virol.* **26**, 15–20.
Submitted by: J. S. Porterfield

Ivertebrate species: *Aedes aegypti*
Designation of lines: KOM3, KOM6, KOM8
Primary explant from which line was derived: Minced trypsinized larvae
Investigator who established the line: Z. Marhoul
Date primary culture, from which line originated, was set up: March,
    May, and June 1974, respectively
Number of subcultures to date: 50, 42, and 40, respectively, as of April
    1975
Morphology: Mainly epithelial-like cells

Karyology: $2n$
Culture medium: Kitamura (1965) medium with 10% Leibowitz (1963)
 L15 medium plus 10% fetal calf serum
Growth characteristics of line: All three lines are split 1:5 at weekly
 intervals; treated with 0.05% pronase in PBS for 1 minute
History of line: All lines frozen in liquid N_2 at fifth passage
Reference: Unpublished
Submitted by: Z. Marhoul

Invertebrate species: *Aedes aegypti*
Designation of line: RML-12 *Aedes aegypti*
Primary explant from which line was derived: Newly hatched larvae
Investigator who established the line: U. K. M. Bhat
Date primary culture, from which line originated, was set up: August
 1974
Number of subcultures to date: 15 as of April 1975
Morphology: Epithelial-like cells
Karyology: 20% diploid and the rest tetraploid or polyploid at the
 eleventh passage
Culture medium: Mitsuhashi and Maramorosch (1964) medium plus
 20% FBS
Growth characteristics of line: Two sublines were established from one
 primary culture—one consisting of cells growing only as sheets at-
 tached to the substrate, the other of hollow multicellular vesicles
 which remain suspended in the medium
Virus susceptibility: West Nile, Japanese B encephalitis, and Dengue-2
History of line: Established at Rocky Mountain laboratory
Reference: Unpublished
Submitted by: C. E. Yunker

Invertebrate species: *Aedes aegypti*
Common name: Yellow fever mosquito
Designation of line: Mos 63
Primary explant from which line was derived: Minced larvae
Investigators who established the line: M. G. R. Varma and Mary
 Pudney
Date primary culture, fròm which line originated, was set up: September
 9, 1972
Number of subcultures to date: 128 as of June 1975
Morphology: Epithelial-like cells
Culture medium: Leibovitz (1963) L-15 medium
Growth characteristics of line: Split 1:12 weekly

Reference: Unpublished
Submitted by: Mary Pudney and M. G. R. Varma

Invertebrate species: *Aedes albopictus*
Common name: Mosquito
Designation of line: ATC-15
Primary explant from which line was derived: Minced trypsinized larvae
Investigator who established the line: K. R. P. Singh
Number of subcultures to date: 112 as of July 1969
Morphology: Epithelial-like cells
Karyology: Diploid
Culture medium: Mitsuhashi and Maramorosch (1964) medium; or lactalbumin hydrolysate in Hanks' BSS plus 10% FCS, and 0.1% bovine plasma albumin; pH 6.8
Growth characteristics of line: Forms tightly attached monolayers within 2–3 days after transfer
Virus susceptibility: Supports growth of many mosquito-borne arboviruses but only a few tick-borne arboviruses. Batai, Calovo, CE, Chenuda, CHIK, CHP, CTF, DEN-1, -2, -3, -4; EEE, EHD, IVS, JE, Kimesovo, Kunjin, Lepovnik, Nodamura, NJVS, RR, SF, Sindbis, SLE, TAH, Tribec, VEF, WN, YF
Reference: Singh, K. R. P. (1967). Cell cultures derived from larvae of *Aedes albopictus* (Skuse) and *Aedes aegypti* (L.). *Curr. Sci.* **36**, 506–508.
Comments: Useful in studies of B-group arboviruses, especially plaque assays
Submitted by: C. E. Yunker

Invertebrate species: *Aedes albopictus*
Designation of line: RML-11 subline of Singh's ATC-15 line
Primary explant from which line was derived: Newly hatched larvae
Investigator who established the line: K. R. P. Singh
Number of subcultures to date: 191 (clonal population)
Morphology: Epithelioid
Karyology: Clonal population: $2n = 6$. Majority polyploid (mostly low multiples of diploid) in one hundred thirty-fifth passage
Culture medium: Lactalbumin hydrolysate in Hanks' BSS plus 1% bovine plasma albumin (fraction V); pH. 6.8
Virus susceptibility: Similar to parent line except Colorado tick fever virus propagates to higher levels in this subline

History of line: Received from Yale Arbovirus Research Unit in forty-ninth passage. Adapted to grow in medium free of serum at about eighty-fifth passage
Reference: Singh, K. R. P. (1967). Cell cultures derived from larvae of *Aedes albopictus* (Skuse) and *Aedes aegypti* (L.). *Curr. Sci.* **36**, 506–508.
Comments: Exists in three distinctive populations, one of which is clonal
Submitted by: C. E. Yunker

Invertebrate species: *Aedes albopictus*
Common name: Mosquito
Primary explant from which line was derived: Ovarian tissue of adult female mosquitoes
Investigators who established the line: S. H. Hsu and M. H. Huang
Date primary culture, from which line originated, was set up: September 1974
Number of subcultures to date: 34 as of March 1975
Morphology: Three major morphological configurations—fibroblastic or spindle-like, stellate, and round cells. Generally this cell contains a well-defined, ectopically placed nucleus, occasionally two nuclei were seen. Each nucleus contains a well-defined nucleolus.
Karyology: Majority of the cells are diploid with chromosome numbers of six $(2n = 6)$
Virus susceptibility: Japanese encephalitis, West Nile, Kunjin, dengues 1, 2, and 4 of B-group and Western equine, Sindbis and Chikungunya of A-group arboviruses. Others are being tested.
Reference: Unpublished
Submitted by: S. H. Hsu

Invertebrate species: *Aedes malayensis*
Common name: Mosquito
Designation of line: Mos 60
Primary explant from which line was derived: Minced larvae
Investigators who established the line: Mary Pudney and M. G. R. Varma
Date primary culture, from which line originated, was set up: February 4, 1972
Number of subcultures to date: 138 as of June 1975
Morphology: Mixed epithelial-like and fibroblast-like cells, predominantly epithelial-like

Karyology: Diploid
Culture medium: A 1:1 ratio of Mitsuhashi and Maramorosch (1964)
    medium and Varma and Pudney (1969) VP_{12} medium plus 15% FCS
    or Leibovitz (1963) L-15 medium plus 15% FCS
Growth characteristics of line: Split 1:44 weekly
Virus susceptibility: JBE, WN with cytopathic effect
Reference: Varma, M. G. R., Pudney, M., and Leake, C. J. (1974). Cell
    lines from larvae of Aedes (stegomyia) malayensis Colless and
    Aedes (S) pseudoscutellaris (Theobald) and their infection with
    some arboviruses. Trans. Roy. Soc. Trop. Med. Hyg. 68, 374–382.
Submitted by: Mary Pudney and M. G. R. Varma

Invertebrate species: Aedes novo-albopictus
Common name: Mosquito
Designation of lines: ATC-170 and ATC-173
Primary explant from which line was derived: Newly hatched larval
    tissues
Investigator who established the line: U. K. M. Bhat
Date primary culture, from which line originated, was set up: March
    1971
Number of subcultures to date: 12 and 5, respectively, as of May 1971
Culture medium: Mitsuhashi and Maramorosch (1964) medium
Reference: Unpublished

Invertebrate species: Aedes pseudoscutellaris
Common name: Mosquito
Designation of line: Mos 61
Primary explant from which line was derived: Minced larvae
Investigators who established the line: M. G. R. Varma and Mary
    Pudney
Date primary culture, from which line originated, was set up: February
    10, 1972
Number of subcultures to date: 134 as of June 1975
Morphology: Mixed epithelial-like and fibroblast-like cells
Karyology: Polyploid
Culture medium: A 1:1 ratio of Mitsuhashi and Maramorosch (1964)
    and Varma and Pudney (1969) VP_{12} medium plus 15% FCS or
    Leibovitz (1963) L-15 medium plus 15% FCS
Growth characteristics of line: Split 1:26 weekly
Virus susceptibility: Wide range of arbovirus—some with CPE
Reference: Varma, M. G. R., Pudney, M., and Leake, C. J. (1974). Cell
    lines from larvae of Aedes (stegomyia) malayensis Colless and

Aedes (S) *pseudoscutellaris* (Theobald) and their infection with some arboviruses. *Trans. Roy. Soc. Trop. Med. Hyg.* **68,** 374–382. Submitted by: Mary Pudney and M. G. R. Varma

Invertebrate species: *Aedes taeniorhynchus*
Common name: Mosquito
Primary explant from which line was derived: First stage larvae
Investigator who established the line: I. Schneider
Date primary culture, from which line originated, was set up: December 1972
Number of subcultures to date: 75 as of April 1975
Morphology: Epithelial-like
Karyology: 70% diploid ($2n = 6$) ; 30% heteroploid
Culture medium: Mitsuhashi and Maramorosch (1964) medium plus 20% FBS
Growth characteristics of line: Generation time about 50 hours
Reference: Unpublished
Submitted by: I. Schneider

Invertebrate species: *Aedes vexans*
Common name: Mosquito
Primary explant from which line was derived: Minced pupae
Date primary culture, from which line originated, was set up: August 1969
Number of subcultures to date: 18 as of August 1969
Karyology: Chromosome numbers range from 156 to 216 with a mean of 190
Culture medium: Grace (1962) medium plus 10% FBS
Reference: Sweet, B. H., and McHale, J. S. (1970). Characterization of cell lines derived from *Culiseta inornata* and *Aedes vexans* mosquitoes. *Exp. Cell Res.* **61,** 51–63.

Invertebrate species: *Aedes w-albus*
Common name: Mosquito
Designation of lines: ATC-136 and ATC-137
Primary explant from which line was derived: Newly hatched trypsinized larvae
Date primary culture, from which line originated, was set up: July and August 1969
Number of subcultures to date: 37 and 27, respectively, as of publication below

Morphology: ATC-136 is mostly epithelial-like cells and a few fibroblast-like cells. ATC-137 is mostly fibroblast-like cells and a few
epithelial-like cells
Karyology: Mostly diploid in early passages but polyploid cells increased
after 10 passes
Culture medium: Mitsuhashi and Maramorosch (1964) medium
Virus susceptibility: CHIK, CHP, DEN-2, JE
Reference: Singh, K. R. P., and Bhat, U. K. M. (1971). Establishment
of two mosquito cell lines from larval tissues of *Aedes w-albus. Experientia* **27,** 142–143.

Invertebrate species: *Anopheles gambiae*
Common name: Mosquito
Designation of line: Mos 55
Primary explant from which line was derived: Minced trypsinized larvae
Investigators who established the line: Z. Marhoul and M. Pudney
Date primary culture, from which line originated, was set up: August
1970
Number of subcultures to date: 225 as of April 1975
Morphology: Epithelial-like cells of similar size
Culture medium: Originally 1:1 ratio of Kitamura (1965) medium modi-
fied by Varma and Pudney (1969)/VP$_{12}$ medium with 15% heat-
inactivated fetal calf serum. Now it is Kitamura (1965) medium
with 10% Leibovitz (1963) L-15 and 10% fetal calf serum
Growth characteristics of line: Split 1:80 or 1:130
Reference: Marhoul, Z., and Pudney, M. (1972). A mosquito cell line
(Mos 55) from *Anopheles gambiae* larvae. *Trans. Roy. Soc. Trop.
Med.* **66,** 183–184.
Submitted by: Z. Marhoul

Invertebrate species: *Anopheles stephensi*
Common name: Mosquito
Primary explant from which line was derived: Minced larvae
Date primary culture, from which line originated, was set up: March
1968
Number of subcultures to date: 120 as of April 1971
Morphology: Most cells are epithelial in appearance and range from 4–9
μm in diameter and 12–20 μm in length
Karyology: Most cells are diploid
Culture medium: Grace (1962) medium modified by Schneider (1969)
Growth characteristics of line: Population doubling time of 65 hours
Virus susceptibility: CHP, JE

Reference: Schneider, I. (1969). Establishment of three diploid cell lines of *Anopheles stephensi* (Diptera: Culicidae). *J. Cell Biol.* **42**, 603–606.

Invertebrate species: *Anopheles stephensi*
Designation of line: Mos 43
Primary explant from which line was derived: Minced trypsinized first instar larvae
Date primary culture, from which line originated, was set up: April 1969
Number of subcultures to date: 293 as of June 1975
Morphology: Most cells are fibroblast-like
Karyology: Most cells are diploid
Culture medium: A 1:1 ratio of Kitamura (1965) medium modified by Varma and Pudney (1969)/Varma and Pudney (1969) VP_{12} medium
Growth characteristics of line: Population doubling time of 16 hours with a maximum cell density of 8×10^6 cells/ml
Virus susceptibility: Calovo, TAH
Reference: Pudney, M., and Varma, M. G. R. (1971). *Anopheles stephensi* var. *mysorensis:* Establishment of a larval cell line (Mos. 43). *Exp. Parasitol.* **29**, 7–12.

Invertebrate species: *Anopheles stephensi*
Designation of lines: Mos 44, Mos 45, Mos 46
Primary explant from which line was derived: First stage larvae
Investigator who established the line: M. G. R. Varma
Date primary culture, from which line originated, was set up: April 1970
Number of subcultures to date: 36–108 as of June 1975
Reference: Pudney, M., McCarthy, D., and Shortridge, K. F. (1973). Rod-shaped virus-like particles in cultured *Anopheles* cells and in an *Anopheles* laboratory colony. *Proc. Int. Colloq. Invertebr. Tissue Cult., 3rd, 1971* pp. 337–345.

Invertebrate species: *Armigeries subalbatus*
Common name: Mosquito
Primary explant from which line was derived: First stage larvae
Investigator who established the line: I. Schneider
Date primary culture, from which line originated, was set up: September 1973
Number of subcultures to date: 21 as of April 1975
Morphology: Epithelial-like cells
Karyology: Diploid

Culture medium: Mitsuhashi and Maramorosch (1964) medium plus 20% FBS
Growth characteristics of line: Generation time about 60 hours
Reference: Unpublished
Submitted by: I. Schneider

Invertebrate species: *Culex molestus*
Common name: Mosquito
Primary explant from which line was derived: Adult ovaries
Date primary culture, from which line originated, was set up: October 1967
Number of subcultures to date: 87 as of February 1970
Karyology: Most cells are diploid with an indication of heterploidy in a few cells
Culture medium: Kitamura (1970) medium
Growth characteristics of lines: Population doubling time of 30 hours with a maximum population of 1.1×10^6 cells/ml
Reference: Kitamura, S. (1970). Establishment of cell line from *Culex* mosquito. *Kobe J. Med. Sci.* **16**, 41–50.

Invertebrate species: *Culex quinquefasciatus*
Common name: Mosquito
Primary explant from which line was derived: Adult ovaries
Number of subcultures to date: 92 as of December 1969
Morphology: Most cells are diploid
Culture medium: Hsu *et al.* (1970) 721 medium
Reference: Hsu, S. H., Mao, W. H., and Cross, J. H. (1970). Establishment of a line of cells derived from ovarian tissue of *Culex quinquefasciatus* Say. *J. Med. Entomol.* **7**, 703–707.

Invertebrate species: *Culex salinarius*
Common name: Mosquito
Primary explant from which line was derived: First stage larvae
Investigator who established the line: I. Schneider
Date primary culture, from which line originated, was set up: March 1970
Number of subcultures to date: 112 as of April 1975
Morphology: Both epithelial-like and fibroblast-like cells
Karyology: Primarily diploid
Culture medium: Hsu *et al.* (1972) modified 721 medium plus 15% FBS
Growth characteristics of line: Generation time approximately 48 hours

Reference: Schneider, I. (1973). Establishment of cell lines from *Culex tritaeniorhynchus* and *Culex salinarius* (Diptera: Culicidae). *Proc. Int. Colloq. Invertebr. Tissue Cult., 3rd, 1971* pp. 121–134.
Submitted by: I. Schneider

Invertebrate species: *Culex tarsalis*
Common name: Mosquito
Primary explant from which line was derived: Embryos
Investigators who established the line: J. Chao and G. H. Ball
Date primary culture, from which line originated, was set up: February 1972
Number of subcultures to date: 86 as of February 1974, and then frozen
Morphology: Spindle-shaped or fibroblast-like in healthy cultures; round or epithelial-like in poor cultures
Karyology: Diploid and polyploid
Culture medium: Schneider (1966) medium; Hsu *et al.* (1970) 721 medium; or Mitsuhashi and Maramorosch (1964) medium
Growth characteristics of line: Monolayer of spindle-shaped cells can be easily loosened by pipetting. One culture can be split in two every 24 hours
Virus susceptibility: BUT, CE, CV, JC, LOK, MD, SLE, TUR, WEE
History of line: Primary cultures and early transfers were in a mixture of Schneider's, Singh's, and Grace's media
Reference: Unpublished
Submitted by: J. Chao

Invertebrate species: *Culex tritaeniorhynchus summorosus* Dyar
Common name: Mosquito
Primary explant from which line was derived: Ovarian tissue of adult female mosquitoes
Investigator who established the line: S. H. Hsu
Date primary culture, from which line originated, was set up: December 1970
Number of subcultures to date: 426 as of March 1975
Morphology: Principally fibroblast-like, being falciparum or spindle-shaped with an occasional stellate configuration
Karyology: Most cells are diploid
Culture medium: Hsu *et al.* (1972) modified 721 medium
Reference: Hsu, S. H., Li, S. Y., and Cross, J. H. (1972). A cell line derived from ovarian tissue of *Culex tritaeniorhynchus summorosus* Dyar. *J. Med. Entomol.* **9**, 86–91.
Submitted by: S. H. Hsu

Invertebrate species: *Culex tritaeniorhynchus*
Primary explant from which line was derived: First stage larvae
Investigator who established the line: I. Schneider
Date primary culture, from which line originated, was set up: April 1970
Number of subcultures to date: 187 as of April 1975
Morphology: Epithelial-like cells
Karyology: Primary diploid
Culture medium: Hsu *et al.* (1972) modified 721 medium plus 15%
    FBS
Growth characteristics of line: Generation time is 20 hours
Reference: Schneider, I. (1973). Establishment of cell lines from *Culex*
    tritaeniorhynchus and *Culex salinarius* (Diptera:Culicidae). *Proc.*
    Int. Colloq. Invertebr. Tissue Cult., 3rd, 1971 pp. 121–134.
Submitted by: I. Schneider

Invertebrate species: *Culiseta inornata*
Common name: Mosquito
Primary explant from which line was derived: Minced adult
Number of subcultures to date: 27 as of August 1969
Karyology: Extremely polyploid, at least $50n$–$70n$
Culture medium: Grace (1962) medium plus 10% FBS
Reference: Sweet, B. H., and McHale, J. S. (1970). Characterization
    of cell lines derived from *Culiseta inornata* and *Aedes vexans* mos-
    quitoes. *Exp. Cell Res.* **61**, 51–63.

Invertebrate species: *Drosophila immigrans*
Common name: Fruitfly
Primary explant from which line was derived: Late embryos and first
    stage larvae
Investigators who established the line: I. Schneider and P. Chakrabartty
Date primary culture, from which line originated, was set up: August
    1973
Number of subcultures to date: 21 as of April 1975
Morphology: Epithelial-like
Karyology: Primarily diploid
Culture medium: Schneider (1966) medium plus 15% FBS
Growth characteristics of line: population doubling time is 72 hours
Reference: Unpublished
Submitted by: I. Schneider

Invertebrate species: *Drosophila melanogaster*
Common name: Fruit fly

Primary explant from which line was derived: Homogenized embryos

Number of subcultures to date: 43 as of publication below

Karyology: Model chromosome number is diploid with evidence of heteroploidy

Culture medium: Horikawa *et al.* (1966) H-6 medium

Growth characteristics of line: Maximum population of 1×10^6 cells/ml. Line is no longer in existence

Reference: Horikawa, M., Ling, L., and Fox, A. S. (1966). Long-term culture of embryonic cells of *Drosophila melanogaster*. *Nature (London)* **210**, 183–185.

Invertebrate species: *Drosophila melanogaster*

Designation of lines: 68C and 68K

Primary explant from which line was derived: 6- to 12-hour embryos

Investigators who established the line: G. Echalier and A. Ohanessian

Date primary culture, from which line originated, was set up: February 1968

Number of subcultures to date: 300+ as of April 1975

Morphology: 68C cells are fibroblast-like; 68K cells are roundish and separate

Karyology: 68C male karyotype, 68K female karyotype haplo IV

Culture medium: Echalier D22 medium plus 10–20% embryonic calf serum. One subline from the 68K line was adapted to grow in serum-free D22 medium

Growth characteristics of line: The cell cycle for 68K is estimated to be 18 hours at 25°C

Virus susceptibility for line 68K: Sigma, Sindbis, WN

Reference: Echalier, G., and Ohanessian, A. (1970). *In vitro* culture of *Drosophila melanogaster* embryonic cells. *In Vitro* **6**, 162–172.

Comments: 68C and 68K were the first established cell lines from *D. melanogaster*

Submitted by: G. Echalier and A. Ohanessian

Invertebrate species: *Drosophila melanogaster*

Designation of line: 67j25D

Primary explant from which line was derived: Homogenized 12-hour embryos

Investigators who established the line: V. T. Kakpakov and V. A. Gvozdev

Date primary culture, from which line originated, was set up: October 1967

Number of subcultures to date: 400 as of February 1975

Morphology: Rounded cells are small about 12 μm in diameter

Karyology: The karyotype of line is fundamentally diploid

Culture medium: Gvozdev and Kakpakov (1968) C-15 and C-45 medium. The composition of medium C-45 is not published

Growth characteristics of line: Population doubling time near 24 hours with a maximum population per ml medium of 2.5×10^7 cells

History of line: 67j25D was established using the C-15 medium supplemented with 15% bovine fetal serum and 10% pupal extract. The interval between the primary culture and the first subculture is two months on the average

Reference: Kakpakov, V. T., Gvozdev, V. A., Platova, T. P., and Polukarova, L. G. (1969). *In vitro* establishment of embryonic cell lines of *Drosophila melanogaster. Genetika* **5,** 67–75.

Submitted by: V. T. Kapakov

Invertebrate species: *Drosophila melanogaster*

Designation of line: 70123

Primary explant from which line was derived: Embryo

Investigator who established the line: V. T. Kakpakov

Karyology: The heterochromatic region of one of the two X chromosomes is enlarged

Reference: Unpublished

Submitted by: V. T. Kakpakov

Invertebrate species: *Drosophila melanogaster*

Designation of lines: 69I, 69J, 69D

Primary explant from which line was derived: 6- to 12-hour-old embryos

Investigator who established the line: C. Richard-Molard

Date primary culture, from which line originated, was set up: 1969

Number of subcultures to date: 300+

Morphology: All cells are roundish and separate

Karyology: 69I, aneuploide (90% of cells); 69J, diploide haplo I (5% of cells are aneuploides); 69D, diploide haplo I

Culture medium: Echalier D22 medium supplemented with 10–20% embryonic calf serum

Growth characteristics of line: Cell cycle is estimated to be 18 hours at 25°C

Virus susceptibility: 69D is sensitive to σ *Drosophila* virus called P- and P+. 69I and 69J are sensitive only to σ *Drosophila* virus called P+

Reference: Richard-Molard, C. (1975). *Arch. Virol.* **47,** 139–145.

Comments: 69I and 69J are homozygous for the P^r allele of the gene ref(2)P. 69D is homozygous for the P^o allele
Submitted by: Christine Richard-Molard

Invertebrate species: *Drosophila melanogaster*
Designation of line: Line 1
Primary explant from which line was derived: Late embryos and first stage larvae
Investigator who established the line: I. Schneider
Date primary culture, from which line originated, was set up: August 1969
Number of subcultures to date: 206 as of April 1975
Morphology: Approx. 50% round; 50% fibroblast-like
Karyology: 60% diploid, 40% heteroploid
Culture medium: Schneider (1966) medium plus 15% FBS
Growth characteristics of line: Generation time approx. 24 hours
Reference: Schneider, I. (1972). Cell lines derived from late embryonic stages of *Drosophila melanogaster*. *J. Embryol. Exp. Morphol.* **27**, 353–365.
Submitted by: I. Schneider

Invertebrate species: *Drosophila melanogaster*
Designation of line: Line 2
Primary explant from which line was derived: Late embryos and first stage larvae
Investigator who established the line: I. Schneider
Date primary culture, from which line originated, was set up: December 1969
Number of subcultures to date: 198 as of April 1975
Morphology: Epithelial-like cells
Karyology: 50% diploid, 50% heteroploid
Culture medium: Schneider (1966) medium plus 10% FBS
Growth characteristics of line: Generation time approx. 16 hours
Virus susceptibility: VS
Reference: Schneider, I. (1972). Cell lines from late embryonic stages of *Drosophila melanogaster*. *J. Embryol. Exp. Morphol.* **27**, 353–365.
Submitted by: I. Schneider

Invertebrate species: *Drosophila melanogaster*
Designation of line: Line 3
Primary explant from which line was derived: Late embryos and first stage larvae

Investigator who established the line: I. Schneider
Date primary culture, from which line originated, was set up: February
    1970
Number of subcultures to date: 177 as of April 1975
Morphology: Epithelial-like, fibroblast-like, and round
Karyology: Diploid $2n = 8$
Culture medium: Schneider (1966) medium plus 15% FBS
Growth characteristics of line: Generation time approx. 24 hours
Reference Schneider, I. (1972). Cell lines from late embryonic stages
    of *Drosophila melanogaster*. *J. Embryol. Exp. Morphol.* **27**, 353–365.
Submitted by: I. Schneider

Invertebrate species: *Drosophila melanogaster*
Primary explant from which line was derived: Imaginal discs
Investigator who established the line: I. Schneider
Date primary culture, from which line originated, was set up: February
    1970
Number of subcultures to date: 31 as of April 1971
Reference: Unpublished

Invertebrate species: *Drosophila melanogaster*
Designation of line: GM_1
Primary explant from which line was derived: Embryonic tissues
Investigators who established the line: G. Mosna and S. Dolfini
Date primary culture, from which line originated, was set up: February
    27, 1970
Number of subcultures to date: 272 as of April 2, 1975
Morphology: Epithelial-like cells
Karyology: $X + Y$ (fragment), normal autosomes II and III, only one
    chromosome IV
Culture medium: Echalier and Ohanessian (1970) D-20 medium
History of line: Date of the first subculture was October 31, 1970. Since
    September 1972, the cells also grow in absence of serum
Reference: Mosna, G., and Dolfini, S. (1972). Morphological and chro-
    mosomal characterization of three new continuous cell lines of *Dro-
    sophila melanogaster*. *Chromosoma* **38**, 1–9.
Submitted by: S. Faccio Dolfini

Invertebrate species: *Drosophila melanogaster*
Designation of line: GM_2
Primary explant from which line was derived: Embryonic tissues
Investigators who established the line: G. Mosna and S. Dolfini

Date primary culture, from which line originated, was set up: February 13, 1970
Number of subcultures to date: 268 as of April 2, 1975
Morphology: Epithelial-like cells
Karyology: Tetraploid cells
Culture medium: Echalier and Ohanessian (1970) D-20 medium
History of line: Date of the first subculture was February 14, 1970. Since April 1972, the cells also grow in absence of serum
Reference: Mosna, G., and Dolfini, S. (1972). Morphological and chromosomal characterization of three new continuous cell lines of *Drosophila melanogaster*. *Chromosoma* **38**, 1–9.
Submitted by: S. Faccio Dolfini

Invertebrate species: *Drosophila melanogaster*
Designation of line: GM_3
Primary explant from which line was derived: Embryonic tissues
Investigators who established the line: G. Mosna and S. Dolfini
Date primary culture, from which line originated, was set up: September 30, 1970
Number of subcultures to date: 225 as of April 2, 1975
Morphology: Epithelial-like cells
Karyology: X + Y (with a deletion), normal autosomes II and III, only one chromosome IV
Culture medium: Echalier and Ohanessian (1970) D-20 medium
History of line: Date of the first subculture was March 15, 1971. Since May 1972, the cells also grow in absence of serum
Reference: Mosna, G., and Dolfini, S. (1972). Morphological and chromosomal characterization of three new continuous cell lines of Drosophila melanogaster. *Chromosoma* **38**, 1–9.
Submitted by: S. Faccio Dolfini

Invertebrate species: *Drosophila melanogaster*
Designation of line: 11P80
Primary explant from which line was derived: From embryos of the stock T(Y;3)P80
Investigator who established the line: C. Halfer
Date primary culture, from which line originated, was set up: March 15, 1974
Number of subcultures to date: 46 as of April 2, 1975
Morphology: Roundish and separate cells
Karyology: Tetraploid, with only 2 X chromosomes and variable number of IV

Culture medium: Echalier and Ohanessian (1970) D-20 medium

History of line: Date of the first subculture was April 18, 1974

Reference: Unpublished

Comments: The cell line is derived from a stock characterized by a translocation between the Y and third chromosome; the translocation is no longer present

Submitted by: Carlota Halfer

Invertebrate species: *Drosophila melanogaster*

Designation of lines: 75A, 75B, 75C, 75D1, 75D2, 75E, 75F, 75L

Primary explant from which line was derived: 6- to 12-hour-old embryos

Investigator who established the line: A. Ohanessian

Date primary culture, from which originated, was set up: October 1974

Number of subcultures to date: About 10 as of April 1975

Culture medium: Line 75D grows in Shields and Sang (1970) medium supplemented with 10% embryonic calf serum. Other lines are maintained in DS medium which consists of 50% Shields and Sang medium and 50% Echalier D22 medium with 10% embryonic calf serum

Virus susceptibility: 75B and 75D1 are sensitive to VSV

Reference: Unpublished

Submitted by: A. Ohanessian

Invertebrate species: *Musca domestica*

Common name: House fly

Primary explant from which line was derived: Embryos

Karyology: Cells are diploid

Culture medium: Eide and Chang (1969) X-2 medium

Reference: Eide, P. E. (1975). Establishment of a cell line from long-term primary embryonic house fly cell cultures. *J. Insect Physiol.* **21**, 1431–1438.

C. ORTHOPTERA

Invertebrate species: *Blabera fusca*

Common name: Cockroach

Designation of lines: Two separate cell lines

Primary explant from which line was derived: Minced trypsinized embryos

Date primary culture, from which line originated, was set up: March 1965

Culture medium: Landureau (1966) medium

Reference: Landureau, J. C. (1968). Cultures *in vitro* de cellules embryonnaires de Blattes (Insectes Dictyopteres). II. Obtention de lignées cellulaires á multiplication continué. *Exp. Cell Res.* **50,** 323–337.

Invertebrate species: *Blattella germanica*
Common name: German cockroach
Date primary culture, from which line originated, was set up: March 1965
Karyology: Most cells are diploid
Culture medium: Landureau (1966) medium
Reference: Landureau, J. C. (1966). Cultures in vitro de cellules embryonnaires de Blattes (Insectes Dictyopteres). *Exp. Cell Res.* **41,** 545–566.

Invertebrate species: *Blattella germanica*
Designation of line: UM-BTE-1
Primary explant from which line was derived: Embryos; segmented germ band
Investigators who established the line: T. J. Kurtti and M. A. Brooks
Date primary culture, from which line originated, was set up: February 17, 1972
Number of subcultures to date: 98 as of February 25, 1975
Morphology: When transferred to fresh medium the cells attach and flatten themselves to the culture flask; in metabolized medium they are round to spindle-shaped
Karyology: Polyploid; typical *B. germanica* chromosomes
Culture medium: Landureau and Jollès (1969) S19 medium modified by Kurtti (1974)
Growth characteristics of line: Has a weekly subculture interval; population doubling time is 5 days; initial seeding density is $3–5 \times 10^5$ cells/ml; cells dispersed by pipetting
Virus susceptibility: Unknown
History of line: Embryos from 5-day-old ootheca were dissociated with a 0.05% trypsin solution for 5 minutes. A 50-week adaptation period was encountered before the line could be regularly subcultured
Reference: Kurtti, T. J. (1974). The development of insect tissue culture systems for studying intracellular symbiotes. Ph. D. Dissertation, Department of Entomology, Fisheries and Wildlife, University of Minnesota, Minneapolis.
Comments: The cells can be cultured as an attached monolayer or in suspension (Wheaton Celstir system)
Submitted by: T. J. Kurtti

Invertebrate species: *Blattella germanica*
Designation of line: UM-BGE-2
Primary explant from which line was derived: Embryos; dorsal closure; histogenesis
Investigators who established the line: T. J. Kurtti and M. A. Brooks
Date primary culture, from which line originated, was set up: January 27, 1972
Number of subcultures to date: 76 as of February 25, 1975
Morphology: Round to multipolar cells
Karyology: Diploid; *B. germanica* chromosomes
Culture medium: Landureau and Jollès (1969) S19 medium modified by Kurtti (1974)
Growth characteristics of line: Weekly subculture interval; one-to-two or -three dilution of the cells at each subculture
Virus susceptibility: Unknown
History of line: Embryos from a 7-day-old ootheca were dissociated with a 0.05% saline solution of trypsin. A 40–50 week adaptation period was encountered before the cells could be regularly subcultured
Reference: Kurtti, T. J. (1974). The development of insect tissue culture systems for studying intracellular symbiotes. Ph.D. Dissertation, Department of Entomology, Fisheries and Wildlife, University of Minnesota, Minneapolis.
Submitted by: T. J. Kurtti

Invertebrate species: *Blattella germanica*
Designation of line: UM-BGE-4
Primary explant from which line was derived: Embryos; germ band and histogenesis
Investigators who established the line: T. J. Kurtti and M. A. Brooks
Date primary culture, from which line originated, was set up: March 12, 1972
Number of subcultures to date: 72 as of February 25, 1975
Morphology: Vesicles; cells colonize as hollow spheres
Karyology: Diploid and tetraploid; *B. germanica* chromosomes
Culture medium: Landureau and Jollès (1969) S19 medium modified by Kurtti (1974)
Growth characteristics of line: Weekly subculture interval; 1:2 dilution of the cells
Virus susceptibility: Unknown
History of line: Vesicles from several primary cultures (6 weeks old) of cells from 5- and 7-day-old ootheca were pooled; 50-week adaptation period

Reference: Kurtti, T. J. (1974). The development of insect tissue culture systems for studying intracellular symbiotes. Ph.D. Dissertation, Department of Entomology, Fisheries and Wildlife, University of Minnesota, Minneapolis.

Comments: The vesicles cannot be dissociated by solutions of trypsin or EDTA; gap and septate junctions abundant between opposed cells

Submitted by: T. J. Kurtti

Invertebrate species: *Blattella germanica*

Designation of line: UM-BGE-5 (two sublines, alpha and beta, were developed)

Primary explant from which line was derived: Embryos; organogenesis

Investigators who established the line: T. J. Kurtti and M. A. Brooks

Date primary culture, from which line originated, was set up: August 27, 1971

Number of subcultures to date: 57, alpha subline; 32, beta subline; as of February 25, 1975

Morphology: alpha (round cells), no attachment to the culture flask; beta (multipolar cells), attached monolayer

Karyology: Not determined

Culture medium: Landureau and Jollès (1969) S19 medium modified by Kurtti (1974)

Growth characteristics of line: alpha—biweekly subculture interval; beta—triweekly subculture interval

Virus susceptibility: Unknown

History of line: Embryos from a 1-day-old ootheca were dissociated with a 0.5% solution of trypsin for 20 minutes. A 90-week adaptation period was observed

Reference: Kurtti, T. J. (1974). The development of insect tissue culture systems for studying intracellular symbiotes. Ph.D. Dissertation, Department of Entomology, Fisheries and Wildlife, University of Minnesota, Minneapolis.

Comments: The alpha subline can be suspended by pipetting whereas the beta subline cannot (except with EDTA)

Submitted by: T. J. Kurtti

Invertebrate species: *Leucophaea maderae*

Common name: Cockroach

Designation of line: LM 42

Primary explant from which line was derived: Nymphal dorsal vessel

Investigators who established the line: C. Vago and J. M. Quiot

Date primary culture, from which line originated, was set up: 1968

Number of subcultures to date: 31 as of September 1970
Morphology: Very elongated cells attached to surface
Culture medium: Vago and Quiot (1969) 72 S.F.M. medium

Invertebrate species: *Leucophaea maderae*
Designation of line: LM 75
Primary explant from which line was derived: Nymphal dorsal vessel
Date primary culture, from which line originated, was set up: October
    1969
Number of subcultures to date: 67 as of May 1971
Morphology: Spherical and fibroblast-like cells, 30–50 μm in diameter
Culture medium: Vago and Quiot (1969) 72 S.F.M. medium modified
    by J. M. Quiot (personal communication, 1971)

Invertebrate species: *Leucophaea maderae*
Designation of line: LM 112
Primary explant from which line was derived: Nymphal ovaries
Date primary culture, from which line originated, was set up: July 1970
Number of subcultures to date: 32 as of June 1971
Morphology: Spherical and spindle-shaped cells (20–40 μm diameter) in
    suspension and rarely attached to surface
Culture medium: Vago and Quiot (1969) 72 S.F.M. medium modified
    by J. M. Quiot (personal communication, 1971)

Invertebrate species: *Periplaneta americana*
Common name: American cockroach
Designation of line: EPa
Primary explant from which line was derived: Minced trypsinized
    embryos
Date primary culture, from which line originated, was set up: September
    1965
Number of subcultures to date: 200+ as of March 1971
Karyology: Fundamentally euploid
Culture medium: Landureau (1966) medium
Reference: Landureau, J. C. (1968). Cultures *in vitro* de cellules em-
    bryonnaires de Blattes (Insectes Dictyopteres). II. Obtention de
    lignées cellulaires à multiplication continué. *Exp. Cell Res.* **50,**
    323–337.
Comments: Chitinase is produced by these cells

Invertebrate species: *Periplaneta americana*
Primary explant from which line was derived: Minced trypsinized
    embryos

Date primary culture, from which line originated, was set up: September 1965
Number of subcultures to date: 32 as of March 1971
Karyology: Polyploid with many cells approximately $16n$
Reference: Landureau, J. C. (1968). Cultures *in vitro* de cellules embryonnaires de Blattes (Insectes Dictyopteres). II. Obtention de lignées cellulaires à multiplication continué. *Exp. Cell Res.* **50**, 323–337.

Invertebrate species: *Periplaneta americana*
Designation of line: HPa 33
Primary explant from which line was derived: Hemocytes from nymphal male
Investigator who established the line: J. C. Landureau

Invertebrate species: *Periplaneta americana*
Designation of line: HPa 34
Primary explant from which line was derived: Hemocytes of adult female
Investigator who established the line: J. C. Landureau

D. HOMOPTERA

Invertebrate species: *Aceratagallia sanguinolenta*
Common name: Clover leafhopper
Designation of line: AS-1
Primary explant from which line was derived: Embryos
Morphology: Consists almost exclusively of epithelial-like cells
Culture medium: Chiu and Black (1967) medium
Virus susceptibility: WTV, PYDV
Reference: Chiu, R., and Black, L. M. (1969). Assay of wound tumor virus by the fluorescent cell counting technique. *Virology* **37**, 667–677.

Invertebrate species: *Agallia constricta*
Common name: Leafhopper
Designation of line: AC20
Primary explant from which line was derived: Minced embryos
Date primary culture, from which line originated, was set up: December 1965
Morphology: Predominantly epithelial-like cells
Culture medium: Chiu and Black (1967) medium or Hirumi and Maramorosch (1964) medium
Growth characteristics of line: Population doubling time of 72 hours
Virus susceptibility: WTV, PYDV

Reference: Chiu, R., and Black, L. M. (1967). Monolayer cultures of insect cell lines and their inoculation with a plant virus. *Nature (London)* **215**, 1076–1078.

Invertebrate species: *Agallia quadripunctata*
Common name: Leafhopper
Primary explant from which line was derived: Minced embryos
Culture medium: Chiu and Black (1967) medium
Reference: Chiu, R., and Black, L. M. (1967). Monolayer cultures of insect cell lines and their inoculation with a plant virus. *Nature (London)* **215**, 1076–1078.

Invertebrate species: *Agalliopsis novella*
Investigator who established the line: I. Windsor
Reference: Windsor, I. (1972). Clover clubleaf: A possible rickettsial disease of plants. Ph.D. Thesis, University of Illinois, Urbana.

Invertebrate species: *Colladonus montanus*
Common name: Leafhopper
Designation of lines: Three separate cell lines
Primary explant from which line was derived: Fragmented trypsinized embryos
Morphology: Monolayers of large epithelial-like cells, some of which are elongate and spindle-shaped. Also present are larger cells with many processes and giant cells 3–4 times the size of usual cell types
Culture medium: Chiu and Black (1967) medium
Reference: Richardson, J., and Jensen, D. D. (1971). Tissue culture of monolayer cell lines of *Colladonus montanus* (Homoptera:Cicadellidae) a vector of the causal agent of western X-disease of peach. *Ann. Entomol. Soc. Amer.* **64**, 722–729.

Invertebrate species: *Dalbulus maidis*
Common name: Corn leafhopper
Designation of line: DMIIB
Primary explant from which line was derived: Embryos
Investigators who established the line: A. McIntosh, E. Parmegiani, and R. Shamy
Date primary culture, from which line originated, was set up: September 28, 1973
Number of subcultures to date: 20 as of April 1975
Morphology: Mainly epithelial-like
Karyology: Diploid at passage 10

Culture medium: Equal parts of Morgan *et al.* (1950) TC 199 with Hanks' salts and Melnick's (1955) monkey kidney medium "A" plus 10% FBS

Growth characteristics of line: Line is proliferating better than when first initiated. 1:2 splits are made weekly

Virus susceptibility: Not tested. Contains an icosahedral virus in its cytoplasm

Reference: Unpublished

Submitted by: A. H. McIntosh

Invertebrate species: *Macrostelles fascifrons*

Common name: Leafhopper

Investigators who established the line: H. Hirumi

Culture medium: Equal parts of Morgan *et al.* (1950) TC 199 with Hanks' salts and Melnick's (1955) monkey kidney medium "A" plus 10% FBS

Reference: McIntosh, A. H., Maramorosch, K., and Rechtoris, C. (1973). Adaptation of an insect cell line (*Agallia constricta*) in a mammalian cell culture medium. *In Vitro* **8,** 375–378.

Invertebrate species: *Macrosteles sexnotatus*

Common name: Leafhopper

Primary explant from which line was derived: Embryos

Number of subcultures to date: 31 as of June 1971

Morphology: Cells of epithelial-like and fibroblast-like morphology

Reference: Peters, D., and Spaansen, C. H. (1973). In vitro culture of *Macrosteles sexnotatus* embryonic cells. *Proc. Int. Colloq. Invertebr. Tissue Cult., 3rd, 1971* pp. 67–74.

Invertebrate species: *Nephotettix apicalis*

Common name: Leafhopper

Primary explant from which line was derived: Embryonic tissue

Investigator who established the line: J. Mitsuhashi

Date primary culture, from which line originated, was set up: May 1970

Number of subcultures to date: 23 as of March 1971

Reference: Unpublished

Invertebrate species: *Nephotettix cincticeps*

Common name: Leafhopper

Primary explant from which line was derived: Embryonic tissues

Investigator who established the line: J. Mitsuhashi

Date primary culture from which line originated was set up: February 1970
Number of subcultures to date: 52 as of March 1971
Reference: Unpublished

E. HEMIPTERA

Invertebrate species: *Triatoma infestans*
Common name: Blood-sucking bug
Designation of line: BTC-32
Primary explant from which line was derived: Embryonic tissues
Investigators who established the line: Mary Pudney and D. Lanar
Date primary culture, from which line originated, was set up: May 15, 1972
Number of subcultures to date: 93 as of June 1975
Morphology: Epithelial-type cells of three sizes
Karyology: Diploid
Culture medium: Leibovitz (1963) L-15 medium plus 10% tryptose phosphate broth and 10% inactivated FCS
Growth characteristics of line: Split 1:7 weekly
Reference: Unpublished
Submitted by: Mary Pudney and M. G. R. Varma

IV. Arachnida

ACARINA

Invertebrate species: *Boophilus microplus*
Common name: Tick
Designation of line: TTC-256
Primary explant from which line was derived: Embryos
Number of subcultures to date: 14 as of publication below
Karyology: Most cells are diploid with 21 chromosomes
Culture medium: Hanks' balanced salt solution with 0.5% lactalbumin hydrolysate plus 10% FCS
Reference: Pudney, M., Varma, M. G. R., and Leake, C. J. (1973). Culture of embryonic cells from the tick *Boophilus microplus* (Ixodidae). *J. Med. Entomol.* **10**, 493–496.

Invertebrate species: *Rhipicephalus appendiculatus*
Common name: Hard tick

Designation of lines: TTC-219, TTC-243, TTC-257
Primary explant from which line was derived: Developing adult tissues
Number of subcultures to date: 39–103 as of June 1975
Karyology: Polyploid
Culture medium: Leibovitz (1963) L-15 medium plus 10% tryptose phosphate broth and 10% FCS
Virus susceptibility: TTC-243 supports replication of WN, Louping ill, Langat, and Quaranfil viruses
Reference: Varma, M. G. R., Pudney, M., and Leake, C. J. (1975). The establishment of three cell lines from the tick *Rhipicephalus appendiculatus* (Acari:Ixodidae) and their infection with some arboviruses. *J. Med. Entomol.* **11**, 698–706.

Invertebrate species: *Rhipicephalus sanguineus* (Latreille)
Common name: Brown dog tick
Primary explant from which line was derived: Minced 12-day-old eggs
Investigators who established the line: S. H. Hsu and H. H. Huang
Date primary culture, from which line originated, was set up: May 28, 1974
Number of subcultures to date: 9 as of April 1975
Morphology: The major cell type is fibroblastic-spindle and stellate cells, and a few round cells are also noted
Karyology: Most cells appear to be of female origin having 22 chromosomes
Culture medium: Equal volume of Leibovitz (1963) and Mitsuhashi and Maramorosch (1964) medium supplemented with 10% tryptose phosphate and 15% heat-inactivated FCS with antibiotics
Growth characteristics of line: This cell line was established in May 1974. Since the growth rate is relatively slow, the subculture interval of about 3–4 weeks. To date, it has only been subcultured for 9 passages
Reference: Unpublished
Submitted by: S. H. Hsu

V. Mollusca

GASTROPODA

Invertebrate species: *Helix aspersa*
Common name: European brown snail

Primary explant from which line was derived: Foot
Culture medium: Vago and Chastang (1958) *H. aspersa* medium
Reference: Vago, C., and Chastang, S. (1958). Obtention de lignées cellulaires en culture de tissus d'invertèbres. *Experientia* **14**, 110–111.

Invertebrate species: *Biomphalaria glabrata*
Common name: Snail
Primary explant from which line was derived: Embryos
Investigator who established the line: E. L. Hansen
Number of subcultures to date: 34+ as of publication below
Reference: Hansen, E. L. (1975). Effect of serum on tissue culture from *Biomphalaria glabrata* (Mollusca). *In Vitro* **10**, 348 (abstr.).

VI. Alphabetical Listing of Invertebrate Tissue Culture Media

Chiu and Black (1967) insect tissue culture medium (mg/100 ml)

D-Glucose	400	Lactalbumin hydrolysate	650
NaCl	105	Yeastolate	500
KCl	80	Penicillin	10000 U
$MgSO_4 \cdot 7H_2O$	185	Streptomycin	10000 μg
$CaCl_2$	30	Neomycin	5000 μg
KH_2PO_4	30	Fungizone	250 μg
$NaHCO_3$	35	FBS	17.5–20.0 ml

Echalier and Ohanessian (1970) D-20 insect tissue culture medium

Glutamic acid	7.35 gm	Neutralize with 10 N KOH and add H_2O to 100 ml, use 54 ml of this solution	
Glycine	3.74 gm		
Glutamic acid	7.35 gm	Neutralize with 10 N NaOH and add H_2O to 100 ml, use 94 ml of this solution	
Glycine	3.74 gm		
$MgCl_2 \cdot 6H_2O$	1.0 gm	Sodium acetate $\cdot 3H_2O$	25 mg
$MgSO_4 \cdot 7H_2O$	3.7 gm	Glucose	2 gm
$NaH_2PO_4 \cdot 2H_2O$	0.47 gm	Lactalbumin hydrolysate	15 gm
$CaCl_2$	0.89 gm	Grace's (1962) vitamins	
Yeastolate	1.5 gm	H_2O to 1000 ml	
Malic acid	670 mg	Adjust pH to 6.7	
Succinic acid	60 mg	FBS 10–20% supplement	

Eide and Chang (1969) X-2 insect tissue culture medium (mg/liter)

NaCl	4533.0	DPN	50.0
KCl	1073.0	Nicotinamide HCl	119.0
$CaCl_2 \cdot 2H_2O$	860.0	Thymidine	200.0
$MgCl_2 \cdot 6H_2O$	300.0	ATP	300.0
$NaH_2PO_4 \cdot 2H_2O$	400.0	Biotin	0.01
$NaHCO_3$	350.0	D-Ca pantothenate	0.02
$CH_3CO_2^-, Na^+$	250.0	L-Carnitine HCl	0.01
Dextrose	11550.0	Choline chloride	0.20
Fructose	2000.0	Folic acid	0.02
Succinic acid	600.0	i-Inositol	0.02
Fumaric acid	200.0	Niacin	0.02
Malic acid	660.0	p-Aminobenzoic acid	0.02
α-Ketoglutaric acid	320.0	Pyridoxine	0.02
Glutathione (reduced)	200.0	Riboflavin	0.02
Ascorbic acid	100.0	Thiamine HCl	0.02
Lactalbumin hydrolysate	17500.0	Fetal calf serum	20%

Goodwin et al. (1973) IPL-25 insect tissue culture medium (mg/liter)

$NaH_2PO_4 \cdot H_2O$	1160.0	L-Isoleucine	750.0
$NaHCO_3$	350.0	L-Leucine	250.0
KCl	1200.0	L-Lysine HCl	700.0
$CaCl_2$	500.0	L-Methionine	1000.0
$MgSO_4 \cdot 7H_2O$	880.0	L-Proline	500.0
Choline chloride	20.0	L-Phenylalanine	1000.0
Sucrose (gm)	16.5	DL-Serine	400.0
Glucose	1000.0	L-Tyrosine	250.0
Glycerol	2700.0	L-Tryptophan	100.0
Malic acid	53.6	L-Threonine	200.0
α-Ketoglutaric acid	29.6	L-Valine	500.0
Succinic acid	4.8	Thiamine HCl	0.08
Fumaric acid	4.4	Riboflavin	0.08
L-Arginine HCl	800.0	Calcium pantothenate	0.008
L-Aspartic acid	1300.0	Pyridoxine HCl	0.40
L-Asparagine	1300.0	p-Aminobenzoic acid	0.32
β-Alanine	300.0	Folic acid	0.08
L-Cystine	100.0	Niacin	0.16
L-Glutamic acid	1500.0	i-Inositol	0.40
L-Glutamine	1000.0	Biotin	0.16
L-Glycine	200.0	Cyanocobalamine	0.24
L-Histidine	200.0		

Goodwin (1975) IPL-40 insect tissue culture medium (mg/liter)

Same as IPL-25 except KCl is 1900.0 mg/liter, glucose is 2500.0 mg/liter, and 1000.0 mg maltose is added

Goodwin (1975) IPL-45 insect tissue culture medium (mg/liter)

Same as IPL-25 except sucrose is 20.0 mg/liter, maltose is 1000.0 mg/liter, glucose is 5000.0 mg/liter, and 800.0 mg hydroxy-L-proline is added

Grace (1962) insect tissue culture medium (mg/100 ml)

NaH$_2$PO$_4$·2H$_2$O	114	L-Tryptophan	10
NaHCO$_3$	35	L-Glutamine	60
KCl	224	L-Threonine	17.5
CaCl$_2$	100	L-Valine	10
MgCl$_2$·6H$_2$O	228	Sucrose (gm)	2.668
MgCO$_4$·7H$_2$O	278	Fructose	40
L-Arginine HCl	70	Glucose	70
L-Aspartic acid	35	Malic	67
L-Asparagine	35	α-Ketoglutaric	37
L-Alanine	22.5	Succinic	6
β-Alanine	20	Fumaric	5.5
L-Cystine HCl	2.5	Thiamine HCl	0.002
L-Glutamic acid	60	Riboflavin	0.002
L-Glycine	65	Ca pantothenate	0.002
L-Histidine	250	Pyridoxine HCl	0.002
L-Isoleucine	5	p-Aminobenzoic acid	0.002
L-Leucine	7.5	Folic acid	0.002
L-Lysine HCl	62.5	Niacine	0.002
L-Methionine	5	i-Inositol	0.002
L-Proline	35	Biotin	0.001
L-Phenylalanine	15	Choline chloride	0.02
DL-Serine	110	Penicillin G, Na salt	3
L-Tyrosine	5	Streptomycin sulfate	10

Grace (1962) insect tissue culture medium modified by Schneider (1969) (gm/liter)

NaCl	3.0	Sucrose	16.0
KCl	1.1	Glucose	1.0
MgCl$_2$·6H$_2$O	1.14	Trehalose	0.4
MgSO$_4$·7H$_2$O	0.4	Cholesterol	2.0 mg
NaHCO$_3$	0.35	Phenol red	0.01%
CaCl$_2$	0.40	FBS	15%

Amino acids as in Grace's (1962) medium
Supplemented with 1% 10X solution of abbreviated NCTC 135
    that lacks sugars, amino acids, and salts

Grozdev and Kakpakov (1968) S-15 insect tissue culture medium (mg/100 ml)

$NaH_2PO_4 \cdot 2H_2O$	50.0	Niacinamide	0.01
$NaHCO_3$	35.0	Vitamin B_{12}	0.002
KCl	156.0	Vitamin A	0.002
$CaCl_2$	50.0	Thiamine	0.002
$MgCl_2 \cdot 6H_2O$	250.0	Riboflavin	0.002
NaCl	400.0	Calcium pantothenate	0.002
Sucrose	500.0	Pyridoxine	0.002
Glucose	500.0	p-Aminobenzoic acid	0.002
Malic acid	67.0	Folic acid	0.002
Succinic acid	6.0	Biotin	0.002
Sodium acetate	2.5	Inositol	0.002
Lactalbumin hydrolysate	1750.0	Choline chloride	0.002
Yeast extract	150.0	Tris-aminomethane	300.0
L-Tryptophan	10.0	Phenol red	1.0
L-Cysteine	2.5	Penicillin	10.0
Glutathione	0.5	Streptomycin	10.0
Nicotinamide dinucleotide	0.5	FBS	15%
Ascorbic acid	10.0		

Hink (1970) TNM-PH insect tissue culture medium

Grace's (1962) medium	90.0 ml	TC yeastolate	0.3 gm
Fetal bovine serum (FBS)	8.0 ml	Lactalbumin hydrolysate	0.3 gm
Chicken egg ultrafiltrate	8.0 ml	Bovine plasma albumin, crystallized	0.5 gm

Hirumi and Maramorosch (1964) insect tissue culture medium

This medium consists of a mixture of 10 ml modified Vago's (1963) No. 22 *Bombyx mori* medium, 20 ml Morgan's (1950) TC 199, and 6 ml of FBS, with 100 units per ml of penicillin and streptomycin.

Modified No. 22 (Vago, 1963) (mg/100 ml)

$NaH_2PO_4 \cdot H_2O$	120	Glucose	70
$MgCl_2 \cdot 6H_2O$	300	Sucrose	40
$MgSO_4 \cdot 7H_2O$	400	Fructose	40
KCl	300	Lactalbumin hydrolysate	1000
$CaCl_2 \cdot 2H_2O$	100	Phenol red (0.5%)	0.5 ml

Horikawa *et al.* (1966) H-6 insect tissue culture medium (mg/liter)

$NaH_2PO_4 \cdot 2H_2O$	200	Lactalbumin hydrolysate	7500
$NaHCO_3$	350	L-Tryptophan	80
KCl	200	L-Cystine HCl	20
$CaCl_2 \cdot 2H_2O$	20	Yeast extract	1200
$MgCl_2 \cdot 6H_2O$	100	Phenol red	10
NaCl	7000	Penicillin G	30
Glucose	5500	Streptomycin sulfate	100
Sucrose	5500	Calf serum	10%

Hsu *et al.* (1970) 721 insect tissue culture medium (mg/100 ml)

KCl	80	Lactalbumin hydrolysate	2000
NaCl	450	Bacto-peptone	500
CaCl$_2$	35	TC-Yeastolate	200
MgSO$_4 \cdot$7H$_2$O	40	*l*-Malic acid	60
KH$_2$PO$_4$	25	α-Ketoglutaric acid	40
NaHCO$_3$	100	Succinic acid	6
D-Glucose	160	Fumaric acid	6
Sucrose	600	Medium No. 199 (1X)	20%
		FBS	10%

Hsu *et al.* (1972) modification of 721 insect tissue culture medium (mg/100 ml)

KCl	28	α-Ketoglutaric acid	14
NaCl	157.5	Succinic acid	2.1
MgSO$_4 \cdot$7H$_2$O	14	Fumaric acid	2.1
KH$_2$PO$_4$	8.75	Lactalbumin hydrolysate	865
CaCl$_2$	12.25	TC-Yeastolate	320
NaHCO$_3$	35	Bacto-peptone	175
D-Glucose	56	Medium No. 199 (10X)	5.0 ml
Sucrose	210	Fetal bovine serum	10.0 ml
l-Malic acid	21	Whole egg ultrafiltrate	1.0 ml

Kitamura (1965) insect tissue culture medium (gm/100 ml)

NaCl	0.65	Sucrose	1.0
KCl	0.05	Lactalbumin hydrolysate	2.0
CaCl$_2$	0.01	TC 199	various amounts
KH$_2$PO$_4$	0.01	Calf serum	various amounts
NaHCO$_3$	0.01		

Kitamura (1965) insect tissue culture medium modified by Varma and Pudney (1969) (mg/100 ml)

NaCl	650	D-Glucose	400
KCl	50	Lactalbumin hydrolysate	650
CaCl$_2 \cdot$2H$_2$O	10	Yeastolate	500
KH$_2$PO$_4$	10	FBS	20%
NaHCO$_3$	10	Penicillin	1000 U/ml
		Streptomycin	1 mg/ml

Kitamura (1970) insect tissue culture medium (mg)

NaCl	650	Glucose	200
KCl	50	Lactalbumin hydrolysate	1000
$CaCl_2 \cdot 2H_2O$	10	H_2O	100 ml
KH_2PO_4	10	TC 199	2 parts
$NaHCO_3$	10	Calf serum	1 part

Landureau (1966) insect tissue culture medium (mM/liter)

L-Arginine	11.5	NaCl	113
L-Aspartic acid	1.5	KCl	12
L-Glutamic acid	13.5	$CaCl_2$	4.4
α-Alanine	1.35	$MgSO_4 \cdot 7H_2O$	5.6
β-Alanine	0.5	$NaHCO_3$	4.3
L-Cysteine HCl	0.83	H_3PO_3	10
L-Glutamine	4.1	Glucose	5.6
L-Glycine	23.5	Trehalose	25
L-Histidine	2.6	α-Ketoglutaric acid	2.5
L-Leucine	1.91	Citric acid	0.8
L-Lysine HCl	0.88	Fumaric acid	0.5
L-Methionine	3.35	Malic acid	5.0
L-Proline	6.53	Succinic acid	0.5
L-Serine	0.76	Yeast extract	0.5 gm/liter
L-Threonine	1.68	Lactalbumin hydrolysate	3.5 gm/liter
L-Tyrosine	2.0	Fetal calf serum	10%
L-Valine	1.28	Grace's (1962) vitamins	

Landureau and Grellet (1972) S20 insect tissue culture medium (mM/liter)

$CaCl_2$	2.0	Proline	5.5
KCl	14.0	Serine	0.3
$MgSO_4 \cdot 7H_2O$	5.0	Threonine	0.84
$MnSO_4 \cdot H_2O$	0.19	Tryptophane	0.5
NaCl	145.0	Tyrosine	0.5
H_3PO_3	11.0	Valine	0.8
Glucose	22.2	Folic acid	0.00005
Arginine HCl	3.8	d-Biotin	0.00004
Aspartic acid	3.8	Choline HCl	1.3
Glutamic acid	10.2	Cyanocobalamine	0.00003
Cysteine HCl	4.1	Inositol	0.0003
Glycine	10.0	Nicotinamide	0.0003
Histidine	1.3	Pyridoxine HCl	0.0002
Isoleucine	1.0	Riboflavin	0.0005
Leucine	1.5	Thiamine HCl	0.002
Lysine HCl	1.0	Ca panthothenate	0.001
Methionine	1.7	FBS	0-20%
Phenylalanine	0.6		

Landureau and Jollès (1969) S19 insect tissue culture medium (mg/liter)

CaCl₂	490	L-Arginine HCl	800
KCl	1050	L-Cysteine HCl	260
MgSO₄·7H₂O	1260	L-Glutamine	300
MnSO₄·H₂O	65	Glycocolle	750
NaCl	8500	L-Histidine	300
NaHCO₃	360	L-Isoleucine	120
H₃PO₃	900	L-Leucine	250
Glucose	3000	L-Lysine HCl	160
Folic Acid	0.01	L-Methionine	500
d-Biotin	0.01	L-Phenylalanine	200
Choline HCl	0.4	L-Proline	750
Inositol	0.05	L-Serine	80
Nicotinamide	0.03	L-Threonine	200
Ca pantothenate	0.1	L-Tryptophane	200
Pyridoxine HCl	0.03	L-Tyrosine	180
Riboflavin	0.05	L-Valine	150
Thiamine HCl	0.01	Plasma protein fraction V	4000
L-Aspartic acid	250	α₂-Macroglobulin	50
L-Glutamic acid	1500	Penicillin G	50
α-L-Alanine	120	Streptomycin	70

Leibovitz (1963) L-15 cell culture medium (mg/liter)

CaCl₂	140.0	L-Lysine	75.0
KCl	400.0	DL-Methionine	150.0
KH₂PO₄	60.0	DL-Phenylalanine	250.0
MgCl₂·6H₂O	200.0	L-Serine	200.0
MgSO₄·7H₂O	200.0	DL-Threonine	600.0
NaCl	8000.0	L-Tryptophane	20.0
Na₂HPO₄	190.0	L-Tyrosine	300.0
Galactose	900.0	DL-Valine	200.0
Sodium pyruvate	550.0	DL-Ca pantothenate	1.0
DL-α-Alanine	450.0	Choline chloride	1.0
L-Arginine	500.0	Folic acid	1.0
L-Asparagine	250.0	Inositol	2.0
L-Cysteine	120.0	Nicotinamide	1.0
L-Glutamine	300.0	Pyridoxine HCl	1.0
Glycine	200.0	Riboflavin 5'-phosphate	0.1
L-Histidine	250.0	Thiamine monophosphate	1.0
L-Leucine	125.0	Phenol red	10.0

Mitsuhashi (1967) CSM-2F insect tissue culture medium (mg/100 ml)

NaH$_2$PO$_4$·H$_2$O	50	Lactalbumin hydrolysate	520
MgCl$_2$·6H$_2$O	120	Bacto-peptone	520
MgSO$_4$·7H$_2$O	160	TC-Yeastolate	200
KCl	120	Choline chloride	40
CaCl$_2$·2H$_2$O	40	TC 199	20 ml
Glucose	80	Fetal bovine serum	20 ml
Fructose	80	Dihydrostreptomycin sulphate	10

Mitsuhashi and Maramorosch (1964) insect tissue culture medium (mg/100 ml)

NaH$_2$PO$_4$·H$_2$O	20	D-Glucose	400
MgCl$_2$·6H$_2$O	10	Lactalbumin hydrolysate	650
KCl	20	Yeastolate	500
CaCl$_2$·2H$_2$O	20	FBS	20%
NaCl	700	Penicillin	100 U/ml
NaHCO$_3$	12	Streptomycin	100 μg/ml

Mitsuhashi (1972) MGM-431 insect tissue culture medium (mg/liter)

NaH$_2$PO$_4$·2H$_2$O	958	L-Phenylalanine	125
NaHCO$_3$	292	DL-Serine	917
KCl	1875	L-Tyrosine	42
MgSO$_4$·7H$_2$O	2333	L-Tryptophan	83
MgCl$_2$·6H$_2$O	1917	L-Threonine	146
CaCl$_2$	833	L-Valine	83
Sucrose	22080	Malic acid	558
Fructose	417	Succinic acid	50
Glucose	3333	Fumaric acid	46
L-α-Alanine	263	α-Ketoglutaric acid	308
β-Alanine	167	Folic acid	0.16
L-Arginine hydrochloride	583	Biotin	0.08
L-Asparagine	293	Thiamine hydrochloride	0.16
L-Aspartic acid	293	Riboflavin	0.16
L-Cystine	21	Calcium pantothenate	0.16
L-Glutamic acid	500	Pyridoxine hydrochloride	0.16
L-Glutamine	500	p-Aminobenzoic acid	0.16
Glycine	542	Niacin	0.16
L-Histidine	2083	L-Inositol	0.16
L-Isoleucine	42	Choline chloride	1.6
L-Leucine	63	Polyvinylpyrrolidone K-90	500
L-Lysine hydrochloride	521	Fetuin	20
L-Methionine	42	FBS	100 (ml)
L-Proline	292		

J. M. Quiot (personal communication, 1975) Gm15 insect tissue culture medium (mg/100 ml)

$NaH_2PO_4 \cdot H_2O$	106	Thiamine HCl	0.02
KCl	300	Riboflavin	0.02
$CaCl_2 \cdot 2H_2O$	100	Ca pantothenate	0.02
$MgCl_2 \cdot 6H_2O$	300	Pyridoxine HCl	0.02
$MgSO_4 \cdot 7H_2O$	400	p-Aminobenzoic acid	0.02
Fructose	40	Folic acid	0.02
Glucose	1000	Niacin	0.02
Bovine albumin Fraction V	100	i-Inositol	0.02
Lactalbumin hydrolysate	500	Biotin	0.01
α-Ketoglutaric acid	35	Choline chloride	0.20
Citric acid	17	Streptomycin sulfate	10
Fumaric acid	6	Penicillin G, Na salt	20,000 U
Malic acid	65	Calf serum	10 ml
Succinic acid	6		

Schneider (1964) insect tissue culture medium (mg/100 ml)

$CaCl_2$	60	α-Ketoglutaric acid	35
$MgSO_4 \cdot 7H_2O$	370	Fumaric acid	6
KCl	160	Malic acid	60
KH_2PO_4	60	Succinic acid	6
NaCl	210	Lactalbumin hydrolysate	1000
$NaHCO_3$	70	TC-Yeastolate	100
Glucose	200	Penicillin	50 U/ml
Trehalose	200	Streptomycin	50 U/ml

Schneider (1966) modification of Schneider (1964) insect tissue culture medium (mg/100 ml)

$CaCl_2$	60	L-Glutamic acid	80
$MgSo_4 \cdot 7H_2O$	370	L-Glutamine	180
KCl	160	L-Glycine	25
KH_2PO_4	45	L-Histidine	40
NaCl	210	L-Isoleucine	15
$NaHCO_3$	40	L-Leucine	15
Na_2HPO_4	70	L-Lysine	165
Glucose	200	L-Methionine	15
Trehalose	200	L-Proline	170
α-Ketoglutaric acid	35	L-Serine	25
Fumaric acid	6	L-Threonine	35
Malic acid	60	L-Tryptophane	10
Succinic acid	6	L-Tyrosine	50
β-Alanine	50	L-Valine	30
L-Aspartic acid	40	TC-Yeastolate	200
L-Arginine	60	Penicillin	50 U/ml
L-Cysteine	6	Streptomycin	50 U/ml
L-Cystine	2		

Shields and Sang (1970) insect tissue culture medium (mg/100 ml)

$MgSO_4 \cdot 7H_2O$	513	L-Glycine	50
$CaCl_2 \cdot 6H_2O$	174	L-α-Alanine	165
KCl	313	L-Valine	42
NaCl	86	L-Methionine	12
$NaH_2PO_4 \cdot 2H_2O$	88	L-Isoleucine	27
$KHCO_3$	18	L-Leucine	40
Monosodium malate $\cdot 2H_2O$	95	L-Tyrosine	26
Monosodium α-ketoglutarate	42	L-Phenylalanine	24
Disodium fumarate	8	L-β-Alanine	10
Disodium succinate $6H_2O$	14	L-Histidine	55
Monosodium L-glutamate	246	L-Tryptophan	10
L-Aspartic acid	15	L-Arginine	50
L-Serine	35	L-Lysine	68
L-Asparagine	30	L-Cystine	20
L-Glutamine	60	L-Cysteine	80
L-Proline	40	Glutathione	0.5
L-Threonine	50	Glucose	460
TC-Yeastolate	200	FBS	10%

Trager (1935) Solution A (gm/liter)

Maltose	20.538	K_2HPO_4	0.2613
NaCl	0.8766	*Bombyx mori* hemolymph	10%
$MgCl_2 \cdot 6H_2O$	0.2033	Penicillin	100 U/ml
$CaCl_2$	0.111	Streptomycin	100 μg/ml
$NaH_2PO_4 \cdot H_2O$	0.207		

Vago and Chastang (1958) insect tissue culture medium (gm/liter)

KCl	3.0	Casein hydrolysate	0.5
$CaCl_2$	0.5	Glutamine	0.1
NaH_2PO_4	1.0	Yeast extract	2.0 ml
$MgSO_4 \cdot 7H_2O$	3.5	Choline	0.002
$MgCl_2 \cdot 6H_2O$	3.0	Penicillin	200,000 IU
Glucose	1.0	Streptomycin	0.05
$NaHCO_3$	to pH 6.4	*Bombyx mori* hemolymph	10%

Vago and Chastang (1958) *Helix aspersa* tissue culture medium (gm/liter)

NaCl	6.5	Casein hydrolysate	0.5
KCl	0.14	Glutamine	0.1
$CaCl_2$	0.12	Yeast extract	2.0 ml
NaH_2PO_4	0.01	Penicillin	200,000 IU
$NaHCO_3$	0.2	Streptomycin	0.05
Glucose	1.0	*H. aspersa* blood	10%

Vago and Quiot (1969) 72 S.F.M. insect tissue culture medium modified
by J. M. Quiot (personal communication, 1971) (gm/liter)

L-Proline	0.67	L-Lysine	0.09
L-Cysteine	0.26	L-Threonine	0.08
L-Glycine	0.65	L-Isoleucine	0.04
L-Phenylalanine	0.10	TC 199 organic fraction	2.38
L-Methionine	0.44	TC 199 mineral fraction	8.89
L-Valine	0.05	Bacitracin	0.10
L-Leucine	0.01	Penicillin	0.12
L-Tryptophan	0.16	Streptomycin	0.05
L-Histidine	0.26	Colimycin	0.025
L-Tyrosine	0.10	$NaHCO_3$(5%) to pH	6.8
L-Arginine	0.76	FCS	15%

Varma and Pudney (1969) VP_{12} insect tissue culture medium (mg)

NaCl	390	Inositol	40
$NaH_2PO_4 \cdot 2H_2O$	55	Lactalbumin hydrolysate	500
$MgCl_2 \cdot 6H_2O$	110	Bovine plasma albumin,	100
$MgSO_4 \cdot 7H_2O$	120	fraction V	
KCl	55	5% Glutamine	0.6 ml
$CaCl_2 \cdot 2H_2O$	40	Basal medium Eagle vitamin	2.0 ml
D-Glucose	200	mixture (100X)	
$NaHCO_3$	50	H_2O	97.4 ml
Choline chloride	25	FBS	10%

Yunker et al. (1967) insect tissue culture medium

Grace's (1962) medium	79%
FBS	10%
Whole chicken egg ultrafiltrate	10%
Bovine plasma albumin, fraction V	1%

Yunker et al. (1967) insect tissue culture medium modified by Hink and
Ignoffo (1970)

Grace's (1962) medium	90.0 ml
FBS	10.0 ml
Egg ultrafiltrate	10.0 ml
Bovine plasma albumin, crystallized	1.0 gm

References to Media Section

Chiu, R. J., and Black, L. M. (1967). *Nature (London)* **215**, 1076–1078.
Echalier, G., and Ohanessian, A. (1970). *In Vitro* **6**, 162–172.
Eide, P. E., and Chang, T. H. (1969). *Exp. Cell Res.* **54**, 302–308.

Goodwin, R. H. (1975). *In Vitro* (in press).

Goodwin, R. H., Vaughn, J. L., Adams, J. R., and Louloudes, S. J. (1973). *Misc. Publ. Entomol. Soc. Amer.* **9**, 66–72.

Grace, T. D. C. (1962). *Nature (London)* **195**, 788–789.

Gvozdev, V. A., and Kakpakov, V. T. (1968). *Genetika* **4**, 226–235.

Hink, W. F. (1970). *Nature (London)* **226**, 466–467.

Hink, W. F., and Ignoffo C. M. (1970). *Exp. Cell Res.* **60**, 307–309.

Hirumi, H., and Maramorosch, K. (1964). *Exp. Cell Res.* **36**, 625–631.

Horikawa, M., Ling, L.-N., and Fox, A. S. (1966). *Nature (London)* **210**, 183–185.

Hsu, S. H., Mao, W. H., and Cross, J. H. (1970). *J. Med. Entomol.* **7**, 703–707.

Hsu, S. H., Li, S. Y., and Cross, J. H. (1972). *J. Med. Entomol.* **9**, 86–91.

Kitamura, S. (1965). *Kobe J. Med. Sci.* **11**, 23–30.

Kitamura, S. (1970). *Kobe J. Med. Sci.* **16**, 41–50.

Kurtti, T. J. (1974). Ph.D. Dissertation, University of Minnesota Minneapolis.

Landureau, J. C. (1966). *Exp. Cell Res.* **41**, 545–556.

Landureau, J. C., and Grellet, P. (1972). *C. R. Acad. Sci.* **274**, 1372–1375.

Landureau, J. C., and Jollès, P. (1969). *Exp. Cell Res.* **54**, 391–398.

Leibovitz, A. (1963). *Amer. J. Hyg.* **78**, 173–180.

Melnick, J. L. (1955). *Ann. N. Y. Acad. Sci.* **61**, 754–772.

Mitsuhashi, J. (1967). *Nature (London)* **215**, 863–864.

Mitsuhashi, J. (1972). *Appl. Entomol. Zool.* **8**, 64–72.

Mitsuhashi, J., and Maramorosch, K. (1964). *Contrib. Boyce Thompson Inst.* **22**, 435–460.

Morgan, J. F., Morton, H. J., and Parker, R. C. (1950). *Proc. Soc. Exptl. Biol. Med.* **73**, 1–8.

Schneider, I. (1964). *J. Exp. Zool.* **156**, 91–104.

Schneider, I. (1966). *J. Embryol. Exp. Morphol.* **15**, 271–279.

Schneider, I. (1969). *J. Cell Biol.* **42**, 603–606.

Shields, G., and Sang, J. H. (1970). *J. Embryol. Exp. Morphol.* **23**, 53–69.

Trager, W. (1935). *J. Exp. Med.* **61**, 501–513.

Vago, C. (1963). *Ann. Epiphyties* **14**, 7.

Vago, C., and Chastang, S. (1958). *Experientia* **14**, 110–111.

Vago, C., and Quiot, J. M. (1969). *Ann. Zool. Ecol. Anim.* **1**, 231.

Varma, M. G. R., and Pudney, M. (1969). *J. Med. Entomol.* **6**, 432–439.

Yunker, C. E., Vaughn, J. L., and Cory, J. (1967). *Science* **155**, 1565–1566.

Index

A

Aceratagallia sanguinolenta
  cell line, 353
  continuous cultures, 309
  culture medium, 353
  morphology, 353
  virus susceptibility, 311, 353
  wound tumor virus in, 310–311
Acetate, 149
2,3-Acetonide of β-ecdysone, activity of,
  165
Achatina marginata, in vitro culture of
  embryos, 63
Acholeplasma, 256, 259
Acholeplasma granularum, 255
Acholeplasma laidlawii, 255, 256, 257, 259
Acid fuchsin, affinity of *Papilio xuthus*
  cell lines to, 26
Acid phosphatase, 264
Actinomycin D, virus persistence and,
  222–223
Acyrthosiphon pisum, culture, 313–314
Adaptation period, length of, 45
Adenosine triphosphate (ATP), sporoplasm culture and, 191
Adrenal cortex, similarity to prothoracic
  glands, 106
Aedes, 202
  arenaviruses in, 227
  bunyaviruses in, 203, 204–205, 208, 209
  Chikungunya virus in, 203
  flaviviruses in, 205
  orbiviruses in, 205
  rhabdoviruses in, 206
  vitellogenic synthesis in, 124
Aedes aegypti, 186, 194, 195, 202, 297
  alphaviruses in, 203, 208, 225
  arbovirus persistence in, 219, 220–221,
    222
  bunyaviruses in, 204, 208
  cell lines, 330–333

identification, 55
  intraspecies, 264
Chilo iridescent virus in, 290
cloning, 4
colony formation, 6
culture media, 330, 331, 332, 333
extrinsic cell contamination of, 264
failure to agglutinate with concanavalin
    A, 9
flaviviruses in, 209, 218
growth characteristics, 330, 331, 332, 333
karyology, 330, 332, 333
lack of syncytia induction in, 240
morphology, 330, 332, 333
mosquito iridescent virus in, 270, 287,
    290, 291
  assay of, 271–275, 286–289
  infection and replication of, 275–283
mycoplasmal contamination in, 257
orbiviruses in, 205, 206
rhabdoviruses in, 206, 211, 213
Sericesthis iridescent virus in, 290
Toxoplasma gondii in, 228, 229
viral contamination of, 242, 246, 247,
    248, 254
virus susceptibility, 330, 331, 332, 333
Aedes albopictus, 297
  alphaviruses in, 203, 207, 208, 214
  arbovirus persistence in, 219–221, 222–
    223
  bunyaviruses in, 203, 204, 207, 208, 209,
    215–218
  cell density of, 10, 11
  cell lines, 202, 334–335
    intraspecies, 264
  cloning, 4
  colony formation, 6
  culture media, 334
  cytopathic effect in, 224–227
  extrinsic cell contamination of, 264
  failure to agglutinate with con-
    canavalin A, 8, 9, 10

371

activity of, 165
imaginal discs and, 137–138

S

St. Louis encephalitis virus, 205
Salivary glands, puffing pattern in, 162
Salt-marsh caterpillar, see Estigmene
  acrea
Salts
  in Biomphalaria glabrata hemolymph,
  94
  in mollusk hemolymph, 83
Samia, spermatogenesis, 123
Samia cynthia, 194
  cell lines, 327–328
  culture medium, 328
  ecdysones and, 104
  karyology, 327
  morphology, 327
  spermatocysts, 142, 143
Sandfly fever Neapolitan virus, 204–205
Sandfly fever Sicilian virus, 205
Sarcophaga, wing disc changes in, 119
Sarcophaga test abdomen, 107, 114
Saw thistle yellow virus, 314
Schistocerca, juvenile hormone secretion
  in, 125
Schistocerca gregaria
  integument, 139
  oenocytes, 153
Schistosoma mansoni, 58, 69, 75
  developmental studies, 59, 64–65
Schneider's medium, 96, 366
Sedimentation profiles, of mosquito
  iridescent virus, 280, 283
Semliki Forest (SF) virus, 203, 208
  attenuation, 221
  persistence and interferon, 219
  ultrastructure, 214–215
Sepia officinalis
  reproductive endocrinology, 72
  in vitro organ culture, 65
Sericesthis iridescent virus, (SIV), 270,
  283, 288–289
  assay of, 283, 285
  mitochondria and, 290
Serine, cell line utilization and requirement for, 31, 33–34, 36
Serum, see also Fetal bovine serum

in culture media, 95–96
Silk glands, response to ecdysone, 124
Silkmoth, spermatogenesis, 133
Silkworm, see Bombyx mori
Silverwater virus, 205
Sindbis virus, 203, 208, 215
  interferon, 219
Sixspotted leafhopper, see Macrosteles
  fascifrons
Slug moth, see Monema flavescens
Snail, see also Helix aspersa, Biomphalaria glabrata
  oxygen consumption of organ slices in,
  62
Sodium citrate, for chromosome preparations, 83
Sodium deoxycholate (SDC), virus sensitivity to, 202
Soldado virus, 206
Spermatocysts, hormonal control of development, 142–143
Spermatogenesis
  early observation, 133
  hormonal effects on, 68, 123, 143, 150
Spin-filter vessel, 301–302
Spodoptera frugiperda, 297
  cell lines, 328–329
  cloning, 4
  colony formation, 6
  culture media, 328
  growth characteristics, 328–329
  karyology, 328
  morphology, 328
  mycoplasmal contamination in, 257
  NPV replication in, 297
  virus susceptibility, 328, 329
  virus yield from, 299
Spores
  alkaline treatment of, 190, 191
  germination in Nosema, 190
  to infect cultured cells, 190
  phagocytosis by insect cells, 191
Sporoplasms
  culture of, 191
  eversion of, 191
Sporozoites, oocytes and, 188
Spruce budworm, see Choristoneura
  fumiferana
Stagnicola emarginata, organ culture, 71
Stainability

A 6
B 7
C 8
D 9
E 0
F 1
G 2
H 3
I 4
J 5